Robust Computational Techniques for Boundary Layers

APPLIED MATHEMATICS
Editor: R.J. Knops

This series presents texts and monographs at graduate and research level
covering a wide variety of topics of current research interest in modern and
traditional applied mathematics, in numerical analysis and computation.

(Full details concerning this series, and more information on titles in
preparation are available from the publisher.)

Applied Mathematics 16

Robust Computational Techniques for Boundary Layers

P. A. Farrell
Kent State University

A. F. Hegarty
Trinity College, Dublin

J. J. H. Miller
University of Limerick

E. O'Riordan
Dublin City University

G. I. Shishkin
Russian Academy of Sciences

CRC Press
Taylor & Francis Group
Boca Raton London New York

CRC Press is an imprint of the
Taylor & Francis Group, an **informa** business

A CHAPMAN & HALL BOOK

CRC Press
Taylor & Francis Group
6000 Broken Sound Parkway NW, Suite 300
Boca Raton, FL 33487-2742

First issued in paperback 2019

© 2000 by Taylor & Francis Group, LLC
CRC Press is an imprint of Taylor & Francis Group, an Informa business

No claim to original U.S. Government works

ISBN-13: 978-1-58488-192-6 (hbk)
ISBN-13: 978-0-367-39878-1 (pbk)

Visit the Taylor & Francis Web site at
http://www.taylorandfrancis.com

and the CRC Press Web site at
http://www.crcpress.com

Library of Congress Cataloging-in-Publication Data

Robust computational techniques for boundary layers / P.A. Farrell...[et al.].
      p. cm. (Applied mathematics and mathematical computation ; 13)
      Includes bibliographical references and index.
      ISBN 1-58488-192-5
      1. Boundary layer—Mathematics. 2. Numerical calculations. I. Farrell, P. A. II. Series.
QA913 .R62 2000
532'.051—dc21
                                                           99-086205
                                                                      CIP

Library of Congress Card Number 99-086205

To Avril, Kathy, Mary, Lida and Pamela

Contents

Preface

In this book we construct numerical methods for solving problems involving differential equations, which have non–smooth solutions with singularities related to boundary layers. Such problems arise in many physical phenomena, and from a mathematical perspective they are singularly perturbed. We construct robust layer–resolving methods for generating numerical approximations to their solution. We say that a numerical method is layer–resolving if it generates, at each point of the solution domain, numerical approximations of the exact solution, and its derivatives, for all values of the singular perturbation parameter from finite to arbitrarily small values. Moreover, these numerical approximations must satisfy pointwise error bounds, independent of the singular perturbation parameter, and be computable with an amount of work, which is also independent of this parameter. The numerical approximations are said to be robust if they inherit the stability properties of the exact solution by preserving the monotonicity of the original problem. It is worth observing that the techniques we use here in the construction of numerical methods for problems with boundary layers are also applicable to problems with a much wider class of singularities.

Robust layer–resolving methods are often needed in real applications. For example in fluid dynamics, if separation and reattachment phenomena are to be reliably modelled, it is essential to determine, for all values of the Reynolds number, pointwise–accurate approximations at all points both within and outside the boundary layers, not only to the flow variables but also to their derivatives. In developing our numerical methods and experimental techniques we always have in mind that our ultimate goal is the numerical modelling of real flow problems, which involve practical issues about, for example, lift, drag and transition phenomena.

Unfortunately, most contemporary numerical methods are not robust and layer–resolving. In particular, the requirement of monotonicity severely limits the usefulness of many known finite element and finite difference methods, and so we are forced to construct non–standard methods using new techniques. At this point it should be observed that robust layer–resolving methods are of little relevance to a problem with a relatively smooth solution or to a singular perturbation problem if there is no interest in capturing boundary layer effects in its solution. In these cases standard numerical methods are usually sufficient.

In the first eight chapters of the book we construct new numerical methods
for broad classes of linear problems in one and two dimensions. We also discuss
the shortcomings, considering our requirement for robust and layer–resolving
behaviour, of some standard numerical methods. For all of these new methods
it is possible to prove, by existing rigorous theoretical error analysis, that these
are robust and layer–resolving. We illustrate this rigorous theoretical approach
for some one–dimensional problems, but, for most of the problems considered in
these chapters, we prefer to demonstrate that our new methods are robust and
layer–resolving by thorough numerical experimentation. On the other hand, for
many linear and most nonlinear problems, current theoretical error analysis is not
sufficiently advanced to verify theoretically that our new methods are robust and
layer–resolving. Therefore we are forced to develop new computational techniques
to provide an experimental parameter–uniform error analysis. In chapter 8, we
describe several new a posteriori experimental techniques for computing realistic
approximations to the error parameters p and C_p of any numerical method, which
is known to be parameter–uniform with some (unknown) order of convergence p
and error constant C_p. To validate these experimental techniques we apply them
to some model problems with known theoretical error estimates and analytic
solutions. We then compare the predicted upper bounds for the error given by
our experimental techniques with the exact errors derived from the analytic
solutions.

In chapter 9 we consider the potentially more accurate numerical methods
based on centred finite difference operators for linear problems in two dimensions. For such non–monotone methods there is no rigorous theoretical error
analysis to determine whether or not a method is parameter–uniform. For this
reason we estimate the error in the approximations generated by non–monotone
methods using the approximations generated previously by monotone methods
and the experimental parameter–uniform error analysis developed in chapter 8.
Then, in chapter 10, we introduce nonlinear singular perturbation problems re-
stricting the discussion, for the sake of simplicity, to semilinear reaction–diffusion
problems. We consider some of the new issues that arise due to the nonlinearity,
in particular the need for parameter–uniform continuation algorithms for solving the associated nonlinear finite difference methods. We construct monotone
numerical methods for these problems and we verify experimentally, using the
experimental parameter–uniform error analysis of chapter 8, that they are robust
and layer–resolving.

In chapters 1 to 10 we develop techniques for constructing robust layer–
resolving methods, and, in addition, we develop further experimental techniques
for justifying computationally that in practice the approximations generated by
these methods display the required robust and layer–resolving convergence. In
chapters 11 and 12 we demonstrate the true power of the techniques discussed
in chapters 1 to 10 by first applying them to the construction of a new numerical

method for solving a non–trivial classical problem in fluid dynamics and then by using them to verify that the approximations generated by this new method converge in a robust and layer–resolving manner.

In the preparatory chapter 11 we introduce the problem of laminar flow of an incompressible fluid past a semi–infinite flat plate. Then we describe a simplification of the Navier–Stokes equations in this case which leads to the simpler Prandtl problem. Because we need it in the next chapter, we devote the remainder of chapter 11 to the construction of an approximation of known accuracy to the self–similar solution of this problem, for all values of the Reynolds number. We obtain the required approximation by using Blasius' semi–analytic technique to reduce the problem of finding this self–similar solution to that of solving a nonlinear third order ordinary differential equation. Since this problem itself is singularly perturbed, we construct a new numerical method to generate numerical approximations to the solution and its derivatives. Since preliminary theoretical error analysis shows that this method is robust and layer–resolving, we apply our experimental parameter–uniform error analysis to demonstrate computationally that this is indeed the case.

In chapter 12 we construct a new numerical method for solving the Prandtl problem. Because there is no rigorous theoretical error analysis for the convergence of the numerical approximations generated by this numerical method to the exact solution and its derivatives, we investigate its performance in two different and independent ways. In the first, we carry out an experimental parameter–uniform error analysis by replacing the unknown exact solution in the expression for the error by the numerical solution on the finest available mesh. This provides us with computed maximum pointwise errors and computed orders of convergence. The results suggest that our new method is a robust layer–resolving method for the Prandtl problem. Our second way of investigating the performance of our new method is to compare the numerical approximations generated by this new method with the semi–analytic approximations of known accuracy constructed in the previous chapter. The results of this procedure are in agreement with those produced by the first way, and they allow us to infer once again that our method is robust and layer–resolving. We end the chapter by showing how we can use our new numerical method to generate benchmark solutions, which may then be used to test the performance of any other numerical method for solving the Prandtl problem or laminar flow problems involving the Navier–Stokes equations, at least in the most interesting case of a large Reynolds number.

The main points of the book can be summarized as follows. We construct robust layer–resolving methods for a wide class of mathematical problems related to linear convection–diffusion and nonlinear reaction–diffusion phenomena and for the classical problem of flow past a plate. We say that we have constructed a robust layer–resolving method if we first specify the finite difference operator and mesh and then we establish that the numerical approximations generated are robust and layer–resolving, that is we determine the existence of positive parameter–uniform constants p and C_p. To find such constants we use two different approaches. The first is theoretical and is applicable to a restricted class of mathematical problems. Furthermore, it yields weak results in the sense that its predicted order of convergence p is usually too small and the error constant C is rarely discussed. The second approach is experimental. It is concerned with concrete representative problems from wide problem classes, and it yields realistic estimates of both p and C. We remark that for most problems of fluid dynamics only the experimental method is feasible.

The book is written in a style which is suitable, not only for mathematicians, but also for engineers and scientists. The fundamental ideas are explained through physical insight, model problems and computational experiments.To illuminate the often subtle issues, we make extensive use of tables and graphics of actual computations. Although the theorems proved in the book are, for simplicity, confined to linear problems in one dimension, similar rigorous theoretical results have been established for all of the robust layer–resolving methods in the first eight chapters. An introduction to this underlying theory, for problems in one and two dimensions, is contained in the book by Miller et al. (1996). A comprehensive account of the theory for linear problems is given in the book by Shishkin (1992b). The theory is presented there in a highly condensed style and the book is currently available only in Russian. The main conclusion of this theoretical work is that standard monotone finite difference operators on piecewise–uniform meshes suffice for the construction of robust layer–resolving methods for a wide class of linear problems. In the present book we show, using the experimental parameter–uniform error analysis developed here, that the same is true for the classical Prandtl boundary layer problem.

In various places in the book we give details of the linear solvers used in the computations in order that readers can implement our methods and reproduce the numerical results. Since we encounter no special restrictions using standard solvers, in conjunction with the monotone methods advocated in this book, we do not consider it necessary to discuss optimal solvers. It should be observed that we are content with a robust layer–resolving method of any accuracy, however small, because we are more interested in the robustness, simplicity and wide applicability of our methods, than in their optimality.

The authors wish to thank their colleagues and research students for many fruitful discussions over the years. Those who helped directly with this book include L. J. Crane, P. W. Hemker, A. Ansari, H. MacMullen, and especially A. Musgrave. For LaTeX support we thank D. O'Riordan.

For decisive financial and material support, throughout the writing of this book, the authors thank especially INCA – the Institute for Numerical Computation and Analysis, Dublin. In addition we are grateful to the Department of Mathematics, Trinity College Dublin, the School of Mathematical Sciences, Dublin City University, the Institute for Mathematics and Mechanics, Russian Academy of Sciences, Ekaterinburg, the Department of Mathematics and Computer Science, Kent State University, the Department of Mathematics and Statistics, University of Limerick, and Forbairt, Dublin, for financial and material support. In particular, the research of G. I. Shishkin was supported in part by the Russian Foundation for Fundamental Investigations Grant No. 98–01–00362, the International Collaboration Programme of Forbairt No. IC/97/057, the Enterprise Ireland Basic Research Grant SC-98–612, the Provost's Academic Development Fund of Trinity College and INCA, Dublin. The research of P. A. Farrell and G. I. Shishkin was supported in part by the Computational Mathematics Program of the National Science Foundation, under grant DMS 9627244.

CHAPTER 1

Introduction to numerical methods for problems with boundary layers

1.1 The location and width of a boundary layer

We begin by considering the simple linear convection–diffusion problem

$$\varepsilon u_\varepsilon''(x) + u_\varepsilon'(x) = 0 \quad \text{for all } x \in \Omega = (0,1), \tag{1.1a}$$

$$u_\varepsilon(0) = 0, \quad u_\varepsilon(1) = 50, \tag{1.1b}$$

where ε is a small positive parameter called the singular perturbation parameter and $0 < \varepsilon \leq 1$.

Note that once we allow the parameter ε in (1.1a) to become small we enter the mathematical realm of singular perturbation problems. The solution of a singularly perturbed problem of the form (1.1) normally has a smooth and a singular component; its singular component is called a boundary layer function. The boundary layer corresponding to this boundary layer function is the subdomain in which the magnitude of the singular component is not negligible. The physical properties associated with a solution containing a boundary layer function are reflected by the mathematical properties of the solution of a singularly perturbed differential equation. In this book we consider singularly perturbed differential equations with the small parameter ε multiplying some or all of the terms involving the highest order derivatives.

The exact solution of problem (1.1) is

$$u_\varepsilon(x) = 50\left(\frac{1 - e^{-x/\varepsilon}}{1 - e^{-1/\varepsilon}}\right). \tag{1.1c}$$

When the parameter ε is small, for example $\varepsilon = 0.001$, the solution u_ε has the form shown in Fig. 1.1 and it is clear that there is a boundary layer at the boundary point $x = 0$. The corresponding boundary layer function is the exponential function $e^{-x/\varepsilon}$.

The techniques used in this book to construct an effective numerical method for solving a problem with a boundary layer function in its solution require *a priori* knowledge about the location and width of the boundary layer.

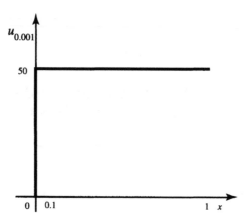

FIG. 1.1. Solution $u_{0.001}(x)$ of problem (1.1) with $\varepsilon = 0.001$ for $x \in [0, 1]$.

The location of a boundary layer is often easy to determine, but it is not obvious how we should define its width w. To illustrate this we consider a typical boundary layer function $e^{-x/\varepsilon}$, the graph of which is shown in Fig. 1.2. Obviously this boundary layer is located at the boundary point $x = 0$, but there are several possible ways to define its width. For example, we could take w to be the smallest value for which we have

$$\sup_{x \geq w} e^{-x/\varepsilon} \leq 10^{-8}.$$

With this definition the width of the boundary layer corresponding to this boundary layer function is $w = 8\varepsilon \ln 10 = O(\varepsilon)$. A more intrinsic definition is to take w to be the smallest value for which we have

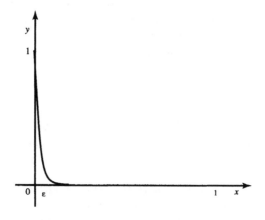

FIG. 1.2. Typical boundary layer function $e^{-x/\varepsilon}$.

$$\sup_{x \geq w} e^{-x/\varepsilon} \leq \varepsilon.$$

In this case the width of the boundary layer is $w = \varepsilon \ln 1/\varepsilon$. Another possibility is to take w to be the smallest value for which, for each integer k satisfying $0 \leq k \leq 3$, we have

$$\sup_{x \geq w} \left| \frac{d^k}{dx^k} e^{-x/\varepsilon} \right| \leq \varepsilon.$$

With this definition the width of the boundary layer is $w = 4\varepsilon \ln 1/\varepsilon$ which, apart from the factor 4, is the same as in the previous case. This last definition turns out to be appropriate for a wide class of singular perturbation problems, if we are interested in studying properties of, not only the solution, but also its derivatives. We remark that, with this definition, the magnitude of the function $e^{-x/\varepsilon}$ and that of its first three derivatives, is exponentially small outside the boundary layer. We also note that the standard proofs to derive theoretical error bounds require that the solution has three continuous derivatives.

1.2   Norms for boundary layer functions

In Fig. 1.1 we observe that, except near the point $x = 0$, the solution u_ε of problem (1.1) is almost equal to the constant function $c(x) \equiv 50$, which is plotted in Fig. 1.3. At first sight, there seems to be no difference in the figures, but if we blow-up the scale of the x-axis in the small subinterval $[0, 10^{-3}]$, as in Fig. 1.4, we see that the two functions are entirely different near the origin, because, as x approaches the origin, one of them has the constant value 50 and the other tends to zero. The point of this comparison is to demonstrate that, in the study of boundary layer functions, it is dangerous to rely solely on the graphical output

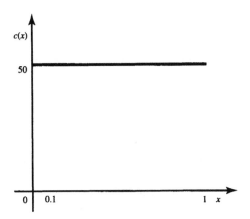

FIG. 1.3. Constant function $c(x) = 50$ for $x \in [0, 1]$.

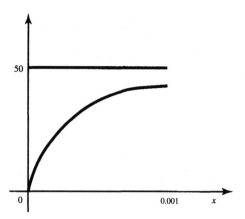

FIG. 1.4. Solution $u_{0.001}(x)$ of problem (1.1) and constant function $c(x) = 50$ for $x \in [0, 0.001]$.

displayed on our computer screens, however high the resolution. When ε is small enough the boundary layer will, generally speaking, not be resolved, and the presence of the boundary layer will not be detected.

We can overcome this difficulty by using a suitable mathematical tool for measuring functions, namely a norm, which enables us to distinguish between different functions. There are many available norms, not all of which are suitable for capturing boundary layers. Therefore, in the remainder of this section, we discuss the effectiveness of several possible norms.

The first norm that we consider as a candidate is the L_1 norm, which is defined for any function f on the interval $[0, 1]$, by

$$\|f\|_1 = \int_0^1 |f(x)| \, dx.$$

In this norm the difference between the solution u_ε of (1.1) and the constant function 50 is

$$\|u_\varepsilon - 50\|_1 = \|50\Big(\frac{e^{-1/\varepsilon} - e^{-x/\varepsilon}}{1 - e^{-1/\varepsilon}}\Big)\|_1 = 50\Big(\varepsilon - \frac{e^{-1/\varepsilon}}{1 - e^{-1/\varepsilon}}\Big).$$

Thus, for $\varepsilon = 0.001$, we have $\|u_{0.001} - 50\|_1 \cong 0.05$, and so the L_1 norm detects little difference between the functions $u_{0.001}$ and 50. The L_1 norm is also an ineffective tool for distinguishing differences in the solution u_ε for different values of the parameter ε; for example, the difference between $u_{0.001}$ and $u_{0.000001}$ in the L_1 norm is

$$\|u_{0.001} - u_{0.000001}\|_1 \cong 50(10^{-3} - 10^{-6}) \cong 0.05.$$

Likewise this norm is not good at detecting differences between solutions of singularly perturbed problems with the same differential equation but different boundary conditions. For example, consider the problem

$$\varepsilon \hat{u}_\varepsilon''(x) + \hat{u}_\varepsilon'(x) = 0, \quad \text{for all } x \in (0,1), \qquad (1.2a)$$
$$\hat{u}_\varepsilon(0) = 100, \quad \hat{u}_\varepsilon(1) = 50, \qquad (1.2b)$$

whose solution is

$$\hat{u}_\varepsilon(x) = 100 - u_\varepsilon(x). \qquad (1.2c)$$

It is clear that u_ε and $\hat{u}_\varepsilon$ are completely different in nature, because u_ε is monotonically increasing while $\hat{u}_\varepsilon$ is monotonically decreasing, but their difference in the L_1 norm is small. namely

$$\|\hat{u}_{0.001} - u_{0.001}\|_1 = 2\|50 - u_{0.001}\|_1 \cong 0.1.$$

We conclude that the L_1-norm is not an appropriate norm for the study of problems with boundary layers.

We now consider another possible norm, the L_2 norm, which is defined for any function f by

$$\|f\|_2 = \sqrt{\int_0^1 |f(x)|^2 \, dx}.$$

In this norm the difference between u_ε and the constant function 50 is

$$\|u_\varepsilon - 50\|_2 = \|50 \frac{e^{-1/\varepsilon} - e^{-x/\varepsilon}}{1 - e^{-1/\varepsilon}}\|_2 \cong 35\sqrt{\varepsilon}$$

and so, when $\varepsilon = 0.000001$, for example, we have $\|u_{0.000001} - 50\|_2 \cong 0.035$, which is again disappointing. Furthermore,

$$\|u_{0.001} - u_{0.000001}\|_2 \cong 1, \quad \text{with} \quad \|u_{0.001}\|_2 \cong \|u_{0.000001}\|_2 \cong 50$$

which shows that in this norm the relative difference between the two functions is just 2%, despite the fact that these two functions have significantly different behaviour near $x = 0$. We conclude that the L_2 norm is also not a good choice for our purposes.

It is clear that the unsatisfactory behaviour of the L_1 and L_2 norms arises from the fact that they involve averaging over the whole interval and therefore local phenomena are often not detected.

We therefore try a norm which does not involve averaging. The most promising candidate is the L_∞ or maximum norm , which is defined by

$$\|f\|_\infty = \max_{x\in[0,1]} |f(x)|.$$

In this norm we have, for all values of ε,

$$\|u_\varepsilon - 50\|_\infty = 50,$$

and the difference between the two functions considered before is

$$\|u_{0.001} - u_{0.000001}\|_\infty \cong 49.6.$$

We see that differences between distinct functions are now detected, irrespective of how small ε is. This suggests that the maximum norm is an appropriate norm for the study of boundary layer phenomena.

Finally, we consider a typical energy norm for a second order ordinary differential equation defined by

$$\||v\||^2 \equiv \varepsilon\|v'\|_2^2 + \|v\|_2^2. \tag{1.3}$$

Caution is required with the use of such an energy norm, as the following example shows. For simplicity, we consider the reaction–diffusion problem

$$\varepsilon u_\varepsilon'' - u_\varepsilon = f(x), \quad x \in \Omega = (0,1), \quad u_\varepsilon(0) = 1, \ u_\varepsilon(1) = 1, \tag{1.4}$$

for which a finite element method is ideally suited, because the differential equation is self–adjoint. In fact it can be shown (see, for example, Bagaev and Shaidurov (1998)) that for the reaction–diffusion problem (1.4) and the standard finite element method with piecewise linear basis functions on a uniform mesh, we have in this norm the ε–uniform error estimate

$$\||\bar{U}_\varepsilon - u_\varepsilon\|| \leq CN^{-1/2}$$

where $\bar{U}_\varepsilon$ is the finite element approximation, C is a constant independent of ε and N is the number of basis functions. At first sight this seems to indicate that this standard finite element method gives satisfactory numerical approximations to the solution u_ε of (1.4) no matter how small ε is. But this is certainly not the case. In fact, the situation is that, unless we choose the number of uniform mesh intervals to be inversely dependent on ε, the boundary layers in this problem are not resolved by this finite element method. Indeed the boundary layer functions in the solution u_ε of (1.4) are $e^{-x/\sqrt{\varepsilon}}$ and $e^{-(1-x)/\sqrt{\varepsilon}}$ and in the energy norm we have

$$|||e^{-x/\sqrt{\varepsilon}}||| = O(\sqrt{\varepsilon}), \quad |||e^{-(1-x)/\sqrt{\varepsilon}}||| = O(\sqrt{\varepsilon}).$$

This shows that the boundary layer functions are not detected in the energy norm when ε is small. On the other hand, in the maximum norm we have

$$||e^{-x/\sqrt{\varepsilon}}||_\infty = ||e^{-(1-x)/\sqrt{\varepsilon}}||_\infty = 1,$$

which shows that these boundary layer functions are captured in the maximum norm regardless of how small ε is.

In other problems, the energy norm does capture the boundary layer function. For example, for the convection–diffusion problem (1.1), the boundary layer function is $e^{-x/\varepsilon}$ and we have

$$|||e^{-x/\varepsilon}||| = O(1).$$

Consider also the two–dimensional convection–diffusion problem

$$\varepsilon u_{xx}(x,y) + \varepsilon u_{yy}(x,y) + u_x(x,y) + u_y(x,y) = f(x,y), \quad (x,y) \in (0,1) \times (0,1)$$

for which the boundary layer functions are $e^{-x/\varepsilon}$, $e^{-y/\varepsilon}$ and $e^{-x/\varepsilon}e^{-y/\varepsilon}$, and the associated energy norm is defined by

$$|||v|||^2 \equiv \varepsilon||v_x||_2^2 + \varepsilon||v_y||_2^2 + ||v||_2^2.$$

In this problem, the boundary layer functions $e^{-x/\varepsilon}$, $e^{-y/\varepsilon}$ are captured in this norm, but for small values of ε the corner boundary layer function $e^{-x/\varepsilon}e^{-y/\varepsilon}$ is not, because

$$|||e^{-x/\varepsilon}e^{-y/\varepsilon}||| = O(\sqrt{\varepsilon}).$$

On the other hand, these boundary and corner boundary layer functions are detected in the maximum norm, since for all values of ε

$$||e^{-x/\varepsilon}e^{-y/\varepsilon}||_\infty = 1.$$

The above examples show that, if we use such an energy norm, we need to determine in advance whether or not the energy norm captures the boundary layer functions in the solution of the problem. In order to avoid the need for such *a priori* information, we use the maximum norm throughout this book. Since we use only this norm, we drop the subscript and let $|| \cdot ||$ denote the maximum norm over the whole solution domain. In the future, if we want to emphasize the domain $\overline{\Omega}$ over which the maximum is taken for a particular problem, we may indicate this by a subscript, for example $|| \cdot ||_{\overline{\Omega}}$.

1.3   Numerical methods

Because the exact solution of a non–trivial problem involving a singularly perturbed differential equation is usually unknown, we need to compute numerical
approximations to it. The domain of these numerical solutions may be either the
whole domain $\overline{\Omega}$ of the exact solution or a discrete set of mesh points or nodes
$\overline{\Omega}^N \subset \overline{\Omega}$, where N is the discretization parameter of the numerical method,
which, for example, corresponds to the number of mesh points in a finite differ-
ence method or the number of nodes in a finite element method.

To be certain of capturing a boundary layer with a numerical method, it is
essential that the approximate solutions generated by the numerical method are
defined globally at each point of the domain of the exact solution. In cases where
the numerical solution is defined only at the mesh points, for example at the
mesh points of a finite difference method, we extend it to the whole domain by
a simple interpolation process such as piecewise linear interpolation. We insist
on a simple process, because we want our techniques to be capable of extension
to complicated problems in many dimensions. For the same reason we consider
only standard finite element methods, and so we do not allow the enrichment of
the finite element subspaces by non–piecewise–polynomial basis functions.

From the discussion in the previous section we know that we need to measure
the error in the maximum norm, which means that we have to decide whether
to use the discrete or global maximum norm with the maximum taken over the
discrete set of mesh points $\overline{\Omega}^N$ or the complete domain $\overline{\Omega}$ respectively. Because
we want our numerical methods to generate solutions defined on the whole of $\overline{\Omega}$,
the global maximum norm is the natural and appropriate norm to use. We refer
to a numerical approximation with a small error in the global maximum norm
as a globally pointwise accurate approximation.

While the solution u_ε of problem (1.1) takes a maximum value of 50, many
of the other problems studied in this book are constructed so that $\|u_\varepsilon\| = 1$ for
all values of ε. The reason for this is that with this normalisation we do not need
to distinguish between the absolute and relative error in a numerical solution.

When the mesh is not necessarily uniform it is appropriate to measure the
size of the error in a numerical approximation in terms of an inverse power of
the discretization parameter N of the numerical method. Hence it is natural also
to measure the width of a boundary layer in terms of the parameter N. In view
of this we define the computational width of a boundary layer as follows: given a
boundary layer function $f(x)$, and a discrete measuring function $g(N)$ such that
$g(N) \to 0$ as $N \to \infty$, then the computational width $W(N)$ of the boundary
layer corresponding to the boundary layer function $f(x)$ relative to $g(N)$ is the
smallest value W for which we have

$$\sup_{x \geq W} |f(x)| \leq g(N).$$

For example, if the boundary layer function is $f(x) = e^{-x/\varepsilon}$ and the measuring function $g(N) = N^{-1}$, then the computational width of the corresponding boundary layer is $W = \varepsilon \ln N$, while if $g(N) = N^{-2}$ the computational width is $W = 2\varepsilon \ln N$. This concept of the computational width of a boundary layer is used later in motivating the construction of a non–uniform mesh fitted to a boundary layer, because it provides a measure of smallness in terms of the discretization parameter associated with the mesh.

When we apply a finite difference or a finite element method to a singular perturbation problem, with boundary layers in its solution, the question of the choice of the mesh $\overline{\Omega}^N$ immediately arises. For a numerical method to capture a boundary layer its numerical approximations must be influenced by its presence. Since we allow the use of only a simple interpolant in the case of a finite difference method, and only piecewise polynomial basis functions in a finite element method, we must place some of the mesh points in the interior of the boundary layer. This guarantees that, if the numerical method is nodally accurate, some information about the exact solution in the boundary layer is contained in the numerical solution. In one dimension the simplest of meshes is a uniform mesh $\overline{\Omega}_u^N = \{x_i\}_0^N$, with $N + 1$ equally spaced mesh points, that is $x_{i+1} - x_i = 1/N$ for all i. But with this mesh none of the mesh points is in the boundary layer, unless N is of order $1/\varepsilon$. Therefore, if ε is small, an unreasonably large number of mesh points is required, especially for problems in more than one dimension. It follows that a uniform mesh is usually not a good choice if the exact solution has a boundary layer.

We then have to decide on what kind of non–uniform mesh to use. It turns out that the simplest effective choice is a piecewise–uniform mesh which is fitted to the boundary layer. Such a mesh is composed of two uniform meshes: a fine mesh in the boundary layer and a coarse mesh outside the boundary layer. The location of the transition point between the fine and coarse mesh is a function of the singular perturbation parameter ε and the discretization parameter N. The correct distribution of the mesh points is to have an equal or comparable number of mesh points in the fine and coarse meshes. Likewise, for problems in more than one dimension, the natural and appropriate distribution of mesh points along the direction normal to a boundary layer is a comparable number of the available mesh points both inside and outside the boundary layer. We explain the fitting of meshes to boundary layers more precisely in later chapters.

1.4 Robust layer–resolving methods

It may not be too difficult to construct a numerical method for a given single value of the singular perturbation parameter ε, say $\varepsilon = 10^{-6}$, but the resulting numerical method may not be suitable for other values of the parameter, say $\varepsilon = 10^{-4}$ or $\varepsilon = 10^{-10}$. In this book we undertake the significantly more difficult task of constructing numerical methods that generate numerical solutions which converge uniformly for *all* values of the parameter ε in the range $(0, 1]$ and that require a parameter–uniform amount of computational work to compute each numerical solution. Such numerical methods are called parameter–uniform or ε–uniform methods and we use the terms parameter–uniform and ε–uniform interchangeably. If a method is ε–uniform the error between the exact solution u_ε and the numerical solution U_ε satisfies an estimate of the following form: for some positive integer N_0, all integers $N \geq N_0$ and all $\varepsilon \in (0, 1]$, we have $\|\overline{U}_\varepsilon - u_\varepsilon\|_{\overline{\Omega}} \leq CN^{-p}$, where C, N_0 and p are positive constants independent of ε and N. Here $\overline{U}_\varepsilon$ denotes the piecewise linear interpolant on the whole domain $\overline{\Omega}$ of the mesh function U_ε defined on the mesh $\overline{\Omega}^N$ and $\|\cdot\|_{\overline{\Omega}}$ denotes the maximum norm over the whole domain $\overline{\Omega}$. Note that we require that the estimate holds at each point of the domain, and not just at the mesh points. It follows that, at each point of $\overline{\Omega}$, $\overline{U}_\varepsilon$ is an ε–uniform continuous analytic approximation to the exact solution u_ε.

The consideration of appropriate simple examples soon convinces us that, in the construction of a robust layer–resolving method, it is essential in the above definition of an ε–uniform method to use the maximum norm with the maximum taken over every point of the solution domain. For example, it is easy to construct a finite element method (see the comment in §1.2), which is ε–uniform in the energy norm, but which fails to detect the presence of a boundary layer. Another simple example, discussed in §2.5, gives exact values at every mesh point of a uniform mesh, but still fails to detect the presence of a boundary layer. To overcome this difficulty, we further insist that the numerical solutions are defined at each point of the domain of the exact solution and that the maximum norm is taken over the whole of this domain. This does ensure that we capture boundary layers, no matter how thin they are, but the performance of numerical methods with all of these properties can still be erratic, as we show in chapters 4 and 9. Therefore, to ensure that our numerical methods not only capture boundary layers but are also robust, we impose one further requirement, namely that they are monotone methods.

In summary, the four key properties that we require of a robust–layer–resolving numerical method are that it is:
- global; that is, defined at each point of the domain of the exact solution
- pointwise–accurate; that is, the error is measured in the global maximum norm
- parameter–uniform; that is, the numerical solutions converge ε–uniformly and can be computed with an ε–uniform amount of computational effort
- monotone; that is, the discrete operator is a monotone operator.

We formalise the above in the following definition of a robust layer–resolving method. Where necessary the terminology is defined more precisely in later chapters.

Definition 1.1 *Let (P_ε) be a family of mathematical problems parameterized by a singular perturbation parameter ε, where ε satisfies $0 < \varepsilon \leq 1$. Assume that each problem in (P_ε) has a unique solution denoted by u_ε, and that each u_ε is approximated by a sequence of numerical solutions $\{(U_\varepsilon, \overline{\Omega}^N)\}_{N=1}^\infty$ obtained using a monotone numerical method (P_ε^N), where U_ε is defined on the mesh $\overline{\Omega}^N$ and N is a discretization parameter. Let $\overline{U}_\varepsilon$ denote the piecewise linear interpolant over $\overline{\Omega}$ of the discrete solution U_ε. Then (P_ε^N) is said to be a robust layer–resolving method if the numerical solutions are computable with an ε–uniform amount of computational work and converge ε–uniformly, in the sense that there exist a positive integer N_0, and positive numbers C and p, where N_0, C and p are independent of N and ε, such that for all $N \geq N_0$*

$$\sup_{0 < \varepsilon \leq 1} \|\overline{U}_\varepsilon - u_\varepsilon\|_{\overline{\Omega}} \leq C N^{-p}. \tag{1.5}$$

The error parameters p and C in the above definition are called respectively the ε–uniform order of convergence and the ε–uniform error constant of the numerical method (P_ε^N).

For the convection diffusion problem (1.1), it is easy to show that the kth order derivative of the solution for all $k \geq 0$ satisfies the bound

$$\left| \frac{d^k}{dx^k} u_\varepsilon(x) \right| \leq C(1 + \varepsilon^{-k} e^{-x/\varepsilon}) \text{ for all } x \in \overline{\Omega}$$

where C is independent of ε. In subsequent chapters we see that when ε–explicit bounds of this kind are known, then ε–uniform numerical methods can be analysed theoretically and error bounds of the form (1.5) can be established. In such cases therefore it is possible to establish theoretically that the method is a robust layer–resolving method.

1.5   Some notation

We introduce the following notation for finite difference operators, which is used throughout the book. Let V be any mesh function, then in one dimension, the forward, backward and centred finite difference operators D^+, D^- and D^0 are defined respectively by

$$D^+ V(x_i) \equiv \frac{V(x_{i+1}) - V(x_i)}{x_{i+1} - x_i}, \tag{1.6a}$$

$$D^- V(x_i) \equiv \frac{V(x_i) - V(x_{i-1})}{x_i - x_{i-1}}, \tag{1.6b}$$

$$D^0 V(x_i) \equiv \frac{V(x_{i+1}) - V(x_{i-1})}{x_{i+1} - x_{i-1}}, \tag{1.6c}$$

and the second order centred difference operator δ^2 is defined by

$$\delta^2 V(x_i) \equiv \frac{2(D^+ V(x_i) - D^- V(x_i))}{x_{i+1} - x_{i-1}}. \tag{1.6d}$$

The corresponding partial finite difference operators in two dimensions are

$$D_x^+ V(x_i, y_j) = \frac{V(x_{i+1}, y_j) - V(x_i, y_j)}{x_{i+1} - x_i}, \tag{1.6e}$$

$$D_x^- V(x_i, y_j) = \frac{V(x_i, y_j) - V(x_{i-1}, y_j)}{x_i - x_{i-1}}, \tag{1.6f}$$

$$D_x^0 V(x_i, y_j) = \frac{V(x_{i+1}, y_j) - V(x_{i-1}, y_j)}{x_{i+1} - x_{i-1}}, \tag{1.6g}$$

$$\delta_x^2 V(x_i, y_j) \equiv \frac{2(D_x^+ V(x_i, y_j) - D_x^- V(x_i, y_j))}{x_{i+1} - x_{i-1}} \tag{1.6h}$$

with analogous definitions of D_y^+, D_y^-, D_y^0 and δ_y^2.

Let u_ε be the solution of (P_ε) and U_ε the solution of (P_ε^N) then, on an arbitrary mesh $\overline{\Omega}^N = \{x_i\}_{i=0}^N$, the exact error $E_{\varepsilon,\text{exact}}^N$ at the mesh points is defined by

$$E_{\varepsilon,\text{exact}}^N = \|U_\varepsilon - u_\varepsilon\|_{\overline{\Omega}^N} \tag{1.7a}$$

and the corresponding ε–uniform exact error at the mesh points is defined by

$$E_{\text{exact}}^N = \max_\varepsilon E_{\varepsilon,\text{exact}}^N. \tag{1.7b}$$

Similarly, the global exact error $\bar{E}_{\varepsilon,\text{exact}}^N$ and the corresponding ε–uniform global exact error $\bar{E}_{\text{exact}}^N$ are defined respectively by

$$\bar{E}_{\varepsilon,\text{exact}}^N = \|\overline{U}_\varepsilon - u_\varepsilon\|_{\overline{\Omega}} \tag{1.7c}$$

and

$$\bar{E}_{\text{exact}}^N = \max_\varepsilon \bar{E}_{\varepsilon,\text{exact}}^N. \tag{1.7d}$$

We assume always that C, sometimes subscripted, is a generic constant independent of ε and N.

Numerical methods on uniform meshes

2.1 Convection–diffusion problems in one dimension

All of the finite difference methods in this chapter are defined on a uniform mesh. We use these methods to solve a variety of linear convection–diffusion test problems in one dimension. Because these test problems have constant data, we can find explicit solutions in simple closed form of both the original problem and of the discrete problem corresponding to the numerical method. This enables us to compute the error exactly and hence to examine both analytically and numerically the convergence properties of the numerical methods for these test problems. The main aim of this chapter is to show that we cannot hope to construct a layer–resolving method if we confine ourselves to a uniform mesh. The main conclusion to be drawn from the chapter is that uniform meshes are not appropriate for the construction of robust layer–resolving methods.

We introduce two classes of singularly perturbed linear convection–diffusion problems in one dimension on the unit interval $\Omega = (0, 1)$. The first class involves only Dirichlet boundary conditions.

Problem Class 2.1.   Linear convection–diffusion in one dimension with Dirichlet boundary conditions.

$$L_\varepsilon u_\varepsilon \equiv \varepsilon u_\varepsilon'' + a(x)u_\varepsilon' = f(x) \quad x \in \Omega, \tag{2.1a}$$
$$u_\varepsilon(0) = A, \quad u_\varepsilon(1) = B, \tag{2.1b}$$

where

$$a, f \in C^2(\Omega), \quad a(x) \geq \alpha > 0, \quad x \in \overline{\Omega}. \tag{2.1c}$$

A maximum or minimum principle is a useful tool for deriving *a priori* bounds on the solutions of the differential equations and their derivatives. The reader is referred to Protter and Weinberger (1984) for a comprehensive discussion of these comparison principles. The differential operator L_ε in problems from Problem Class 2.1 satisfies the following minimum principle.

Theorem 2.1 *Let L_ε be the differential operator in (2.1a) and $v \in C^2(\overline{\Omega})$. If $v(0) \geq 0$, $v(1) \geq 0$ and $L_\varepsilon v(x) \leq 0$ for all $x \in \Omega$, then $v(x) \geq 0$ for all $x \in \overline{\Omega}$.*

Proof The proof is by contradiction. Assume that there exists a point $p \in \overline{\Omega}$ such that $v(p) < 0$. It follows from the hypotheses that $p \notin \{0,1\}$. Define the auxiliary function $w = v e^{\alpha x/(2\varepsilon)}$ and note that $w(p) < 0$. Choose $q \in \Omega$ such that $w(q) = \min_\Omega w(x) < 0$. Therefore, from the definition of q, we have $w'(q) = 0$ and $w''(q) \geq 0$. But then

$$L_\varepsilon v(q) = (\varepsilon w''(q) + (a(q) - \alpha)w'(q) - \frac{\alpha}{2\varepsilon}(a(q) - \alpha/2)w(q))e^{-q\alpha/(2\varepsilon)} > 0$$

which is a contradiction. $\square$

We now consider a typical problem from Problem Class 2.1 with constant data $a = \alpha$ and f. In this case the exact solution of problem (2.1a-b) is

$$u_\varepsilon(x) = u_\varepsilon(0) + x\frac{f}{\alpha} - (u_\varepsilon(0) - u_\varepsilon(1) + \frac{f}{\alpha})\left(\frac{1 - e^{-\alpha x/\varepsilon}}{1 - e^{-\alpha/\varepsilon}}\right).$$

It follows that for $x \neq 0$

$$\lim_{\varepsilon \to 0} u_\varepsilon(x) = (x - 1)\frac{f}{\alpha} + u_\varepsilon(1),$$

which is the solution of the reduced problem

$$\alpha v_0'(x) = f, \ x \in \Omega, \quad \text{and} \quad v_0(1) = u_\varepsilon(1).$$

Note that, in general, $v_0(0) \neq u_\varepsilon(0)$, for $\varepsilon > 0$ and so a boundary layer occurs at the boundary point $x = 0$. Differentiating the above representation of u_ε, we see that u_ε and its derivatives, for all integers $k \geq 0$, satisfy the bounds

$$|u_\varepsilon^{(k)}(x)| \leq C(1 + \varepsilon^{-k}e^{-\alpha x/\varepsilon}), \quad \text{for all} \quad x \in \overline{\Omega},$$

where C is a constant independent of ε. Moreover, when a and f depend on x and are sufficiently smooth, we show in the next chapter that the derivatives up to order k, depending on the smoothness of the data, satisfy similar bounds. From these it follows that, outside the small open neighbourhood $(0, k\alpha^{-1}\varepsilon\ln(1/\varepsilon))$ of the boundary point $x = 0$, in other words outside the boundary layer, the solution and its derivatives are ε–uniformly bounded in the sense that, for all k,

$$\sup_{x \geq k\alpha^{-1}\varepsilon\ln(1/\varepsilon)} |u_\varepsilon^{(k)}(x)| \leq C,$$

where C is independent of ε. This may be seen by observing that for all $x \geq k\alpha^{-1}\varepsilon\ln(1/\varepsilon)$

$$\varepsilon^{-k}e^{-\alpha x/\varepsilon} \leq \varepsilon^{-k}e^{-\ln(1/\varepsilon)^k} = 1.$$

On the other hand, for $x \leq \varepsilon$, the derivatives grow without bound as the parameter ε tends to zero. In particular, at the boundary point $x = 0$ itself, we see that

$$|u_\varepsilon^{(k)}(0+)| \leq C\varepsilon^{-k}.$$

The second class of singularly perturbed convection–diffusion problems is similar to the first, except that a Neumann boundary condition is given at the left–hand boundary point.

Problem Class 2.2.   Linear convection–diffusion in one dimension with Dirichlet and Neumann boundary conditions.

$$L_\varepsilon u_\varepsilon \equiv \varepsilon u_\varepsilon'' + a(x)u_\varepsilon' = f(x), \quad x \in \Omega, \tag{2.2a}$$
$$\varepsilon u_\varepsilon'(0) = A, \quad u_\varepsilon(1) = B, \tag{2.2b}$$

where

$$a, f \in C^2(\Omega), \quad a(x) \geq \alpha > 0, \quad x \in \overline{\Omega}. \tag{2.2c}$$

The differential operator L_ε in problems from Problem Class 2.2 also satisfies a minimum principle, which we state and prove in the next chapter. With constant data $a = \alpha$ and f, the exact solution of problem (2.2a-b) is

$$u_\varepsilon(x) = (x - 1)\frac{f}{\alpha} + \frac{1}{\alpha}\left(\varepsilon u_\varepsilon'(0) - \varepsilon\frac{f}{\alpha}\right)\left(e^{-\alpha/\varepsilon} - e^{-\alpha x/\varepsilon}\right) + u_\varepsilon(1)$$

and it is clear that u_ε and its derivatives satisfy the same bounds as those of problem (2.1a-b). That is, for all integers $k \geq 0$,

$$|u_\varepsilon^{(k)}(x)| \leq C(1 + \varepsilon^{-k}e^{-\alpha x/\varepsilon}), \quad \text{for all} \quad x \in \overline{\Omega},$$

where C is a constant independent of ε. Thus the solutions of problems from the Problem Classes 2.1 and 2.2 have similar boundary layers. However, despite the similar structure of the solutions it turns out that the performance of some standard numerical methods can differ dramatically when they are used to solve problems from these two classes. This is illustrated in subsequent sections of this chapter by solving two simple convection–diffusion problems, one from each

class, using these methods. The first such problem is a member of Problem Class
2.1

$$\varepsilon u_\varepsilon'' + 2u_\varepsilon' = 0 \quad x \in \Omega, \tag{2.3a}$$
$$u_\varepsilon(0) = 1, \quad u_\varepsilon(1) = 0 \tag{2.3b}$$

with the exact solution

$$u_\varepsilon(x) = \frac{e^{-2x/\varepsilon} - e^{-2/\varepsilon}}{1 - e^{-2/\varepsilon}}. \tag{2.3c}$$

The second problem is from Problem Class 2.2

$$\varepsilon u_\varepsilon'' + 2u_\varepsilon' = 0 \quad x \in \Omega, \tag{2.4a}$$
$$\varepsilon u_\varepsilon'(0) = -2, \quad u_\varepsilon(1) = 0 \tag{2.4b}$$

and its exact solution is

$$u_\varepsilon(x) = e^{-2x/\varepsilon} - e^{-2/\varepsilon}. \tag{2.4c}$$

We see from these expressions that the solutions of problems (2.3) and (2.4) are
monotone decreasing as x increases and that in both cases $0 \le u_\varepsilon(x) \le 1$ for all
$x \in \overline{\Omega}$. Graphs of these exact solutions appear in Fig. 2.1 and Fig. 2.4.

We remark in passing that the similar discussions to those for Problem Classes
2.1 and 2.2 in this, and the next two chapters, are valid also for the slightly larger
problem classes corresponding to the more general differential equation

$$\varepsilon u_\varepsilon'' + a(x)u_\varepsilon' - b(x)u_\varepsilon(x) = f(x), \quad x \in \Omega,$$

where $b \in C^2(\Omega)$ and $b(x) \ge \beta \ge 0$.

2.2 Centred finite difference method

In this section we consider the classical centred finite difference method on a
uniform mesh for problems in Problem Class 2.1 by discretizing the differential
operator L_ε.

**Method 2.1. Centred finite difference operator on uniform mesh for
Problem Class 2.1.**

$$\varepsilon \delta^2 U_\varepsilon(x_i) + a(x_i)D^0 U_\varepsilon(x_i) = f(x_i), \quad x_i \in \Omega_u^N, \tag{2.5a}$$
$$U_\varepsilon(0) = u_\varepsilon(0), \quad U_\varepsilon(1) = u_\varepsilon(1), \tag{2.5b}$$

where $\overline{\Omega}_u^N$ *is the uniform mesh*

$$\overline{\Omega}_u^N = \{x_i | x_i = i/N, i = 0, \dots, N\}. \tag{2.5c}$$

We use this method to solve the specific problem (2.3) corresponding to the
two values $\varepsilon = 10^{-2}$ and $\varepsilon = 10^{-5}$. In Fig. 2.1 we compare the exact solutions

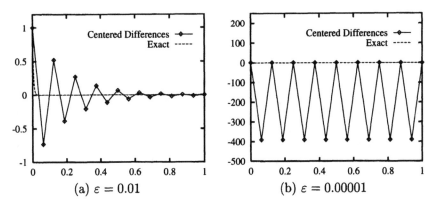

FIG. 2.1. *Comparison of the exact solution of problem* (2.3), *for* $\varepsilon = 0.01$ *and* 0.00001, *and the numerical solution generated by Method 2.1 with* $N = 16$.

with the resulting numerical approximations for $N = 16$. We see that the numerical solutions have large non–physical oscillations, which increase unboundedly as ε decreases. In Fig. 2.1(a) the numerical solution oscillates between positive and negative values. This is in contrast to the behaviour of the exact solution, which decreases monotonically from 1 to 0. Taking account of the coarse vertical scale in Fig. 2.1(b), we see that the exact solution is negligible compared with the numerical solution, which has extremely large oscillations. It is also obvious from Fig. 2.1, that the numerical oscillations increase in magnitude as ε decreases. We conclude from these numerical experiments that Method 2.1 is not satisfactory for Problem Class 2.1.

We now study analytically the development of the oscillations in the numerical solutions using an explicit expression for these solutions. It is not hard to verify that the solution of the discrete problem, obtained when Method 2.1 is applied to problem (2.3), can be written in the form

$$U_\varepsilon(x_i) = \frac{\lambda^i - \lambda^N}{1 - \lambda^N}, \quad \text{where} \quad \lambda = \frac{1 - 1/\varepsilon N}{1 + 1/\varepsilon N}. \tag{2.6}$$

From this it is clear that for $\varepsilon N < 1$, that is for ε sufficiently small compared to $1/N$, we have $\lambda < 0$ and so λ^i is positive or negative depending on whether i is even or odd. Since, for each fixed N, $\lim_{\varepsilon \to 0} \lambda = -1$ we have for N odd

$$\lim_{\varepsilon \to 0} U_\varepsilon(x_i) = \frac{(-1)^i + 1}{2}$$

and so, in this case, U_ε oscillates in a bounded fashion for all ε satisfying $\varepsilon N < 1$. On the other hand, for N even we have

$$\lim_{\varepsilon \to 0} U_\varepsilon(x_i) = \begin{cases} 1 - i/N, & \text{if } i \text{ is even,} \\ -\infty, & \text{otherwise,} \end{cases}$$

which shows that now the oscillations are unbounded as ε decreases. Note that, if the boundary conditions of the problem are reversed, that is if $u_\varepsilon(0) = 0$ and $u_\varepsilon(1) = 1$, and N is even, then the numerical solution oscillates without an upper bound between the values i/N and $+\infty$.

The presence of these large non–physical oscillations in the numerical solutions underlies the well–known fact that centred finite difference operators on uniform meshes are unsuitable for the numerical solution of singular perturbation problems, since they give unacceptable numerical approximations even for the simple one–dimensional problems of Problem Class 2.1.

It is interesting to observe from (2.3c) that the exact solution of problem (2.3) at the mesh point x_i can be written in the form

$$u_\varepsilon(x_i) = \frac{r^i - r^N}{1 - r^N}, \quad \text{where} \quad r = e^{-2/\varepsilon N}$$

which is similar to the expression for $U_\varepsilon(x_i)$ in (2.6). Writing $t = 1/\varepsilon N$ in the expression for λ in (2.6) we obtain $\lambda(t) = \frac{1-t}{1+t}$, which is a Padé approximation to $r(t) = e^{-2t}$. It is well known that the Padé approximation $\lambda(t)$ is a reasonable approximation to $r(t)$ for t close to zero, but for large t we have

$$\lim_{t \to \infty} \left(\lambda(t) - r(t) \right) = -1$$

which indicates again that we can expect difficulties to arise when $\varepsilon \to 0$.

Remark We can obtain a standard finite element method for problems from Problem Class 2.1 by first formulating the following weak form of problem (2.1):

$$\text{Find } u_\varepsilon \in H^1_E(\Omega) \quad \text{such that}$$
$$-(\varepsilon u_\varepsilon', v') + (au_\varepsilon', v) = (f, v), \quad \text{for all} \quad v \in H^1_0(\Omega).$$

Here $(v, w) = \int_0^1 v(x)w(x)\,dx$ is the usual L_2 inner product, $H^1_E(\Omega) = \{v \in H^1(\Omega)|\ v(0) = u_\varepsilon(0), v(1) = u_\varepsilon(1)\}$, and $H^1_0(\Omega) = \{v \in H^1(\Omega)|\ v(0) = v(1) = 0\}$. We then discretize this weak form by choosing a finite dimensional subspace $S^N \subset H^1_E(\Omega)$ which is spanned by a finite set of basis functions $\{\phi_i(x)\}_{i=0}^N$. Typical basis functions are the piecewise linear functions defined by

$$\phi_i(x_j) = \delta_{i,j}, \quad i = 0, \dots, N,$$

where $\bar{\Omega}^N = \{x_j\}_{j=0}^N$ is a finite set of nodes in the domain $\bar{\Omega}$ with $x_0 = 0$ and $x_N = 1$. A finite element approximation $\bar{U}_\varepsilon$ to the exact solution u_ε of the form

$$\bar{U}_\varepsilon(x) = u_\varepsilon(0)\phi_0(x) + \sum_{i=1}^{N-1} U_\varepsilon(x_i)\phi_i(x) + u_\varepsilon(1)\phi_N(x)$$

is obtained by solving the following linear system of $N - 1$ algebraic equations for the $N - 1$ interior nodal values $\{U_\varepsilon(x_i)\}_1^{N-1}$

$$-(\varepsilon \bar{U}'_\varepsilon, \phi'_i) + (a\bar{U}'_\varepsilon, \phi_i) = (f, \phi_i) \quad \text{for all} \quad i = 1, ...N - 1.$$

In the special case of a uniform mesh $\bar{\Omega}^N = \bar{\Omega}_u^N$ and the quadrature rules

$$(a\bar{U}'_\varepsilon, \phi_i) \approx a(x_i)(\bar{U}'_\varepsilon, \phi_i) \quad \text{and} \quad (f, \phi_i) \approx f(x_i)(1, \phi_i)$$

it is not hard to see that the resulting linear system is the same as that for Method 2.1. This shows that this standard finite element method on a uniform mesh also gives unsatisfactory numerical solutions for problems from Problem Class 2.1. It is natural to inquire about the behaviour of the solutions of this finite element method on a more general non–uniform mesh. In this case the numerical solutions are like those obtained from centred finite difference methods on non–uniform meshes, which are discussed in detail in chapter 4.

2.3 Monotone matrices and discrete comparison principles

To understand the source of the numerical oscillations described in the previous section it is helpful to consider the system matrix associated with a numerical method. We identify certain properties of such a matrix, which guarantee that numerical oscillations do not occur in the corresponding numerical solution. In what follows the notation $\mathbf{x} \geq \mathbf{0}$ means that each component of the vector $\mathbf{x}$ is non–negative and similarly $\mathbf{A} \geq \mathbf{0}$ means that each element of the matrix $\mathbf{A}$ is non–negative.

Definition 2.2 *An $n \times n$ real matrix $\mathbf{A}$ is monotone if $\mathbf{Ax} \geq \mathbf{0}$ implies that $\mathbf{x} \geq \mathbf{0}$.*

It is well known that a matrix $\mathbf{A}$ is monotone if and only if $\mathbf{A}$ is nonsingular and $\mathbf{A}^{-1} \geq \mathbf{0}$. In practice, a useful way of proving that a particular matrix is not monotone is to find a vector $\mathbf{y}$ such that $\mathbf{Ay} \geq \mathbf{0}$ and $y_i < 0$ for some i, $1 \leq i \leq n$. Note however that, even if the matrix $\mathbf{A}$ is not monotone, the matrix $-\mathbf{A}$ may be monotone. To take account of this possibility, for a particular matrix $\mathbf{A}$, the following easily verified statement is useful: suppose that there exists a vector $\mathbf{y}$ such that $\mathbf{Ay} \geq \mathbf{0}$ and $y_i y_j < 0$ for some $i \neq j$, $1 \leq i, j \leq n$, then both $\mathbf{A}$ and $-\mathbf{A}$ are not monotone.

Definition 2.3 *An $n \times n$ real matrix $\mathbf{A}$ is an M-matrix if $\mathbf{A}$ is nonsingular, $\mathbf{A}^{-1} \geq \mathbf{0}$ and $a_{i,j} \leq 0$, for all $i \neq j$, $1 \leq i, j \leq n$.*

Clearly the set of M-matrices is a subset of the set of monotone matrices. It is obvious that the hypotheses of the following two lemmas form a set of sufficient conditions for an $n \times n$ matrix to be an M-matrix, and hence monotone.

Lemma 2.4 *A matrix is irreducible if and only if its directed graph is strongly connected.*

Lemma 2.5 *Suppose that the $n \times n$ real matrix $\mathbf{A}$ is irreducibly diagonally dominant and $a_{i,j} \leq 0$, $i \neq j$, $a_{i,i} > 0$, for all $1 \leq i, j \leq n$. Then $\mathbf{A}$ is nonsingular and $\mathbf{A}^{-1} > 0$.*

See Varga (1962, Theorems 1.6 and 3.11) for a proof of these results.

Any finite difference method of the form

$$L^N V(x_i) = \sum_{j=0}^{N} a_{i,j} V(x_j) = f(x_i), \quad x_i \in \Omega^N, \quad V(x_0), V(x_N) \text{ given} \quad (2.7)$$

where $\overline{\Omega}^N = \{x_i\}_{i=0}^{N}$ is an arbitrary mesh, can be written as a linear system

$$\mathbf{A}^N \mathbf{v} = \mathbf{f}$$

where $\mathbf{A}^N = (a_{i,j})$ is the $(N-1) \times (N-1)$ system matrix associated with this finite difference method, $v_i = V(x_i)$ and $f_i = f(x_i) - a_{i,0}V(x_0) - a_{i,N}V(x_N)$. The system matrix associated with a finite difference method obtained from the discretization of a differential equation is often irreducibly diagonally dominant. In such circumstances, to ascertain whether or not this system matrix is an M-matrix, we see from Lemma 2.5 that it suffices to examine the signs of the diagonal and off–diagonal elements.

We illustrate the use of this approach by applying it to the centred finite difference Method 2.1, which has the tridiagonal system matrix $\mathbf{A}^N$ with

$$a_{i,i-1} = \frac{\varepsilon}{h^2}(1 - \rho_i/2), \quad a_{i,i} = \frac{-2\varepsilon}{h^2}, \quad a_{i,i+1} = \frac{\varepsilon}{h^2}(1 + \rho_i/2),$$

where $\rho_i = a(x_i)h/\varepsilon$, $h = 1/N$, $f_i = f(x_i)$ for $1 < i < N - 1$ and

$$f_1 = f(x_1) - \frac{\varepsilon}{h^2}(1 - \rho_1/2)u_\varepsilon(0); \qquad f_{N-1} = f(x_{N-1}) - \frac{\varepsilon}{h^2}(1 + \rho_{N-1}/2)u_\varepsilon(1).$$

If ε is sufficiently small then $\rho_i > 2$, and so the lower and upper diagonal elements are of different sign. Hence this matrix is not an M-matrix for all ε sufficiently small. In fact, we can show that the matrix $\mathbf{A}^N$ is not monotone when Method 2.1 is applied to the specific problem (2.3). Choosing $u_i = U_\varepsilon(x_i)$ for $i = 1, ..., N-1$, we have from (2.6), for ε sufficiently small and N even,

$$u_1 u_2 = U_\varepsilon(x_1)U_\varepsilon(x_2) < 0.$$

and so

$$\mathbf{A}^N \mathbf{u} = \mathbf{f} \geq \mathbf{0}, \quad u_1 u_2 < 0,$$

which shows that both $\mathbf{A}^N$ and $-\mathbf{A}^N$ are not monotone matrices for the specific problem (2.3).

In practice, we often want to construct a numerical method whose system matrix is an M-matrix. We see in the next section that this requirement leads naturally to upwind numerical methods. These methods are characterized by the property that the approximation to the first derivative involves points from only the upwind direction with respect to the flow.

To end this theoretical discussion we introduce the concept of a monotone finite difference method and of a discrete minimum principle, which are useful in establishing theoretical error estimates for finite difference methods.

Definition 2.6 *A finite difference method with the associated system matrix $\mathbf{A}^N$ is monotone if either $\mathbf{A}^N$ or $-\mathbf{A}^N$ is a monotone matrix.*

Definition 2.7 *The finite difference operator L^N in (2.7) satisfies a discrete minimum principle, if, for any mesh function V, the inequalities*

$$\min\{V(x_0), V(x_N)\} \geq 0 \quad \text{and} \quad L^N V(x_i) \leq 0 \quad \text{for all} \quad x_i \in \Omega^N,$$

imply that

$$V(x_i) \geq 0 \text{ for all } x_i \in \overline{\Omega}^N.$$

Note that if L^N satisfies a discrete minimum principle, then it automatically satisfies the following discrete maximum principle: for any mesh function V the inequalities

$$\max\{V(x_0), V(x_N)\} \leq 0 \quad \text{and} \quad L^N V(x_i) \geq 0 \quad \text{for all} \quad x_i \in \Omega^N,$$

imply that $V(x_i) \leq 0$ for all $x_i \in \overline{\Omega}^N$. This can be seen by applying the discrete minimum principle to the mesh function $-V$.

It is obvious from the above definitions that, if $\mathbf{A}^N$ is the system matrix associated with a finite difference method, and $-\mathbf{A}^N$ is a monotone matrix, then its finite difference operator automatically satisfies a discrete minimum principle. A discrete minimum (or maximum) principle is a standard analytical device for deriving theoretical error estimates from local truncation error estimates. Without the existence of such a discrete minimum (or maximum) principle, the theoretical analysis of the pointwise error of a numerical method is significantly more difficult, especially for problems in more than one dimension.

2.4   Upwind finite difference methods

In this section we consider an upwind finite difference method on a uniform mesh, where the first derivative term at a given mesh point is approximated by a discrete derivative, which uses mesh points only in the upwind direction from that mesh point.

Method 2.2. Upwind finite difference operator on uniform mesh for Problem Class 2.1.

$$L_\varepsilon^N U_\varepsilon \equiv \varepsilon \delta^2 U_\varepsilon + a(x_i) D^+ U_\varepsilon = f(x_i), \quad x_i \in \Omega_u^N, \tag{2.8a}$$
$$U_\varepsilon(0) = u_\varepsilon(0), \quad U_\varepsilon(1) = u_\varepsilon(1), \tag{2.8b}$$

where $\overline{\Omega}_u^N$ *is the uniform mesh*

$$\overline{\Omega}_u^N = \{x_i = i/N, i = 0, \dots, N\}. \tag{2.8c}$$

The non–zero entries of the system matrix $\mathbf{A}^N$ associated with this finite difference method are

$$a_{i,i-1} = \frac{\varepsilon}{h^2}, \quad a_{i,i} = -\frac{\varepsilon}{h^2}(2 + \rho_i), \quad a_{i,i+1} = \frac{\varepsilon}{h^2}(1 + \rho_i)$$

where $\rho_i = a(x_i)h/\varepsilon$ as before. It is clear that $\mathbf{A}^N$ is irreducibly diagonally dominant and also that, irrespective of the value of ε, the sign pattern of a typical row of this tridiagonal matrix is $(+, -, +)$. This shows that $-\mathbf{A}^N$ is an M-matrix, and so the finite difference operator L_ε^N in (2.8a) satisfies a discrete minimum principle. It follows that, for any particular problem from Problem Class 2.1 for which all the elements of $\mathbf{f}$ are of one sign, no oscillations between positive and negative values occur in the numerical solution. This is illustrated in Fig. 2.2, where a graph of the discrete solutions with $N = 16$ is given for two values of ε.

A discrete minimum principle can also be established directly, without appealing to the properties of the elements in the system matrix. Analogously to the continuous case, this is usually proved by contradiction as we now show.

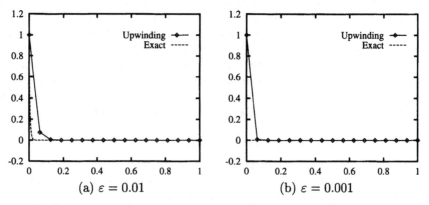

FIG. 2.2. Comparison of the exact solution of problem (2.3) for $\varepsilon = 0.01$ and 0.001, and the numerical solution generated by Method 2.2 with $N = 16$.

Theorem 2.8 *The upwind finite difference operator L_ε^N in (2.8a) satisfies the discrete minimum principle in definition (2.7).*

Proof We suppose that $V(x_k) = \min_i V(x_i) < 0$. Then, from the hypotheses, it is clear that $k \neq 0, N$. Hence the mesh function takes its minimum negative value at some internal mesh point x_k. Since $V(x_k)$ is the minimum value, $D^+V(x_k) \geq 0$ and $\delta^2 V(x_k) \geq 0$, and so to avoid a contradiction we must have $V(x_k) = V(x_{k-1}) = V(x_{k+1}) < 0$. Repeating this argument we eventually conclude that $V(x_0) < 0$, which is a contradiction.   □

An immediate consequence of this result is that the discrete solutions generated by Method 2.2 satisfy the following stability condition.

Lemma 2.9 *The solutions U_ε obtained when Method 2.2 is applied to any problem from Problem Class 2.1 satisfy*

$$|U_\varepsilon(x_i)| \leq |U_\varepsilon(1)| + |U_\varepsilon(0)| + \frac{1}{\alpha}\|f\|.$$

Proof We introduce the two mesh functions

$$\Psi^\pm(x_i) = |U_\varepsilon(0)| + |U_\varepsilon(1)| + \frac{1}{\alpha}\|f\|(1 - x_i) \pm U_\varepsilon(x_i).$$

Note that $\Psi^\pm(x_0), \Psi^\pm(x_N) \geq 0$ and $L_\varepsilon^N \Psi^\pm(x_i) \leq 0$. The proof is then completed in the usual way using the discrete minimum principle given in Theorem 2.8.   □

Spurious numerical oscillations do not occur when the upwind Method 2.2 is applied to any problem with constant data from Problem Class 2.1. We verify this now for the specific problem (2.3) by examining an explicit form of the solution. The solution of the discrete problem obtained when Method 2.2 is applied to problem (2.3) has a similar form to that of the centred finite difference method, namely $U_\varepsilon(x_i) = \frac{\lambda^i - \lambda^N}{1 - \lambda^N}$, where now $\lambda = \frac{1}{1 + 2/\varepsilon N}$. It is clear that $0 < \lambda < 1$, and that $U_\varepsilon(x_i)$ is monotone decreasing with increasing x_i, which implies that no oscillations occur. However, despite the absence of numerical oscillations, this method is not satisfactory, in the sense that it is not an ε–uniform method. To see this, we consider the error at the first mesh point $x_1 = 1/N$. This is given by

$$U_\varepsilon(x_1) - u_\varepsilon(x_1) = \frac{\lambda - \lambda^N}{1 - \lambda^N} - \frac{r - r^N}{1 - r^N}, \quad \text{where } r = e^{-2/N\varepsilon}.$$

Taking $\varepsilon N = 1$ and letting $N \to \infty$, we see that this error tends to the non–zero quantity

$$1/3 - e^{-2} = 0.197998,$$

which proves that Method 2.2 is not ε–uniform for problem (2.3) and that the maximum pointwise error is about 20% no matter how large N is.

We now examine more systematically the behaviour of the errors obtained from further numerical experiments. It is obvious that, in general, it is impossible to compute $E_{\varepsilon,\text{exact}}^N$ and E_{exact}^N in (1.7), because an explicit expression for the exact solution u_ε is often not known. However, for the simple one–dimensional problems (2.3) and (2.4) a simple closed form of u_ε is available and so we can find the quantities $E_{\varepsilon,\text{exact}}^N$ and E_{exact}^N in these two cases.

In this book, we compute the numerical solutions typically for the values $\varepsilon = 2^{-i}, i = 1, \ldots, 18$ and values of N between 8 and 512, and we then tabulate the corresponding exact errors $E_{\varepsilon,\text{exact}}^N$ and E_{exact}^N for these values of ε and N. Since we are primarily interested in the ε–uniform error E_{exact}^N, for the sake of economy we sometimes reduce the range of ε given in the tables, if we observe that the error $E_{\varepsilon,\text{exact}}^N$ stabilizes, with respect to ε, for all N between 8 and 512. This stabilization occurs when, to machine accuracy, we are solving effectively the reduced differential equation, that is the equation obtained by setting $\varepsilon = 0$.

We remark that, if we want to show that a method is not globally ε–uniform, in the first instance we examine whether or not it is ε–uniform at the mesh points. If it turns out that it is ε–uniform at the mesh points, we then consider its behaviour at points of the domain between the mesh points.

For Method 2.2 applied to problem (2.3) the exact maximum pointwise errors $E_{\varepsilon,\text{exact}}^N$ and E_{exact}^N are given in Table 2.1. We see that the maximum value of the error in each column of Table 2.1 is not less that 0.197897 and it is clear that the method is not ε–uniform at the mesh points for this problem. In other words, for any value of N, there is always some value of ε for which the error is at least 0.197897. These values lie along the diagonal corresponding to

Table 2.1 *Maximum pointwise errors $E_{\varepsilon,\text{exact}}^N$ and E_{exact}^N generated by Method 2.2 applied to problem (2.3) for various values of ε and N.*

	Number of intervals N						
ε	8	16	32	64	128	256	512
2^{-1}	0.065802	0.036415	0.019245	0.009930	0.005042	0.002541	0.001275
2^{-2}	0.130372	0.075930	0.041465	0.021752	0.011153	0.005649	0.002843
2^{-3}	**0.197897**	0.132113	0.076564	0.041720	0.021865	0.011206	0.005674
2^{-4}	0.181682	**0.197998**	0.132121	0.076565	0.041721	0.021865	0.011206
2^{-5}	0.110776	0.181684	**0.197998**	0.132121	0.076565	0.041721	0.021865
2^{-6}	0.058823	0.110776	0.181684	**0.197998**	0.132121	0.076565	0.041721
2^{-7}	0.030303	0.058823	0.110776	0.181684	**0.197998**	0.132121	0.076565
2^{-8}	0.015385	0.030303	0.058823	0.110776	0.181684	**0.197998**	0.132121
2^{-9}	0.007752	0.015385	0.030303	0.058823	0.110776	0.181684	**0.197998**
2^{-10}	0.003891	0.007752	0.015385	0.030303	0.058823	0.110776	0.181684
2^{-12}	0.000976	0.001949	0.003891	0.007752	0.015385	0.030303	0.058823
2^{-14}	0.000244	0.000488	0.000976	0.001949	0.003891	0.007752	0.015385
2^{-16}	0.000061	0.000122	0.000244	0.000488	0.000976	0.001949	0.003891
2^{-18}	0.000015	0.000031	0.000061	0.000122	0.000244	0.000488	0.000976
E_{exact}^N	0.197897	0.197998	0.197998	0.197998	0.197998	0.197998	0.197998

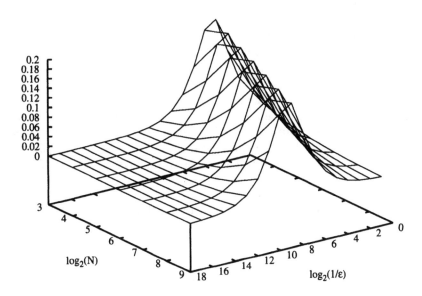

FIG. 2.3. *Maximum pointwise error* $E^N_{\varepsilon,\text{exact}}$ *as a function of* N *and* ε *generated by Method 2.2 applied to problem (2.3).*

$\varepsilon N = 1$, and thus they form a ridge of persistent errors of approximately constant magnitude along this diagonal line, no matter how large N is. A graph of $E^N_{\varepsilon,\text{exact}}$ is shown in Fig. 2.3 and the ridge of persistent error along the diagonal $\varepsilon N = 1$ is immediately apparent. Since the maximum value of the exact solution is unity, it follows that the numerical solution has an ε–uniform error of roughly 20% no matter how large N is.

It is important to observe that, if a numerical approximation is computed on a coarse mesh with $\varepsilon N < 1$, that is to the left of the ridge, which is the usual situation when the singular perturbation parameter ε is small, then the error *increases* as the mesh is refined. This shows that doubling the computational work causes a doubling of the error! Of course, for each fixed ε, in principle, it is always possible to choose N sufficiently large so that $\varepsilon N > 1$. However, such a choice of N is impractical for real problems, especially in more than one dimension. For example, if $\varepsilon = 10^{-3}$ and N is the number of mesh points in each coordinate direction, then in order to have $\varepsilon N > 1$ we need to take more than 10^6 mesh points in two dimensions and 10^9 mesh points in three dimensions, while if $\varepsilon = 10^{-9}$ the corresponding numbers are 10^{18} in two dimensions and

10^{27} in three dimensions. This reinforces our conclusion that Method 2.2 is not a practical method for Problem Class 2.1.

2.5 Fitted operator methods for Problem Class 2.1

We have seen that the numerical methods considered above, using centred and upwind finite difference operators on a uniform mesh, are not layer-resolving for problems involving boundary layers. We now investigate whether a method with a fitted finite difference operator on a uniform mesh gives us better numerical solutions. A finite difference operator is said to be fitted if it is designed to reflect, in some sense, the exact solution or the boundary layer functions. The first such method was proposed by Allen and Southwell (1955) in the engineering literature and the convergence of its numerical solutions was analysed by Il'in (1969) in a one–dimensional linear case.

Method 2.3.   Allen–Southwell–Il'in finite difference operator on uniform mesh for Problem Class 2.1.

$$\varepsilon \sigma_i \delta^2 U_\varepsilon + a(x_i) D^0 U_\varepsilon = f(x_i) \quad x_i \in \Omega_u^N, \tag{2.9a}$$

$$U_\varepsilon(0) = u_\varepsilon(0), \quad U_\varepsilon(1) = u_\varepsilon(1), \tag{2.9b}$$

where the fitting factor

$$\sigma_i = (\rho_i/2) \coth(\rho_i/2), \quad \rho_i = a(x_i)/(N\varepsilon), \tag{2.9c}$$

on the uniform mesh

$$\overline{\Omega}_u^N = \{x_i | x_i = i/N, i = 0, \dots, N, h = 1/N\}. \tag{2.9d}$$

The non–zero elements of the system matrix $\mathbf{A}^N$ associated with this method are

$$a_{i,i-1} = \frac{\varepsilon}{h^2}(\sigma_i - \rho_i/2) = \frac{\varepsilon}{h^2}\Big(\frac{\rho_i}{e^{\rho_i} - 1}\Big) > 0, \quad a_{i,i} = -a_{i,i-1} - a_{i,i+1},$$

$$a_{i,i+1} = \frac{\varepsilon}{h^2}(\sigma_i + \rho_i/2) = \frac{\varepsilon}{h^2}\Big(\frac{\rho_i}{1 - e^{-\rho_i}}\Big) > 0.$$

It is easy to see that the tridiagonal matrix $-\mathbf{A}^N$ is an M-matrix and hence Method 2.3 is a monotone method.

When $\varepsilon = O(1)$, we have $\rho_i \to 0$ and $\sigma_i \to 1$ as $N \to \infty$ which shows that, in this case, Method 2.3 is essentially the same as Method 2.1. On the other hand, when $\varepsilon \to 0$ we have $\rho_i \to \infty$, and so Method 2.3 now reduces to

$$a(x_i) D^+ U_0 = f(x_i), \quad x_i \in \Omega_u^N, \quad U_0(1) = u_\varepsilon(1)$$

which is an appropriate finite difference method for the corresponding reduced problem.

The above choice of the fitting factor σ_i may be motivated in the following way. We apply a fitted three-point finite difference operator to the homogeneous continuous problem with constant coefficients, and then choose the fitting factor σ so that the exact solution of the continuous problem is also a solution of the corresponding discrete problem. In the above case the discrete problem is

$$\varepsilon\sigma\delta^2 U_\varepsilon + aD^0 U_\varepsilon = 0 \quad x_i \in \Omega_u^N, \quad U_\varepsilon(0) = 1, \ U_\varepsilon(1) = 0,$$

the solution of which is

$$U_\varepsilon(x_i) = \frac{\lambda^i - \lambda^N}{1 - \lambda^N}, \quad \text{where } \lambda = \frac{2\sigma - \rho}{2\sigma + \rho}, \quad \text{and } \rho = a/\varepsilon N.$$

The exact solution of the continuous problem

$$\varepsilon u_\varepsilon'' + a u_\varepsilon' = 0 \quad x \in \Omega, \quad u_\varepsilon(0) = 1, \ u_\varepsilon(1) = 0$$

at the mesh points $x_i \in \Omega_u^N$ is

$$u_\varepsilon(x_i) = \frac{r^i - r^N}{1 - r^N}, \quad r = e^{-\rho}, \quad \rho = a/\varepsilon N.$$

Putting $\lambda = r$ ensures that $U_\varepsilon(x_i) = u_\varepsilon(x_i)$ and yields

$$\sigma = (\rho/2)\coth(\rho/2), \quad \rho = a/\varepsilon N.$$

as required.

It follows from this choice of the fitting factor that Method 2.3 is exact at the mesh points for problem (2.3). For more general problems with variable coefficients, it can be shown that, at the mesh points, the method generates solutions which converge to the exact solution irrespective of the value of ε. Indeed, Il'in (1969) proved that, for all problems from Problem Class 2.1, we have the following ε-uniform error estimate at the mesh points for the numerical solutions given by Method 2.3

$$\sup_{0<\varepsilon\le1} \max_{0\le i\le N} |U_\varepsilon(x_i) - u_\varepsilon(x_i)| \le CN^{-1},$$

where C is a constant independent of ε and N. But this does not mean that Method 2.3 is ε-uniform in our global sense for Problem Class 2.1, because it tells us nothing about the ε-uniform convergence of the numerical solutions at points between the mesh points. We show later in this section that we do not have ε-uniform convergence of the piecewise linear interpolants of these numerical approximations at points between the mesh points. This confirms our assertion that Method 2.3 is not an ε-uniform method for Problem Class 2.1 in the global sense of this book.

There are many different ways of generating Method 2.3. One such is through a finite element formulation using the following weak form of Problem Class 2.1

$$-(\varepsilon u'_\varepsilon, v') + (a u'_\varepsilon, v) = (f, v) \quad \text{for all} \quad v \in H^1_0(\Omega).$$

We construct a finite element basis with the exponential basis functions $\{\phi_i(x)\}^N_{i=0}$ first introduced in Hemker (1977), which are defined to be the exact solutions of the constant coefficient differential equations

$$\phi''_i + a(x_i)\phi'_i = 0, \quad \phi_i(x_j) = \delta_{i,j}, \quad x_j = j/N \in \Omega^N_u. \tag{2.10}$$

A finite element approximation

$$\bar{U}_\varepsilon = u_\varepsilon(0)\phi_0(x) + \sum_{i=1}^{N-1} U_\varepsilon(x_i)\phi_i(x) + u_\varepsilon(1)\phi_N(x)$$

is then obtained by solving the discrete system of equations

$$-(\varepsilon \bar{U}'_\varepsilon, \phi_i) + (a\bar{U}'_\varepsilon, \phi_i) = (f, \phi_i) \quad \text{for all} \quad i = 1, ... N - 1. \tag{2.11}$$

Approximating the integrals in (2.11) by the quadrature rules

$$(a\bar{U}'_\varepsilon, \phi_i) \approx a(x_i)(\bar{U}'_\varepsilon, \phi_i) \quad \text{and} \quad (f, \phi_i) \approx f(x_i)(1, \phi_i)$$

produces the same equations as in Method 2.3. On the other hand, if we employ the quadrature rules

$$(a\bar{U}'_\varepsilon, \phi_i) \approx \bar{a}(x_i)(\bar{U}'_\varepsilon, \phi'_i) \quad \text{and} \quad (f, \phi_i) \approx \bar{f}(x_i)(1, \phi_i)$$

and the basis functions ϕ_i defined by

$$\phi''_i + \bar{a}(x_i)\phi'_i = 0, \quad \phi_i(x_j) = \delta_{i,j}, \quad x_j = j/N \in \Omega^N_u$$

where $\bar{a}(x_i) = (a(x_i) + a(x_{i+1}))/2$, then the resulting finite difference method coincides with that proposed by El–Mistikawy and Werle (1978). Note that in the case of constant data a and f, the method of El–Mistikawy and Werle and Method 2.3 coincide, which means that the observations we make below about Method 2.3 apply also to the method of El–Mistikawy and Werle.

In both Berger et al. (1981) and Hegarty et al. (1980), it is proved that, for problems from Problem Class 2.1 with sufficiently smooth data, the following ε-uniform error estimate holds at the mesh points for the solution U_ε of the El–Mistikawy and Werle finite difference method

$$\sup_{0<\varepsilon\leq1} \max_{0\leq i\leq N} |U_\varepsilon(x_i) - u_\varepsilon(x_i)| \leq CN^{-2},$$

where C is a constant independent of ε and N.

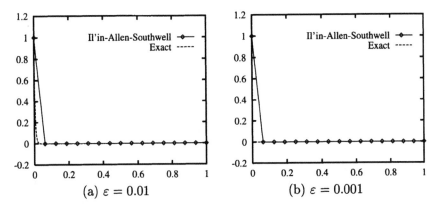

FIG. 2.4. Numerical solution generated by Method 2.3 applied to problem (2.4), for $\varepsilon = 0.01$ and 0.001, with $N = 16$.

Both the El–Mistikawy and Werle finite difference method and Method 2.3 use a uniform mesh, and so, when $\varepsilon N < 1$, none of the mesh points is in the boundary layer. Therefore, again, no information is obtained about the behaviour of the exact solution within the boundary layer from the corresponding numerical solutions or their piecewise linear interpolants. In particular, if $\varepsilon = 10^{-2}$ and $N = 1/16$ it can be seen in Fig. 2.4 that there is no mesh point in the boundary layer. In practice, it is often the case that $\varepsilon \ll 1$, and so, although Il'in's method is ε–uniform at the mesh points, we would have to take $N > 1/\varepsilon \gg 1$ in order to guarantee that there are mesh points in the boundary layer.

We now determine explicitly the global maximum pointwise error when Method 2.3 is applied to problem (2.3) and the resulting finite difference solution at the mesh points is extended to the whole domain by piecewise linear interpolation. Since Method 2.3 is exact at the mesh points for this problem, the maximum error $|\overline{U}_\varepsilon(x) - u_\varepsilon(x)|$ in each subinterval $[x_{i-1}, x_i]$, $1 \le i \le N$ must occur at points x_i^* in the interior of the subinterval, and thus we have

$$(\overline{U}_\varepsilon - u_\varepsilon)'(x_i^*) = 0.$$

We now locate such a point x_1^* in the first subinterval $(0, \frac{1}{N})$. For all $x \in (0, \frac{1}{N})$ we have

$$u_\varepsilon'(x) = -\frac{2}{\varepsilon} \frac{e^{-2x/\varepsilon}}{1 - e^{-2/\varepsilon}} \quad \text{and} \quad \overline{U}_\varepsilon'(x) \equiv \frac{u_\varepsilon(1/N) - 1}{1/N}$$

and so x_1^* satisfies

$$\frac{e^{-2x_1^*/\varepsilon}}{1 - e^{-2/\varepsilon}} = \frac{B(\rho)}{1 - e^{-2/\varepsilon}}, \quad \text{where } B(\rho) = \frac{1 - e^{-\rho}}{\rho} \text{ and } \rho = 2/N\varepsilon.$$

Solving for x_1^*, we get

$$x_1^* = -\frac{\varepsilon}{2} \ln B(\rho).$$

Evaluating the error at the point x_1^* yields

$$\overline{U}_\varepsilon(x_1^*) - u_\varepsilon(x_1^*) = \frac{k(B(\rho))}{1 - e^{-2/\varepsilon}}, \text{ where } k(t) = 1 - t(1 - \ln t).$$

Since $B(0) = 1$ and $B(\rho)$ is positive and monotonically decreasing for $\rho \in [0, \infty)$, it follows that $0 < B(\rho) \leq 1$ for all $\rho \in [0, \infty)$. Furthermore, $k(0) = 1$ and $k(t)$ is positive and monotonically decreasing for $t \in (0, 1)$ and so, for all ε sufficiently small, the maximum global error in the first subinterval is

$$\overline{U}_\varepsilon(x_1^*) - u_\varepsilon(x_1^*) \approx k(B(\rho)) > 0. \tag{2.12}$$

Taking $\varepsilon = 1/N$, for any value of N we have $\rho = 2$, and the right–hand side of (2.12) has the approximate value 0.205. We conclude that the global maximum error is at least 20%. This demonstrates the fact that a numerical method, which is exact at the mesh points, can have an ε–uniform global error in its piecewise linear interpolant of 20% or more. In particular, this is the case for Method 2.3 applied to problem (2.3).

We remark here that, for a singularly perturbed linear ordinary differential equation, it is possible to construct a global ε–uniform method, based on a fitted finite difference operator on a uniform mesh, if we allow the use of a more complicated form of interpolation. For example, consider the exponential basis functions $\{\phi_i(x)\}_{i=0}^N$ defined in (2.10). It is shown in O'Riordan (1984), for example, that for any problem from Problem Class 2.1 the following global ε–uniform error estimate holds

$$\sup_{0 < \varepsilon \leq 1} \| \sum_{i=0}^N U_\varepsilon(x_i)\phi_i - u_\varepsilon \|_\Omega \leq CN^{-1},$$

where $U_\varepsilon(x_i)$ is the numerical solution obtained from Method 2.3. However, extensions of this approach, and other kinds of interpolation, to linear problems in several dimensions and to nonlinear problems, is not easy in practice. Furthermore, we see in chapter 10 that if an inappropriate mesh is used then no interpolation process can be found to generate an ε–uniform method for a class of semilinear problems in one dimension.

In addition to finding an approximation to the solution u_ε of a differential equation, an accurate approximation to the scaled first derivative $\varepsilon u_\varepsilon'$, of the solution is often required. A standard approximation is the scaled discrete first derivative $\varepsilon D^+ U_\varepsilon$. We now examine the corresponding error in a case where explicit analytic expressions can be found for both of these quantities. For problem (2.3) we have

$$\varepsilon u'_\varepsilon(x) = -2 \frac{e^{-2x/\varepsilon}}{1 - e^{-2/\varepsilon}}.$$

Since Method 2.3 applied to this constant coefficient problem is exact at the nodes we have

$$\varepsilon D^+ U_\varepsilon(x_i) = \varepsilon D^+ u_\varepsilon(x_i) = -2 \frac{e^{-2x_i/\varepsilon}}{1 - e^{-2/\varepsilon}} \left(\frac{1 - e^{-\rho}}{\rho} \right), \quad \text{where} \quad \rho = 2/N\varepsilon.$$

Thus the error in these scaled quantities at the point $x = 0$ is

$$\varepsilon |D^+ U_\varepsilon(0) - u'_\varepsilon(0)| = \frac{2(1 - B(\rho))}{1 - e^{-2/\varepsilon}}.$$

Taking $\varepsilon N = 2$, and letting $N \to \infty$ we get

$$\lim_{N \to \infty} \varepsilon |D^+ U_\varepsilon(0) - u'_\varepsilon(0)| = 2e^{-1} \neq 0.$$

In other words $\varepsilon D^+ U_\varepsilon$ does not converge ε–uniformly to $\varepsilon u'_\varepsilon$ as the mesh is refined. We remark that it is possible to construct non–standard discrete approximations to $\varepsilon u'_\varepsilon$, based on the above exponential basis functions, which are ε–uniformly convergent for this one–dimensional problem. But extensions of this technique to partial differential equations and to nonlinear ordinary differential equations are not, in general, feasible.

2.6   Neumann boundary conditions

We have seen in Table 2.1 that a persistent error of about 20% occurs when Method 2.2 is applied to problem (2.3), which has Dirichlet boundary conditions. While this may be acceptable in certain circumstances, it is important to realize that upwind finite difference operators on uniform meshes can produce much larger errors in other equally simple problems. To illustrate this we now consider their application to problem (2.4) from Problem Class 2.2. This problem has a Neumann boundary condition which can be discretized in a standard fashion by the scaled discrete derivative

$$\varepsilon D^+ U_\varepsilon(0).$$

The resulting upwind finite difference method on a uniform mesh for problems in Problem Class 2.2 is defined as follows.

Method 2.4. Upwind finite difference operator on uniform mesh for Problem Class 2.2.

$$L_\varepsilon^N U_\varepsilon \equiv \varepsilon \delta^2 U_\varepsilon + a(x_i) D^+ U_\varepsilon = f(x_i), \quad x_i \in \Omega_u^N, \qquad (2.13a)$$
$$\varepsilon D^+ U_\varepsilon(0) = \varepsilon u_\varepsilon'(0), \quad U_\varepsilon(1) = u_\varepsilon(1), \qquad (2.13b)$$

where

$$\overline{\Omega}_u^N = \{x_i = i/N, i = 0, \dots, N\}. \qquad (2.13c)$$

In Table 2.2 we display the maximum pointwise errors arising when this method is applied to problem (2.4). It is evident that they are unacceptably large. The errors, which are approximately equal to $2/\varepsilon N$, do not stabilize with decreasing values of ε. On the contrary they grow without an upper bound.

This unacceptable behaviour can also be verified analytically in the following way. An explicit expression for the solution of the discrete problem, obtained when Method 2.4 is applied to problem (2.4), is given by

$$U_\varepsilon(x_i) = \frac{2(\lambda^i - \lambda^N)}{\varepsilon N(1 - \lambda)}, \quad \text{where} \quad \lambda = \frac{1}{1 + 2/\varepsilon N}.$$

It is clear that, for each fixed N, $\lim_{\varepsilon \to 0} \lambda = 0$ and so

$$U_\varepsilon(0) = \frac{2(1 - \lambda^N)}{\varepsilon N(1 - \lambda)} \to \infty \quad \text{as } \varepsilon \to 0.$$

On the other hand, the exact solution of problem (2.4) at the boundary point $x = 0$ has the limiting value

$$\lim_{\varepsilon \to 0} u_\varepsilon(0) = 1.$$

Table 2.2 *Maximum pointwise errors $E_{\varepsilon,\text{exact}}^N$ generated by Method 2.4 applied to problem (2.4).*

ε	Number of intervals N						
	8	16	32	64	128	256	512
2^{-2}	0.9925	0.4981	0.2493	0.1247	0.0624	0.0312	0.0156
2^{-4}	4.0000	2.0000	1.0000	0.5000	0.2500	0.1250	0.0625
2^{-6}	16.0000	8.0000	4.0000	2.0000	1.0000	0.5000	0.2500
2^{-8}	64.0000	32.0000	16.0000	8.0000	4.0000	2.0000	1.0000
2^{-10}	256.0000	128.0000	64.0000	32.0000	16.0000	8.0000	4.0000
2^{-12}	1024.0000	512.0000	256.0000	128.0000	64.0000	32.0000	16.0000
2^{-14}	4096.0000	2048.0000	1024.0000	512.0000	256.0000	128.0000	64.0000
2^{-16}	16384.0000	8192.0000	4096.0000	2048.0000	1024.0000	512.0000	256.0000
2^{-18}	65536.0000	32768.0000	16384.0000	8192.0000	4096.0000	2048.0000	1024.0000

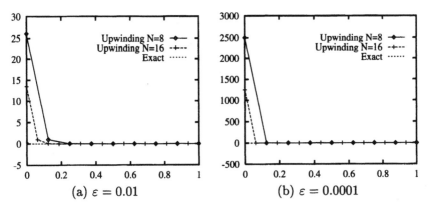

FIG. 2.5. Comparison of the exact solution of problem (2.4), for $\varepsilon = 0.01$ and 0.0001, and the numerical solutions generated by Method 2.4.

It follows that, as ε decreases, the error at the left–hand boundary is unbounded as $\varepsilon \to 0$. This shows that the numerical solutions bear no relation to the exact solution near the boundary point $x = 0$. This unbounded behaviour of the numerical solutions is illustrated in Fig. 2.5, where they are compared with the exact solutions for $\varepsilon = 0.01$ and $\varepsilon = 0.0001$. Note the different vertical scales in the graphs (a) and (b).

Remark In passing we note that it is easier to construct an ε–uniform method for a problem in Problem Class 2.2 with a homogeneous Neumann boundary condition than for the problem considered above. This is due to the fact that the behaviour of the exact solution of such a problem is less singular than that of a problem with an inhomogenous Neumann boundary condition. To illustrate this we consider, for example, the convection–diffusion problem

$$\varepsilon u_\varepsilon''(x) + a(x)u_\varepsilon' = f(x), \quad \varepsilon u_\varepsilon'(0) = A, \quad u_\varepsilon(1) = B.$$

It is not hard to verify that the derivative of its solution can be written in the closed form

$$u_\varepsilon'(x) = u_\varepsilon'(0)e^{-I(x)/\varepsilon} + \frac{1}{\varepsilon}\int_{t=0}^{x} f(t)e^{-(I(x)-I(t))/\varepsilon}\,dt \quad \text{where} \quad I(x) \equiv \int_0^x a(s)\,ds$$

Suppose now that the Neumann boundary condition is homogeneous, that is $u_\varepsilon'(0) = A = 0$. Then the first term on the right–hand side vanishes. It follows that the boundary layer at $x = 0$ is very weak, since $u_\varepsilon'(x)$ is bounded independently of ε. Note, however, that the behaviour of u_ε is still weakly singular near $x = 0$ because $u_\varepsilon''(0) = O(1/\varepsilon)$. In such cases we say that the solution u_ε has a weak boundary layer in a neighbourhood of the point $x = 0$.

2.7   Error estimates in alternative norms

The numerical methods discussed so far produce numerical solutions on uniform meshes which have unacceptable errors in the maximum norm. This leads to the obvious question: do these numerical solutions have better behaviour in some other norm? For a finite difference method the discrete ℓ_1 and ℓ_2 norms are frequently used, while a discrete energy norm is natural for a finite element method. In this section we discuss the measurement of the error in these norms, and we show that these norms are unsuitable measures of the error in the numerical solutions when a boundary layer is present in the exact solution of the problem. This reinforces the discussion in §1.1, where we saw that the analogous continuous norms are unsuitable for measuring various properties of the exact solutions.

We consider first the discrete ℓ_1 and ℓ_2 norms defined respectively for all $\mathbf{V} \in R^{N+1}$ by

$$\|\mathbf{V}\|_{1,d} = \frac{x_1 - x_0}{2}|V_0| + \sum_{i=1}^{N-1} \frac{x_{i+1} - x_{i-1}}{2}|V_i| + \frac{x_N - x_{N-1}}{2}|V_N|, \qquad (2.14)$$

and

$$\|\mathbf{V}\|_{2,d}^2 = \frac{x_1 - x_0}{2}|V_0|^2 + \sum_{i=1}^{N-1} \frac{x_{i+1} - x_{i-1}}{2}|V_i|^2 + \frac{x_N - x_{N-1}}{2}|V_N|^2. \qquad (2.15)$$

The errors in these two norms for problem (2.3) using Method 2.2, corresponding to the maximum norm errors in Table 2.1, are given respectively in Tables 2.3 and 2.4. Unlike Table 2.1 we see that a ridge of constant errors does not appear in either of these two tables. It is important for the reader to understand that the absence of this ridge in these cases is not a gain. On the contrary it shows that the use of the ℓ_1 or ℓ_2 norm to measure the errors leads to the masking of an

Table 2.3 ℓ_1 *norm errors generated by Method 2.2 applied to problem (2.3).*

ε	Number of intervals N						
	8	16	32	64	128	256	512
2^{-1}	0.035368	0.019644	0.010338	0.005302	0.002684	0.001351	0.000677
2^{-2}	**0.048667**	**0.027467**	**0.014517**	0.007453	0.003774	0.001899	0.000953
2^{-3}	0.042783	0.026111	0.014326	**0.007487**	**0.003825**	**0.001933**	**0.000971**
2^{-4}	0.028915	0.021468	0.013063	0.007164	0.003744	0.001912	0.000966
2^{-6}	0.007812	0.007792	0.007229	0.005367	0.003266	0.001791	0.000936
2^{-8}	0.001953	0.001953	0.001953	0.001948	0.001807	0.001342	0.000816
2^{-10}	0.000488	0.000488	0.000488	0.000488	0.000488	0.000487	0.000452
2^{-12}	0.000122	0.000122	0.000122	0.000122	0.000122	0.000122	0.000122
2^{-14}	0.000031	0.000031	0.000031	0.000031	0.000031	0.000031	0.000031
2^{-16}	0.000008	0.000008	0.000008	0.000008	0.000008	0.000008	0.000008
2^{-18}	0.000002	0.000002	0.000002	0.000002	0.000002	0.000002	0.000002

Table 2.4 ℓ_2 *norm errors generated by Method 2.2 applied to problem (2.3).*

ε	Number of intervals N						
	8	16	32	64	128	256	512
2^{-1}	0.042177	0.023142	0.012162	0.006242	0.003163	0.001592	0.000799
2^{-2}	0.067968	0.038337	0.020427	0.010569	0.005381	0.002715	0.001364
2^{-3}	**0.078335**	0.049398	0.027544	0.014609	0.007544	0.003837	0.001936
2^{-4}	0.065810	**0.055437**	0.034934	0.019477	0.010331	0.005335	0.002713
2^{-5}	0.039411	0.046536	**0.039200**	0.024702	0.013772	0.007305	0.003772
2^{-6}	0.020833	0.027868	0.032906	**0.027718**	0.017467	0.009739	0.005165
2^{-7}	0.010719	0.014731	0.019705	0.023268	**0.019600**	0.012351	0.006886
2^{-8}	0.005440	0.007579	0.010417	0.013934	0.016453	**0.013859**	0.008733
2^{-9}	0.002741	0.003847	0.005359	0.007366	0.009853	0.011634	**0.009800**
2^{-10}	0.001376	0.001938	0.002720	0.003790	0.005208	0.006967	0.008226
2^{-12}	0.000345	0.000487	0.000688	0.000969	0.001360	0.001895	0.002604
2^{-14}	0.000086	0.000122	0.000172	0.000244	0.000344	0.000485	0.000680
2^{-16}	0.000022	0.000031	0.000043	0.000061	0.000086	0.000122	0.000172
2^{-18}	0.000005	0.000008	0.000011	0.000015	0.000022	0.000031	0.000043

important attribute of the error, when these methods are used to solve problems with boundary layers in their solutions.

To understand how the ridge of constant errors is masked in the ℓ_1 norm, for this simple problem, we examine the explicit closed form expression for the error in the approximation of the boundary layer function $e^{-x/\varepsilon}$. At a mesh point x_i this is given by

$$\left(U_\varepsilon - u_\varepsilon\right)(x_i) = \lambda^i - e^{-x_i/\varepsilon}, \quad \text{where } \lambda = \frac{1}{1 + 1/\varepsilon N}.$$

Since $(1 + t)^{-1} \geq e^{-t}$, for all $t \geq 0$ we have $\lambda^i \geq e^{-x_i/\varepsilon}$ and so

$$\|U_\varepsilon - u_\varepsilon\|_{1,d} = \frac{1}{N}\sum_{i=1}^{N-1}\left(\lambda^i - e^{-x_i/\varepsilon}\right) = \frac{1}{N}\left(\frac{1 - \lambda^N}{1 - \lambda} - \frac{1 - e^{-1/\varepsilon}}{1 - e^{-1/\varepsilon N}}\right) \leq CN^{-1}$$

where C is independent of ε. This shows that Method 2.2 applied to problem (2.3) is ε–uniform in the discrete ℓ_1 norm. It is easy to show that the same is true in the case of the discrete ℓ_2 norm.

We see in Tables 2.3 and 2.4 that the ε–uniform errors in the ℓ_1 and ℓ_2 norms decrease with increasing N, while in Table 2.1 they are essentially constant. Hence, measuring the errors in the discrete ℓ_1 or ℓ_2 norm, leads to the erroneous conclusion that Method 2.2 is a robust layer–resolving method for problems from Problem Class 2.1. We conclude that use of the discrete ℓ_1 and ℓ_2 norms can give misleading information about the suitability of a specific numerical method for solving problems with boundary layers.

CHAPTER 3

Layer resolving methods for convection diffusion problems in one dimension

3.1 Bakhvalov fitted meshes

The numerical methods constructed in the previous chapter are based on uniform meshes, and as a result serious difficulties arose. In this and subsequent chapters we show that these difficulties can be overcome if we use an appropriate non–uniform mesh. It turns out that the simplest possible kind of non–uniform mesh, namely a piecewise–uniform mesh, is sufficient for our purposes. We do not claim that these piecewise–uniform meshes are optimal in any sense. On the contrary, we want to find the simplest possible way of constructing ε–uniform methods for the problems considered in this book, in the expectation that we can use the same approach to solve more complex problems in the future. For this reason we are not interested in the construction of complicated meshes for one dimensional problems, which are difficult or impossible to apply to the solution of nonlinear problems in several dimensions. In the remainder of this section we give a brief description of various non–uniform meshes that have been proposed for problems with boundary layers. Then, in the next section, we introduce fitted piecewise–uniform meshes.

Bakhvalov (1969) was the first to introduce special non–uniform meshes for solving singularly perturbed boundary value problems. He showed that ε–uniform numerical methods composed of these meshes and standard finite difference operators can be constructed for some classes of problems in one dimension. We illustrate the essential idea of Bakhvalov by constructing such a mesh for problems from Problem Class 2.1, which have a boundary layer near $x = 0$. The mesh is obtained using a mesh generating function $\lambda : \overline{\Omega} \to [-1, 1]$, which is continuous and strictly decreasing, with $\lambda(0) = 1$ and $\lambda(1) = -1$. The mesh points $\{x_i\}$ are defined implicitly by the equations

$$\lambda(x_i) = 1 - \frac{2i}{N}.$$

In particular, if we use the mesh generating function $\lambda(x) = 1 - 2x$, then the resulting mesh is uniform. For problems from Problem Class 2.1, a boundary layer function of the form $e^{-\gamma x/\varepsilon}$, where γ is positive, occurs in the exact solution u_ε. This motivates the use of the Bakhvalov mesh generating function

$$\lambda(t) = \begin{cases} \psi(t), & t \in [0, \tau] \\ \psi'(\tau)(t - \tau) + \psi(\tau), & t \in [\tau, 1] \end{cases}$$

where ψ is defined by $\psi(t) = e^{-\gamma t/\varepsilon}$ and the transition parameter τ, which marks the change from a fine to a coarse mesh, is defined implicitly by the equation

$$\psi'(\tau) = -\frac{1 + \psi(\tau)}{1 - \tau}. \tag{3.1}$$

It is clear that $\lambda(1) = -1$ and that $\lambda \in C^1(\overline{\Omega})$. Once the transition point τ is determined, by solving (3.1), the Bakhvalov mesh points are obtained from

$$x_i = \begin{cases} -\frac{\varepsilon}{\gamma} \ln(1 - \frac{2i}{N}), & i < N/2 \\ \tau + (\psi(\tau) - (1 - \frac{2i}{N}))\frac{(1-\tau)}{1+\psi(\tau)}, & i \geq N/2 \end{cases}$$

This is an elegant representation of the mesh, but there are some practical limitations to this approach. First, (3.1) is a nonlinear equation for the transition point τ, the solution of which cannot be written in closed form. Furthermore, extensions of the Bakhvalov mesh to singularly perturbed partial differential equations have been few (Shishkin (1983)). This is due not only to the complicated construction, but also to the fact that the theoretical techniques required to prove ε–uniform error estimates for the resulting numerical methods are difficult.

With a view to developing higher order methods for two–point boundary value problems, Gartland (1988) examined exponentially graded meshes. With these meshes the solution domain is decomposed into three regions: an inner region $[0, \tau_1]$ where the mesh is graded exponentially in ε, a transition region $[\tau_1, \tau_2]$ where the mesh changes geometrically from fine to coarse, and an outer region $[\tau_2, 1]$ where the mesh is uniform. However, the number of mesh points used in these meshes grows with decreasing ε (albeit only like $O(\ln(1/\varepsilon))$), and so numerical methods composed of these meshes are not ε–uniform. On top of this the design of these meshes is not trivial.

Other results related to the Bakhvalov mesh are given, for example, in Liseiken (1983) and Vulanović (1986). General adaptive mesh refinement methods have been applied to singularly perturbed problems (we refer the reader to Roos et al. (1996) for a discussion), but we know of no strong theoretical results on the convergence of such methods.

In this book we do not discuss exponentially graded meshes. Instead, we show that much simpler piecewise–uniform meshes suffice for the construction of ε–uniform numerical methods for wid classes of problems with boundary layers in their solutions. Because of their simplicity we firmly believe that these meshes will, in due course, prove to be useful for much wider classes of nonlinear problems in several dimensions than are treated in this book.

3.2   Piecewise–uniform fitted meshes

For convection–diffusion problems from Problem Class 2.1 we now introduce simple piecewise–uniform meshes of the form

$$\overline{\Omega}_\varepsilon^N = \{x_i | x_i = 2i\sigma/N, \ i \leq N/2; \ x_i = x_{i-1} + 2(1-\sigma)/N, \ N/2 < i\} \quad (3.2a)$$

where the transition parameter σ, which determines the point of transition from a fine to a coarse mesh, is fitted to the boundary layer by taking

$$\sigma = \min\{\frac{1}{2}, \frac{1}{\alpha}\varepsilon \ln N\}. \quad (3.2b)$$

Note that σ depends on both ε and N and that α is the lower bound on the coefficient a defined in (2.1c). We call a piecewise–uniform mesh with this special choice of σ a piecewise–uniform fitted mesh. A piecewise–uniform mesh of this kind is defined in Shishkin (1988a). It is clear that this piecewise–uniform mesh is only slightly more complex than a uniform mesh, because it is simply two uniform meshes glued together at a carefully chosen transition point. We note that it reduces to a uniform mesh if $\sigma = 1/2$, which occurs whenever $N \geq e^{\alpha/2\varepsilon}$

Combining the standard upwind finite difference operator with this piecewise–uniform fitted mesh leads to the following finite difference method.

Method 3.1.   Upwind finite difference operator on piecewise–uniform fitted mesh for Problem Class 2.1.

$$L_\varepsilon^N U_\varepsilon \equiv \varepsilon \delta^2 U_\varepsilon + a(x_i)D^+ U_\varepsilon = f(x_i), \quad x_i \in \Omega_\varepsilon^N, \quad (3.3a)$$
$$U_\varepsilon(0) = u_\varepsilon(0), \quad U_\varepsilon(1) = u_\varepsilon(1), \quad (3.3b)$$

where $\overline{\Omega}_\varepsilon^N$ is the piecewise-uniform fitted mesh

$$\overline{\Omega}_\varepsilon^N = \{x_i | x_i = 2i\sigma/N, \ i \leq N/2; \ x_i = x_{i-1} + 2(1-\sigma)/N, \ N/2 < i\} \quad (3.3c)$$

with

$$\sigma = \min\{\frac{1}{2}, \frac{1}{\alpha}\varepsilon \ln N\}. \quad (3.3d)$$

Applying this method to problem (2.3), which has the known solution (2.3c), leads to the exact maximum pointwise errors $E_{\varepsilon,\text{exact}}^N$ given in Table 3.1. We see that, for some values of ε, the errors $E_{\varepsilon,\text{exact}}^N$ in this table are larger than the corresponding values in Table 2.1, which were obtained using Method 2.2. However, this does not mean that the numerical solutions given by Method 2.2

Table 3.1 *Maximum pointwise errors $E^N_{\varepsilon,\text{exact}}$ and E^N_{exact} generated by Method 3.1 applied to problem (2.3) for various values of ε and N.*

	Number of intervals N						
ε	8	16	32	64	128	256	512
2^{-2}	0.101617	0.060691	0.037054	0.021752	0.011153	0.005649	0.002843
2^{-4}	0.157096	0.087900	0.047322	0.025794	0.014341	0.008029	0.004485
2^{-6}	0.180862	0.105544	0.057754	0.030755	0.016289	0.008682	0.004679
2^{-8}	0.188122	0.112179	0.062870	0.034200	0.018305	0.009688	0.005099
2^{-10}	0.190018	0.114009	0.064449	0.035445	0.019271	0.010378	0.005531
2^{-12}	0.190497	0.114478	0.064866	0.035791	0.019575	0.010633	0.005737
2^{-14}	0.190617	0.114596	0.064971	0.035880	0.019655	0.010704	0.005801
2^{-16}	0.190647	0.114626	0.064998	0.035903	0.019676	0.010722	0.005818
2^{-18}	**0.190655**	**0.114633**	**0.065004**	**0.035908**	**0.019681**	**0.010727**	**0.005822**
E^N_{exact}	0.190655	0.114633	0.065004	0.035908	0.019681	0.010727	0.005822

are better than those given by Method 3.1. With Method 2.2 there are no mesh points in the boundary layer and so the boundary layer is not represented in the numerical solution. In contrast, in Method 3.1 the piecewise–uniform mesh always has mesh points in the boundary layer; in fact half of the mesh points are there, and so the boundary layer is resolved by Method 3.1. The larger errors in Method 3.1 are the price we pay for the resolution of the boundary layer. Since the method is ε–uniform, we can both resolve the boundary layer and also recover the smaller error by increasing the number of mesh points. These contrasting behaviours may be seen in Fig. 3.1, and they are even more apparent in Fig. 3.2 where the solutions given by several methods are compared in a small region containing the boundary layer; note the different scale of the horizontal axis in Fig. 3.2. Moreover, it is significant that all of the ε–uniform errors E^N_{exact} in Table 3.1 are smaller than the corresponding values in Table 2.1, and that

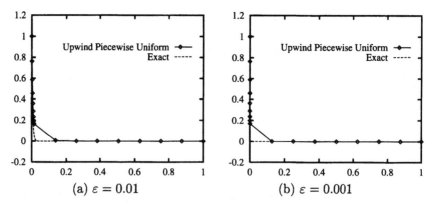

FIG. 3.1. Numerical solution generated by Method 3.1 applied to problem (2.3), for $\varepsilon = 0.01$ and 0.001, with $N = 16$.

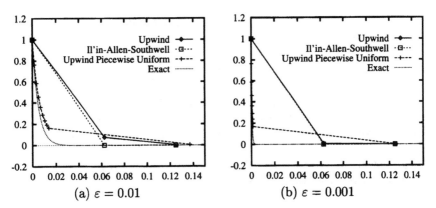

FIG. 3.2. Comparison of the boundary layer in the numerical solutions of problem (2.3), for $\varepsilon = 0.01$ and 0.001, generated by Methods 2.2, 2.3, and 3.1 with $N = 16$.

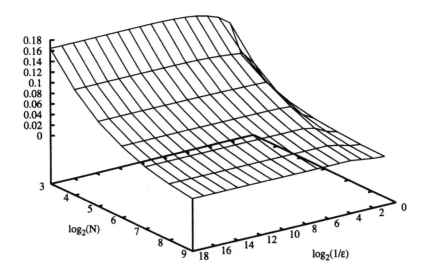

FIG. 3.3. Maximum pointwise error $E^N_{\varepsilon,\text{exact}}$ as a function of N and ε for problem (2.3) generated by Method 3.1.

they decrease rapidly as N increases. In Fig. 3.3, the maximum pointwise errors $E^N_{\varepsilon,\text{exact}}$ at the mesh points for problem (2.3) using Method 3.1 are plotted as a function of N and ε. Note that for all values of ε the error decreases steadily with increasing N. Comparing the graph in this figure with Fig. 2.3, we see that there is no trace of the ridge of persistent error that occurs with Method 2.2.

We now show that a numerical method using a piecewise–uniform fitted mesh and a standard finite difference operator also resolves the boundary layer that occurs in problem (2.4), which has a Neumann boundary condition.

Method 3.2.   Upwind finite difference operator on piecewise–uniform fitted mesh for Problem Class 2.2.

$$L^N_\varepsilon U_\varepsilon \equiv \varepsilon \delta^2 U_\varepsilon + a(x_i)D^+U_\varepsilon = f(x_i), \quad x_i \in \Omega^N_\varepsilon, \tag{3.4a}$$

$$\varepsilon D^+ U_\varepsilon(0) = \varepsilon u'(0), \quad U_\varepsilon(1) = u_\varepsilon(1), \tag{3.4b}$$

where

$$\Omega^N_\varepsilon = \{x_i | x_i = 2i\sigma/N, \ i \le N/2; \ x_i = x_{i-1} + 2(1-\sigma)/N, \ N/2 < i\} \tag{3.4c}$$

with

$$\sigma = \min\{0.5, (1/\alpha)\varepsilon \ln N\}. \tag{3.4d}$$

The exact maximum pointwise errors $E^N_{\varepsilon,\text{exact}}$ and E^N_{exact}, obtained when this method is applied to problem (2.4), are given in Table 3.2. In Fig. 3.4 graphs of the exact solutions for $\varepsilon = 0.01$ and $\varepsilon = 0.0001$ and the numerical solutions with $N = 8$ and $N = 16$ are shown, and in Fig. 3.5 blow–ups of the same graphs are given in the boundary layers. Note the different horizontal scales.

Table 3.2 *Maximum pointwise errors $E^N_{\varepsilon,\text{exact}}$ and E^N_{exact} generated by Method 3.2 applied to problem (2.4) for various values of ε and N.*

	Number of intervals N						
ε	8	16	32	64	128	256	512
2^{-2}	0.579240	0.357192	0.217354	0.124737	0.062382	0.031194	0.015598
2^{-4}	0.718018	0.415225	0.237295	0.135552	0.077208	0.043654	0.024446
2^{-6}	0.779227	0.452500	0.256571	0.144081	0.080440	0.044716	0.024758
2^{-8}	0.798032	0.466054	0.265642	0.149768	0.083672	0.046309	0.025420
2^{-10}	0.803008	0.469836	0.268433	0.151835	0.085193	0.047371	0.026083
2^{-12}	0.804270	0.470810	0.269171	0.152412	0.085661	0.047757	0.026395
2^{-14}	0.804587	0.471055	0.269358	0.152561	0.085784	0.047865	0.026491
2^{-16}	0.804666	0.471116	0.269405	0.152598	0.085816	0.047893	0.026516
2^{-18}	**0.804686**	**0.471132**	**0.269417**	**0.152608**	**0.085824**	**0.047900**	**0.026522**
E^N_{exact}	0.804686	0.471132	0.269417	0.152608	0.085824	0.047900	0.026522

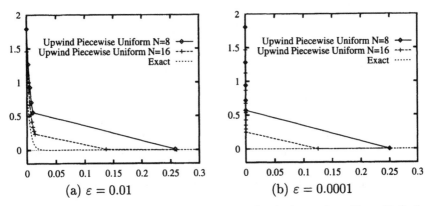

FIG. 3.4. Comparison of the exact and numerical solutions of problem (2.4), for $\varepsilon = 0.01$ and 0.001, generated by Method 3.2 with $N = 8$ and 16.

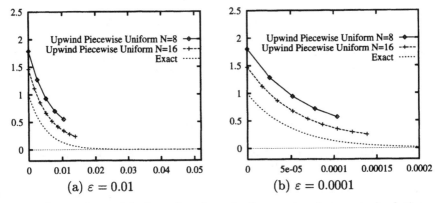

FIG. 3.5. Comparison of the boundary layer in the exact and numerical solutions of problem (2.4), for $\varepsilon = 0.01$ and 0.001, generated by Method 3.2 with $N = 8$ and 16.

The following intuitive argument indicates why appropriately constructed piecewise–uniform fitted meshes are so successful for such problems, and it also motivates our choice of the transition parameter σ. If ε is sufficiently small relative to $1/N$ then, from the definition of σ, we have $\sigma = (\varepsilon/\alpha) \ln N$. It follows that $e^{-\alpha\sigma/\varepsilon} = 1/N$, and so, for all $i \geq N/2$, we have

$$e^{-\alpha x_i/\varepsilon} \leq 1/N.$$

This means that in the coarse mesh region $[\sigma, 1]$ the boundary layer function $e^{-\alpha x/\varepsilon}$ is small in a discrete sense. On the other hand, in the fine mesh region $[0, \sigma]$, it is the mesh width

$$h = \frac{2\sigma}{N} \le \frac{1}{N}$$

that is small in a discrete sense. It is these two properties that allow us to prove, in the next section, that the numerical method is ε–uniform.

3.3 Theoretical results

In this section we prove some theoretical results which demonstrate mathematically that Method 3.1 is an ε–uniform method for Problem Class 2.1. If the reader has little interest in formal mathematical proofs, then this section may be skipped with no loss of new ideas.

In the next two lemmas we establish *a priori* bounds on the exact solution, and its derivatives, of problems in Problem Class 2.1.

Lemma 3.1 *The solution u_ε of any problem from Problem Class 2.1 satisfies the bound*

$$\|u_\varepsilon\| \le \max\{|u_\varepsilon(0)|, |u_\varepsilon(1)|\} + \frac{1}{\alpha}\|f\|$$

Proof Consider the two functions $\psi^\pm(x) = \max\{|u_\varepsilon(0)|, |u_\varepsilon(1)|\} + \frac{1}{\alpha}\|f\|(1 - x) \pm u_\varepsilon(x)$. Note that the functions $\psi^\pm$ are nonnegative at $x = 0, 1$ and that, for all $x \in \Omega$, we have

$$L_\varepsilon \psi^\pm(x) = -\frac{a(x)}{\alpha}\|f\| \pm f(x) \le 0$$

The minimum principle in Theorem 2.1 then implies that $\psi^\pm(x) \ge 0$ for all $x \in \overline{\Omega}$, from which the result follows immediately. $\square$

Lemma 3.2 *The derivatives $u_\varepsilon^{(k)}$ of the solution u_ε of any problem from Problem Class 2.1 satisfy the bounds*

$$\|u_\varepsilon^{(k)}\| \le C\varepsilon^{-k} \max\{\|f\|, \|u_\varepsilon\|\}, \quad k = 1, 2$$

$$\|u_\varepsilon^{(3)}\| \le C\varepsilon^{-3} \max\{\|f\|, \|f'\|, \|u_\varepsilon\|\}$$

where C depends only on $\|a\|$ and $\|a'\|$.

Proof Integrating by parts, we have

$$\int_0^x au_\varepsilon'(t) \, \mathrm{d}t = au_\varepsilon|_0^x - \int_0^x (a'u_\varepsilon)(t) \, \mathrm{d}t$$

and so

$$\left|\int_0^x (f - au_\varepsilon')(t) \, \mathrm{d}t\right| \le \|f\| + C\|u_\varepsilon\|, \tag{3.5}$$

where C depends on $\|a\|$ and $\|a'\|$. By the Mean Value Theorem, there exists a point $z \in (0, \varepsilon)$ such that

$$u'_\varepsilon(z) = \frac{u_\varepsilon(\varepsilon) - u_\varepsilon(0)}{\varepsilon}$$

and so

$$|\varepsilon u'_\varepsilon(z)| \leq 2\|u_\varepsilon\|. \tag{3.6}$$

Integrating the differential equation (2.1a) we get

$$\varepsilon u'_\varepsilon(z) - \varepsilon u'_\varepsilon(0) = \int_0^z (f - au'_\varepsilon)(t) \, dt. \tag{3.7}$$

Combining this with (3.5) and (3.6), it follows that $|\varepsilon u'_\varepsilon(0)| \leq \|f\| + C\|u_\varepsilon\|$. Using (3.7) with $z = x$ we see that $|\varepsilon u'_\varepsilon(x)| \leq \|f\| + C\|u_\varepsilon\|$ for all $x \in \Omega$, and the required result for $k = 1$ follows. Then, from the differential equation, we have

$$\varepsilon u''_\varepsilon = f - au'_\varepsilon \quad \text{and} \quad \varepsilon u'''_\varepsilon = (f - au'_\varepsilon)'$$

from which we obtain successively the required bounds on the second and third derivatives. $\qquad \square$

Note that the above proof requires only that $a, f \in C^1(\Omega)$.

To derive ε–uniform error estimates, we need sharper bounds on the derivatives of the solution u_ε. We derive these using the following decomposition of the solution into smooth and singular components

$$u_\varepsilon = v_\varepsilon + w_\varepsilon \tag{3.8a}$$

where v_ε can be written in the form $v_\varepsilon = v_0 + \varepsilon v_1 + \varepsilon^2 v_2$, and v_0, v_1 and v_2 are defined respectively to be the solutions of the problems

$$av'_0 = f, \quad v_0(1) = u_\varepsilon(1) \tag{3.8b}$$
$$av'_1 = -v''_0, \quad v_1(1) = 0 \tag{3.8c}$$
$$\text{and} \quad L_\varepsilon v_2 = -v''_1, \quad v_2(0) = 0, \ v_2(1) = 0. \tag{3.8d}$$

Thus the smooth component v_ε is the solution of

$$L_\varepsilon v_\varepsilon = f, \quad v_\varepsilon(0) = v_0(0) + \varepsilon v_1(0), \ v_\varepsilon(1) = u_\varepsilon(1) \tag{3.9a}$$

and consequently the singular component w_ε is the solution of the homogeneous problem

$$L_\varepsilon w_\varepsilon = 0, \quad w_\varepsilon(0) = u_\varepsilon(0) - v_\varepsilon(0), \ w_\varepsilon(1) = 0. \tag{3.9b}$$

The next lemma shows that the first two derivatives of v_ε are bounded independently of ε and that the boundary layer function w_ε is exponentially small outside the boundary layer.

Lemma 3.3 *The solution u_ε of any problem from Problem Class 2.1 can be decomposed into the sum*

$$u_\varepsilon = v_\varepsilon + w_\varepsilon$$

where v_ε and w_ε are defined in (3.9a) and (3.9b). Furthermore, the components $v_\varepsilon, w_\varepsilon$ and their derivatives satisfy the bounds

$$\|v_\varepsilon^{(k)}\| \le C(1 + \varepsilon^{2-k}), \quad k = 0, 1, 2, 3$$

$$|w_\varepsilon(x)| \le C e^{-\alpha x/\varepsilon} \quad \text{for all } x \in \overline{\Omega}$$

and

$$\|w_\varepsilon^{(k)}\| \le C \varepsilon^{-k}, \quad k = 1, 2, 3.$$

Proof We observe that v_0, v_1 are independent of ε and that v_2 is the solution of a similar problem to that defining u_ε. For any problem from Problem Class 2.1 we have $v_1'' \in C^0(\Omega)$ and hence, from the previous lemma, for $k = 0, 1, 2$ we have the bounds

$$\|v_\varepsilon^{(k)}\| \le C(1 + \varepsilon^{2-k})$$

and for $k = 3$ we have $\varepsilon v_\varepsilon''' = (f - a v_\varepsilon')'$ which leads to the bound $\|v_\varepsilon^{(3)}\| \le C \varepsilon^{-1}$.

To obtain the required bounds on w_ε and its derivatives we consider the two functions $\psi^\pm(x) = C e^{-\alpha x/\varepsilon} \pm w_\varepsilon(x)$. It is easy to see that $\psi^\pm$ are nonnegative at $x = 0, 1$ and using (3.9b) that

$$L_\varepsilon \psi^\pm = -C \frac{\alpha}{\varepsilon}(a - \alpha) e^{-\alpha x/\varepsilon} \le 0.$$

The minimum principle given in Theorem 2.1 then implies that $\psi^\pm(x) \ge 0$, for all $x \in \overline{\Omega}$, from which the required pointwise bound on w_ε follows. Applying the previous lemma to w_ε then yields the required bounds on the derivatives of w_ε.
□

Remark With a little more work we can obtain the following sharper pointwise bounds on the derivatives of w_ε

$$|w_\varepsilon^{(k)}(x)| \le C \varepsilon^{-k} e^{-\alpha x/\varepsilon}, \quad k = 1, 2, 3. \tag{3.10}$$

To see this we integrate by parts and use the pointwise bounds for w_ε given in Lemma 3.3, to get

$$\left| \int_x^1 (a w_\varepsilon')(t) \, dt \right| = \left| a w_\varepsilon \big|_x^1 - \int_x^1 (a' w_\varepsilon)(t) \, dt \right| \le C e^{-\alpha x/\varepsilon}. \tag{3.11}$$

By the Mean Value Theorem, there exists a point $z \in (1 - \varepsilon, 1)$ such that

$$|w'_\varepsilon(z)| = \left| \frac{w_\varepsilon(1) - w_\varepsilon(1 - \varepsilon)}{\varepsilon} \right| = \left| \frac{w_\varepsilon(1 - \varepsilon)}{\varepsilon} \right|$$

$$\leq C\varepsilon^{-1} e^{-\alpha(1-\varepsilon)/\varepsilon} \leq C\varepsilon^{-1} e^{-\alpha/\varepsilon}. \tag{3.12}$$

Now, integrating the differential equation defining w_ε we have, as in Lemma 3.2,

$$\varepsilon w'_\varepsilon(x) = \varepsilon w'_\varepsilon(1) + \int_x^1 (aw'_\varepsilon)(t) \, dt. \tag{3.13}$$

and using (3.11) and (3.12) we get

$$|\varepsilon w'_\varepsilon(1)| = |-\varepsilon w'_\varepsilon(z) + \int_z^1 (aw'_\varepsilon)(t) \, dt| \leq C e^{-\alpha/\varepsilon}.$$

Using (3.13) and (3.11) it follows that $|\varepsilon w'_\varepsilon(x)| \leq C e^{-\alpha x/\varepsilon}$ for all $x \in \Omega$, which is the required bound for $k = 1$. From the differential equation we have

$$\varepsilon w''_\varepsilon = -aw'_\varepsilon \quad \text{and} \quad \varepsilon w'''_\varepsilon = (-aw'_\varepsilon)'$$

which yield, successively, the required bounds on the second and third derivatives of w_ε.

Just as the solution u_ε of the continuous problem can be decomposed into the sum given in Lemma 3.3, the discrete solution U_ε of Method 3.1 can be decomposed into the analogous sum

$$U_\varepsilon = V_\varepsilon + W_\varepsilon$$

where V_ε and W_ε are respectively the solutions of the problems

$$L^N_\varepsilon V_\varepsilon = f(x_i), \quad x_i \in \Omega^N_\varepsilon, \quad V_\varepsilon(0) = v_\varepsilon(0), \quad V_\varepsilon(1) = v_\varepsilon(1) \tag{3.14}$$

$$L^N_\varepsilon W_\varepsilon = 0, \quad x_i \in \Omega^N_\varepsilon, \quad W_\varepsilon(0) = w_\varepsilon(0), \quad W_\varepsilon(1) = 0. \tag{3.15}$$

Combining these decompositions, we arrive at an analogous decomposition of the error $U_\varepsilon - u_\varepsilon = V_\varepsilon - v_\varepsilon + W_\varepsilon - w_\varepsilon$. Therefore we can estimate the error $U_\varepsilon - u_\varepsilon$ by estimating the errors $V_\varepsilon - v_\varepsilon$ and $W_\varepsilon - w_\varepsilon$ separately. This is done in the next two lemmas, which lead to the required ε–uniform error estimate given in Theorem 3.6.

Lemma 3.4 At each mesh point $x_i \in \overline{\Omega}^N_\varepsilon$ the smooth component of the error satisfies the estimate

$$|(V_\varepsilon - v_\varepsilon)(x_i)| \leq C N^{-1}(1 - x_i)$$

where v_ε is the solution of (3.9a) and V_ε is the solution of (3.14).

Proof This is obtained using the following standard stability and consistency argument. We consider the local truncation error

$$L_\varepsilon^N(V_\varepsilon - v_\varepsilon) = (L_\varepsilon - L_\varepsilon^N)v_\varepsilon = \varepsilon\left(\frac{d^2}{dx^2} - \delta^2\right)v_\varepsilon + a\left(\frac{d}{dx} - D^+\right)v_\varepsilon.$$

Then, by standard local truncation error estimates and Lemma 3.3, we obtain

$$|L_\varepsilon^N(V_\varepsilon - v_\varepsilon)(x_i)| \leq \frac{\varepsilon}{3}(x_{i+1} - x_{i-1})\|v_\varepsilon^{(3)}\| + \frac{a(x_i)}{2}(x_{i+1} - x_i)\|v_\varepsilon^{(2)}\| \leq CN^{-1}.$$

With the two functions $\psi^\pm(x_i) = CN^{-1}(1 - x_i) \pm (V_\varepsilon - v_\varepsilon)(x_i)$, and the discrete minimum principle for L_ε^N given in Theorem 2.8, the proof is completed in the usual way. □

While the above result is not only valid on the piecewise–uniform fitted mesh Ω_ε^N but, also on a uniform mesh, in the next lemma the role of the transition parameter in the fitted piecewise–uniform mesh is, for the first time, crucial.

Lemma 3.5 *For all $N \geq 4$ and at each mesh point $x_i \in \Omega_\varepsilon^N$ the singular component of the error satisfies the estimate*

$$|(W_\varepsilon - w_\varepsilon)(x_i)| \leq CN^{-1}\ln N$$

where w_ε is the solution of (3.9b) and W_ε is the solution of (3.15).

Proof We consider first the case $\sigma = 1/2$, and so $\varepsilon^{-1} \leq C\ln N$ and $h = N^{-1}$. The local truncation error is bounded in the standard way and, using Lemma 3.3, we then obtain

$$|L_\varepsilon^N(W_\varepsilon - w_\varepsilon)(x_i)| \leq C\varepsilon^{-2}(x_{i+1} - x_{i-1})e^{-\alpha x_{i-1}/\varepsilon} \leq C\varepsilon^{-2}N^{-1}e^{-\alpha x_{i-1}/\varepsilon} \tag{3.16}$$

We introduce the two mesh functions

$$\Psi_i^\pm = \frac{Ce^{2\gamma h/\varepsilon}}{\gamma(\alpha - \gamma)}\,\varepsilon^{-1}N^{-1}Y_i \pm (W_\varepsilon - w_\varepsilon)(x_i)$$

where γ is any constant satisfying $0 < \gamma < \alpha$ and

$$Y_i = \frac{\lambda^{N-i} - 1}{\lambda^N - 1}, \quad \lambda = 1 + \frac{\gamma h}{\varepsilon}.$$

It is easy to see that

$$D^+Y_i \leq -\frac{\gamma}{\varepsilon}e^{-\gamma x_{i+1}/\varepsilon} \tag{3.17}$$

and so Y_i decreases monotonically with $0 \leq Y_i \leq 1$. We then have $\Psi_0^\pm > 0$, $\Psi_N^\pm = 0$ and using (3.16), (3.17) and $(\varepsilon\delta^2 + \gamma D^+)Y_i = 0$, we obtain

$$L_\varepsilon^N \Psi_i^\pm = \frac{Ce^{2\gamma h/\varepsilon}}{\gamma(\alpha - \gamma)} \varepsilon^{-1} N^{-1}(a(x_i) - \gamma)D^+ Y_i \pm L_\varepsilon^N (W_\varepsilon - w_\varepsilon)(x_i)$$

$$\leq -C\varepsilon^{-2} N^{-1}\left(\frac{a(x_i) - \gamma}{\alpha - \gamma} e^{-\gamma x_{i-1}/\varepsilon} - e^{-\alpha x_{i-1}/\varepsilon}\right)$$

$$< 0.$$

By the discrete minimum principle we conclude that $\Psi_i^\pm \geq 0$ and so for all $x_i \in \overline{\Omega}_\varepsilon^N$

$$|(W_\varepsilon - w_\varepsilon)(x_i)| \leq \frac{Ce^{2\gamma h/\varepsilon}}{\gamma(\alpha - \gamma)} \varepsilon^{-1} N^{-1} Y_i \leq CN^{-1} \ln N$$

as required.

We now consider the case $\sigma = \frac{\varepsilon}{\alpha} \ln N$ and we give separate proofs in the coarse and fine mesh subintervals. First, suppose that $x_i \in [\sigma, 1]$. Using the triangle inequality we have

$$|(W_\varepsilon - w_\varepsilon)(x_i)| \leq |W_\varepsilon(x_i)| + |w_\varepsilon(x_i)|. \tag{3.18}$$

Using Lemma 3.3 we have

$$|w_\varepsilon(x_i)| \leq Ce^{-\alpha x_i/\varepsilon} \leq Ce^{-\alpha\sigma/\varepsilon} = CN^{-1}. \tag{3.19}$$

To establish a similar bound for $|W_\varepsilon(x_i)|$ we introduce the mesh function Y_i, which is the solution of the constant coefficient finite difference problem

$$(\varepsilon\delta^2 + \alpha D^+)Y_i = 0, \quad 1 \leq i \leq N - 1$$
$$Y_0 = 1, \qquad Y_N = 0.$$

Then

$$Y_i = \begin{cases} 1 + (Y_{N/2} - 1)\ell_i, & \text{if } i \leq N/2 \\ Y_{N/2} r_i, & \text{if } i \geq N/2 \end{cases} \tag{3.20}$$

where

$$\ell_i = \frac{1 - \lambda^{-i}}{1 - \lambda^{-N/2}}, \quad \lambda = 1 + \frac{\alpha h}{\varepsilon}$$

$$r_i = \frac{\Lambda^{N-i} - 1}{\Lambda^{N/2} - 1}, \quad \Lambda = 1 + \frac{\alpha H}{\varepsilon}$$

and $Y_{N/2}$ satisfies

$$(\varepsilon\delta^2 + \alpha D^+)Y_{N/2} = 0. \tag{3.21}$$

We now observe the inequality, for all $N \geq 1$,

$$\left(1 + \frac{2\ln N}{N}\right)^{-N/2} \leq 2N^{-1} \tag{3.22}$$

which is a consequence of the simple inequality $\ln(1 + t) > t(1 - t/2)$ for all $0 < t < 1$ with $t = 2N^{-1}\ln N$.

Since $\lambda = 1 + 2\alpha\sigma\varepsilon^{-1}N^{-1} = 1 + 2N^{-1}\ln N$ we obtain from (3.22)

$$\lambda^{-N/2} \leq 2N^{-1}. \tag{3.23}$$

Then, from (3.20),(3.21) and (3.23), we get, for all $N \geq 4$,

$$0 \leq \frac{\varepsilon}{\alpha}D^-\ell_{N/2} = \frac{\lambda^{-N/2}}{1 - \lambda^{-N/2}} \leq \frac{2N^{-1}}{1 - 2N^{-1}} \leq 4N^{-1}$$

$$-\frac{\varepsilon}{\alpha}D^+r_{N/2} = \frac{\Lambda^{N/2-1}}{\Lambda^{N/2} - 1} \geq \frac{1}{\Lambda}$$

$$0 \leq Y_{N/2} = \frac{D^-\ell_{N/2}}{D^-\ell_{N/2} - (1/2)(\lambda + \Lambda)D^+r_{N/2}} \leq \frac{8\Lambda N^{-1}}{\lambda + \Lambda} \leq 8N^{-1}$$

and

$$D^+Y_i \leq 0, \quad 0 \leq i \leq N - 1.$$

We introduce the two mesh functions

$$\Psi_i^\pm = |W_\varepsilon(0)|Y_i \pm W_\varepsilon(x_i).$$

Then $\Psi_0^\pm \geq 0, \Psi_N^\pm = 0$ and

$$L_\varepsilon^N \Psi_i^\pm = |W_\varepsilon(0)|(a(x_i) - \alpha)D^+Y_i \leq 0$$

It follows from the discrete minimum principle that $\Psi_i^\pm \geq 0$ and so using Lemma 3.3, for all $x_i \in [\sigma, 1]$,

$$|W_\varepsilon(x_i)| \leq |W_\varepsilon(0)|Y_i \leq |w_\varepsilon(0)|Y_{N/2} \leq CN^{-1} \tag{3.24}$$

as required.

It remains to prove the result for $x_i \in [0, \sigma)$. For $i = 0$ there is nothing to prove. For $x_i \in (0, \sigma)$ the proof follows the same lines as for the case $\sigma = 1/2$, except that we use the discrete minimum principle on $[0, \sigma]$ and the already established bound $|W_\varepsilon(x_{N/2})| \leq CN^{-1}$.

We have in this case

$$|L_\varepsilon^N(W_\varepsilon - w_\varepsilon)(x_i)| \le C\sigma\varepsilon^{-2}N^{-1}e^{-\alpha x_{i-1}/\varepsilon}. \qquad (3.25)$$

For all $i, 0 \le i \le N/2$, we introduce the two mesh functions

$$\Psi^\pm(x_i) = \frac{Ce^{2\gamma h/\varepsilon}}{\gamma(\alpha - \gamma)}\sigma\varepsilon^{-1}N^{-1}Y_i + C'N^{-1} \pm (W_\varepsilon - w_\varepsilon)(x_i)$$

where γ is as before and for all $i, 0 \le i \le N/2$

$$Y_i = \frac{\lambda^{N/2-i} - 1}{\lambda^{N/2} - 1} \,, \qquad \lambda = 1 + \frac{\gamma h}{\varepsilon}$$

$$D^+Y_i \le -\frac{\gamma}{\varepsilon}e^{-\gamma x_{i+1}/\varepsilon} < 0\,, \qquad 0 \le Y_i \le 1.$$

Then $\Psi_0^\pm > 0$, $\Psi_{N/2}^\pm \ge 0$ and $L_\varepsilon^N\Psi_i^\pm \le 0$ and the discrete minimum principle for L_ε^N on $(0, \sigma)$ gives $\Psi_i^\pm \ge 0$. Thus, for all $x_i \in (0, \sigma)$, we have

$$|(W_\varepsilon - w_\varepsilon)(x_i)| \le CN^{-1}\ln N$$

as required. □

Remark If instead of the two mesh functions $\Psi^\pm$ in the above argument, we use the simpler functions

$$C\frac{\sigma(\sigma - x_i)}{\varepsilon^2 N} + CN^{-1} \pm (W_\varepsilon - w_\varepsilon)(x_i)$$

we obtain the weaker estimate

$$|W_\varepsilon(x_i) - w_\varepsilon(x_i)| \le CN^{-1}(\ln N)^2$$

rather than the estimate in Lemma 3.5.

We now have the following ε–uniform error estimate.

Theorem 3.6 *Let u_ε be a solution to any problem from Problem Class 2.1 and let U_ε be the corresponding numerical solution generated by Method 3.1. Then, for all $N \ge 4$, we have*

$$\sup_{0<\varepsilon\le 1} \|U_\varepsilon - u_\varepsilon\|_{\Omega_\varepsilon^N} \le CN^{-1}\ln N$$

where C is a constant independent of N and ε.

Proof This follows immediately by combining the two previous lemmas. □

In the remainder of this section we outline the proof of the ε–uniform error estimate for the approximate solutions obtained when Method 3.2 is applied to any problem from Problem Class 2.2. This is analogous to the estimate given in Theorem 3.6 for Method 3.1 and Problem Class 2.1. We highlight only the essential differences in the proof. We begin with the appropriate minimum principle for a problem from Problem Class 2.2.

Theorem 3.7 *Let L_ε be the differential operator in (2.2a) and suppose that $v \in C^2(\overline{\Omega})$. Then, if $v'(0) \leq 0$, $v(1) \geq 0$ and $L_\varepsilon v(x) \leq 0$ for all $x \in \Omega$, it follows that $v(x) \geq 0$ for all $x \in \overline{\Omega}$.*

Proof As in Theorem 2.1, the proof is by contradiction. We assume that there exists a point $p \in \overline{\Omega}$ such that $v(p) = \min_{\overline{\Omega}} v(x) < 0$. From the hypotheses it follows that $p \neq 1$. We define the auxiliary function $w(x) = v(x)e^{\alpha x/(2\varepsilon)}$. Then $w(p) < 0$ and we choose $q \in \overline{\Omega}$ such that $w(q) = \min_{\overline{\Omega}} w(x) < 0$. Again from the hypotheses we see that $q \neq 1$. Suppose now that $q \in \Omega$ then $w'(q) = 0$ and $w''(q) \geq 0$. But

$$L_\varepsilon v(x) = (\varepsilon w''(x) + (a(x) - \alpha)w'(x) - \frac{\alpha}{2\varepsilon}(a(x) - \alpha/2)w(x))e^{-x\alpha/(2\varepsilon)}$$

and so $L_\varepsilon v(q) > 0$, which contradicts the hypotheses. The only remaining possibility is that w attains its minimum at the end–point $x = 0$. In that case $w'(0) \geq 0$ and $w(0) < 0$. But $w'(0) = v'(0) + \frac{\alpha}{2\varepsilon}v(0)$ and, since $w(0) < 0$ implies $v(0) < 0$, we conclude that $v'(0) > 0$ which is a contradiction. □

This minimum principle leads to the following stability result analogous to Lemma 3.1.

Lemma 3.8 *The solution u_ε of any problem from Problem Class 2.2 satisfies the bound*

$$\|u_\varepsilon\| \leq |u_\varepsilon(1)| + \frac{\varepsilon}{\alpha}|u_\varepsilon'(0)| + \frac{1}{\alpha}\|f\|.$$

Proof Using the two functions

$$\psi^\pm(x) = \frac{\varepsilon}{\alpha}|u_\varepsilon'(0)|(e^{-\alpha x/\varepsilon} - e^{-\alpha/\varepsilon}) + |u_\varepsilon(1)| + \frac{1}{\alpha}\|f\|(1-x) \pm u_\varepsilon(x).$$

and the minimum principle for L_ε in Theorem 3.7, the result follows in the usual way. □

We now consider the decomposition

$$u_\varepsilon = v_\varepsilon + w_\varepsilon, \quad v_\varepsilon = v_0 + \varepsilon v_1 + \varepsilon^2 v_2 \tag{3.26a}$$

where the components v_0, v_1 and v_2 are the solutions of the problems

$$av_0' = f, \quad v_0(1) = u_\varepsilon(1), \tag{3.26b}$$
$$av_1' = -v_0'', \quad v_1(1) = 0, \tag{3.26c}$$
$$\text{and} \quad L_\varepsilon v_2 = -v_1'', \quad \varepsilon v_2'(0) = 0, \ v_2(1) = 0. \tag{3.26d}$$

Thus the components v_ε and w_ε are the solutions of the problems

$$L_\varepsilon v_\varepsilon = f, \quad v_\varepsilon'(0) = v_0'(0) + \varepsilon v_1'(0), \ v_\varepsilon(1) = u_\varepsilon(1) \tag{3.27a}$$

and

$$L_\varepsilon w_\varepsilon = 0, \quad w_\varepsilon'(0) = u_\varepsilon'(0) - v_\varepsilon'(0), \ w_\varepsilon(1) = 0. \tag{3.27b}$$

It is not hard to prove that the bounds on the components of the decomposition of the solution and their derivatives, given in Lemma 3.3 for any problem from Problem Class 2.1, also apply to the components v_ε and w_ε defined in (3.27) for any problem from Problem Class 2.2. We omit the proof.

We now state and prove a theorem which establishes a discrete minimum principle for Method 3.2 analogous to Definition 2.7 and Theorem 2.8 for Method 2.2.

Theorem 3.9 *Let L_ε^N be the upwind finite difference operator defined in (2.13) and let Ω^N be an arbitrary mesh of $N+1$ mesh points. If V is any mesh function defined on this mesh such that*

$$D^+V(x_0) \leq 0, \quad V(x_N) \geq 0 \quad and \quad L_\varepsilon^N V \leq 0 \ in \ \Omega^N,$$

then

$$V(x_i) \geq 0 \ for \ all \ x_i \in \overline{\Omega}^N.$$

Proof Assume that $V(x_k) = \min_i V(x_i) < 0$. From the hypotheses we know that $k \neq N$. Suppose now that $k = 0$ and so $V(x_0) = \min_i V(x_i)$. Since $D^+V(x_0) \leq 0$ we have $V(x_1) \leq V(x_0)$ and the only possibility is that $V(x_1) = V(x_0)$. Then $L_\varepsilon^N V(x_1) \leq 0$ implies that $D^+V(x_1) \leq 0$, which in turn implies that $V(x_2) = V(x_1) = V(x_0)$, to avoid a contradiction. Applying the same argument repeatedly we eventually conclude that $V(x_N) = V(x_0) < 0$, which is a contradiction. An analogous argument applies for all k, $0 < k < N$.   □

This discrete minimum principle leads at once to the corresponding discrete stability result for Method 3.2, which is analogous to Lemma 2.9 for Method 2.2.

Lemma 3.10 *The solution U_ε, obtained when Method 3.2 is applied to any problem from Problem Class 2.2, satisfies the bound*

$$|U_\varepsilon(x_i)| \le |U_\varepsilon(1)| + C\varepsilon|D^+U_\varepsilon(0)| + \frac{1}{\alpha}\|f\|.$$

Proof We introduce the two mesh functions

$$\Psi^\pm(x_i) = \varepsilon|D^+U_\varepsilon(0)|\Phi_i + |U_\varepsilon(1)| + \frac{1}{\alpha}\|f\|(1-x_i) \pm U_\varepsilon(x_i),$$

where Φ_i is the solution of the constant coefficient problem

$$\varepsilon\delta^2\Phi_i + \alpha D^+\Phi_i = 0, \quad \varepsilon D^+\Phi_0 = -1, \quad \Phi_N = 0.$$

Then the proof is completed in the usual way using the discrete minimum principle for L_ε^N given in Theorem 3.9. $\qquad\square$

We now state the following ε–uniform error estimate for Method 3.2 analogous to Theorem 3.6 for Method 3.1.

Theorem 3.11 *Let u_ε be the solution of any problem from Problem Class 2.2 and let U_ε be the corresponding numerical solution generated by Method 3.2. Then, for all $N \ge 4$, we have*

$$\sup_{0<\varepsilon\le1} \|U_\varepsilon - u_\varepsilon\|_{\overline{\Omega}_\varepsilon^N} \le CN^{-1}\ln N$$

where C is a constant independent of N and ε.

Proof Lemmas analogous to Lemmas 3.4 and 3.5 are proved by stability and consistency arguments analogous to those used in the proofs of these lemmas. These arguments require two bounds, the first of which is obtained from

$$\varepsilon D^+(V_\varepsilon - v_\varepsilon)(0) = \varepsilon v_\varepsilon'(0) - \varepsilon D^+v_\varepsilon(0) = \frac{\varepsilon}{h}\int_0^h (s-h)v_\varepsilon''(s)\,ds$$

which, using Lemma 3.3, leads to the bound

$$|\varepsilon D^+(V_\varepsilon - v_\varepsilon)(0)| \le C\varepsilon N^{-1}$$

and the second of which follows from

$$\varepsilon D^+(W_\varepsilon - w_\varepsilon)(0) = \frac{\varepsilon}{h}\int_0^h (s-h)w_\varepsilon''(s)\,ds \le \frac{Ch}{\varepsilon}$$

which implies that

$$|\varepsilon D^+(W_\varepsilon - w_\varepsilon)(0)| \le CN^{-1}\ln N.$$

We omit further details. $\qquad\square$

In summary, we see that the simple piecewise–uniform fitted mesh (3.2), in conjunction with the standard upwind finite difference operator (3.3a), yields a monotone numerical method, whose solutions have the required ε–uniform pointwise–accurate behaviour at the mesh points $\overline{\Omega}_\varepsilon^N$. In the next section we show, both computationally and theoretically, that this ε–uniform pointwise–accurate behaviour extends to the entire domain $\overline{\Omega}$, when the discrete solution at the mesh points is extended to the whole domain by piecewise linear interpolation.

3.4   Global accuracy on piecewise–uniform meshes

In this section we compute the exact global ε–uniform maximum pointwise error in the numerical solutions of the specific problem (2.3) obtained with Method 3.1.

We observe that, in general, on any subinterval $[x_{i-1}, x_i]$, the exact maximum pointwise error occurs either at one of the endpoints x_{i-1}, x_i or at an interior point x_i^* such that

$$(\overline{U}_\varepsilon - u_\varepsilon)'(x_i^*) = 0$$

where $\overline{U}_\varepsilon$ denotes the piecewise linear interpolant of the finite difference solution $\{U_\varepsilon(x_i)\}_{i=0}^N$. In the case of a numerical solution of problem (2.3), obtained by Method 3.1, we can determine x_i^* explicitly for each i, $1 \le i \le N$. Consequently, for each ε, we can evaluate the exact global maximum pointwise error $\bar{E}_{\varepsilon,\text{exact}}^N$ from the expression

$$\bar{E}_{\varepsilon,\text{exact}}^N = \max\{E_{\varepsilon,\text{exact}}^N, \max_i |\overline{U}_\varepsilon(x_i^*) - u_\varepsilon(x_i^*)|\}$$

and the exact global ε–uniform maximum pointwise error $\bar{E}_{\text{exact}}^N$ from the expression

$$\bar{E}_{\text{exact}}^N = \max_\varepsilon \bar{E}_{\varepsilon,\text{exact}}^N.$$

The results for problem (2.3) using Method 3.1 are given in Table 3.3. We see that this error has essentially the same magnitude as the exact ε–uniform maximum pointwise error at the mesh points. More precisely, comparing Table 3.3 with Table 3.1, we see that for all $N \ge 128$ the exact global ε–uniform error $\bar{E}_{\text{exact}}^N$ and the exact nodal ε–uniform error E_{exact}^N are approximately equal. This behaviour is in stark contrast to the situation for numerical solutions on a uniform mesh. We saw in §2.5 that the piecewise linear interpolant of the numerical solution, obtained when Method 2.3 is applied to the same problem (2.3) has an error of 20% or more, even though the numerical solution is exact at the mesh points.

We now state and prove a theorem, which extends the nodal ε–uniform error estimates in Theorems 3.6 and 3.11 to global ε–uniform error estimates.

Table 3.3 *Exact maximum global pointwise errors $\bar{E}_{\varepsilon,\text{exact}}^N$ and $\bar{E}_{\text{exact}}^N$ generated by Method 3.1 applied to problem (2.3) for various values of ε and N.*

ε	Number of intervals N						
	8	16	32	64	128	256	512
2^{-2}	0.110465	0.064689	0.038941	0.022394	0.011323	0.005693	0.002854
2^{-4}	0.203972	0.090851	0.048979	0.026481	0.014584	0.008112	0.004512
2^{-6}	0.272095	0.131234	0.059160	0.031363	0.016511	0.008761	0.004705
2^{-8}	0.301089	0.155676	0.071770	0.034718	0.018508	0.009762	0.005124
2^{-10}	0.311136	0.164958	0.079362	0.036106	0.019473	0.010449	0.005554
2^{-12}	0.314311	0.168013	0.082013	0.038271	0.019772	0.010701	0.005760
2^{-14}	0.315267	0.168953	0.082852	0.038984	0.019852	0.010772	0.005824
2^{-16}	0.315546	0.169231	0.083104	0.039204	0.019872	0.010790	0.005841
2^{-18}	**0.315626**	**0.169311**	**0.083178**	**0.039268**	**0.019877**	**0.010795**	**0.005845**
$\bar{E}_{\text{exact}}^N$	0.315626	0.169311	0.083178	0.039268	0.019877	0.010795	0.005845

Theorem 3.12 *Let u_ε be a solution of any problem from Problem Class 2.1 or 2.2 and let U_ε be the corresponding numerical solution generated by Method 3.1 or 3.2, respectively. Then, for all $N \geq 4$, we have*

$$\sup_{0 < \varepsilon \leq 1} \|\bar{U}_\varepsilon - u_\varepsilon\|_{\overline{\Omega}} \leq C N^{-1} \ln N$$

where $\bar{U}_\varepsilon$ is the piecewise linear interpolant of U_ε on $\overline{\Omega}$ and C is a constant independent of N and ε.

Proof We give the proof for Method 3.1 applied to Problem Class 2.1. The proof for the other case is analogous. Let $\bar{u}_\varepsilon$ be the piecewise linear interpolant of the values of u_ε at the mesh points $\overline{\Omega}_\varepsilon^N$, that is,

$$\bar{u}_\varepsilon(x) = \sum_{i=0}^{N} u_\varepsilon(x_i)\phi_i(x),$$

where $\phi_i(x)$ is the piecewise linear function defined by $\phi_i(x_j) = \delta_{i,j}$ for all i, j, $0 \leq i, j \leq N$. Using the triangle inequality we have

$$\|\bar{U}_\varepsilon - u_\varepsilon\| \leq \|\bar{U}_\varepsilon - \bar{u}_\varepsilon\| + \|\bar{u}_\varepsilon - u_\varepsilon\| \tag{3.28}$$

where the first term on the right–hand side is a difference between two interpolants, and the second is an interpolation error. To bound the first term on the right–hand side of (3.28), we note that

$$\bar{U}_\varepsilon - \bar{u}_\varepsilon = \sum_{i=1}^{N-1} (U_\varepsilon - u_\varepsilon)(x_i)\phi_i.$$

Using Theorem 3.6 at each mesh point x_i, we have

$$|(U_\varepsilon - u_\varepsilon)(x_i)| \leq CN^{-1} \ln N.$$

Since $\phi_i(x) \geq 0$ and $\|\sum_{i=1}^{N-1} \phi_i\| \leq 1$, it follows that

$$\|\bar{U}_\varepsilon - \bar{u}_\varepsilon\| \leq CN^{-1} \ln N. \tag{3.29}$$

To bound the second term on the right–hand side of (3.28) we note that for each $i, 0 \leq i \leq N - 1$ and any $f \in C^2(\Omega_i)$, where $\Omega_i = (x_i, x_{i+1})$ and $x \in \overline{\Omega}_i$ we have the following classical estimate for linear interpolation

$$|(\bar{f} - f)(x)| \leq \frac{1}{2}(x_{i+1} - x_i)^2 \|f''\|_{\Omega_i}. \tag{3.30}$$

Corresponding to the decomposition (3.8) of the exact solution u_ε, we decompose the interpolation error into the sum of two components

$$\bar{u}_\varepsilon - u_\varepsilon = \bar{v}_\varepsilon - v_\varepsilon + \bar{w}_\varepsilon - w_\varepsilon$$

where $\bar{v}_\varepsilon$, $\bar{w}_\varepsilon$ are the piecewise linear interpolants of the values of v_ε, w_ε respectively at the mesh points $\overline{\Omega}_\varepsilon^N$. We bound each component separately. From (3.30) and Lemma 3.3 we have

$$\|\bar{v}_\varepsilon - v_\varepsilon\| \leq CN^{-2} \tag{3.31}$$

and, for all $x \in \overline{\Omega}_i$,

$$|(\bar{w}_\varepsilon - w_\varepsilon)(x)| \leq C(x_{i+1} - x_i)^2 \varepsilon^{-2}. \tag{3.32}$$

In the case of a uniform mesh, we have $\sigma = 1/2$ and $\varepsilon^{-1} \leq C \ln N$. It follows from (3.32) that

$$\|\bar{w}_\varepsilon - w_\varepsilon\| \leq C(N^{-1} \ln N)^2.$$

On the other hand, if $\sigma = (\varepsilon/\alpha) \ln N$ it follows from (3.32) that

$$\|\bar{w}_\varepsilon - w_\varepsilon\|_{[0,\sigma]} \leq C\left(\frac{\sigma}{\varepsilon}\right)^2 N^{-2} \leq C(N^{-1} \ln N)^2$$

and from Lemma 3.3 that

$$\|\bar{w}_\varepsilon - w_\varepsilon\|_{[\sigma,1]} \leq 2\|w_\varepsilon\|_{[\sigma,1]} \leq Ce^{-\alpha\sigma/\varepsilon} = CN^{-1}.$$

In all cases therefore

$$\|\bar{w}_\varepsilon - w_\varepsilon\| \leq CN^{-1} \ln N. \tag{3.33}$$

Using (3.31), (3.33) and the triangle inequality completes the proof.   □

Remark   Note that, if we use the transition parameter $\sigma^* = \min\{1/2, 2(\varepsilon/\alpha)\ln N\}$ instead of $\sigma = \min\{1/2, (\varepsilon/\alpha)\ln N\}$, then

$$|w_\varepsilon(x)| \le e^{-\alpha\sigma^*/\varepsilon} = CN^{-2}$$

and so (3.33) is replaced by the stronger ε–uniform interpolation error estimate

$$\|\bar{w}_\varepsilon - w_\varepsilon\|_{\overline{\Omega}} \le CN^{-2}(\ln N)^2.$$

However, because of (3.29), even with σ^* as the transition parameter, there is no improvement of the error estimate in Theorem 3.12.

3.5   Approximation of derivatives

Frequently, users of numerical methods are interested in approximating a scaled first derivative of the solution, and sometimes even a scaled second derivative, rather than the solution itself. In the case of problem (2.4) from Problem Class 2.2, an appropriately scaled first derivative of the solution is

$$\varepsilon u_\varepsilon'(x) = -2e^{-2x/\varepsilon}$$

which is bounded for all $\varepsilon \in (0,1]$, since $\|\varepsilon u_\varepsilon'\| = 2$. A standard approximation of the derivative u_ε' is the simple forward discrete derivative D^+U_ε and we approximate the scaled discrete derivative $\varepsilon D^+U_\varepsilon$ by the scaled derivative $\varepsilon u_\varepsilon'$.

Problems (2.3) and (2.4) are designed so that $\|u_\varepsilon\| = 1$, thus ensuring, as before, that the absolute and relative errors coincide. But for approximations to the scaled first derivatives this is not the case and so, in principle, we need to consider the exact maximum pointwise relative error

$$\max_{0 \le i < N} \frac{\varepsilon|D^+U_\varepsilon(x_i) - u_\varepsilon'(x_i)|}{\varepsilon\|u_\varepsilon'\|}.$$

In practice, for a non–trivial problem, we do not estimate the denominator $\varepsilon\|u_\varepsilon'\| = O(1)$. Instead, we estimate the exact maximum pointwise absolute error

$$E_{\varepsilon,\text{exact}}^N(\varepsilon D^+U_\varepsilon) = \max_{0 \le i < N} \varepsilon|D^+U_\varepsilon(x_i) - u_\varepsilon'(x_i)|$$

and the ε–uniform maximum pointwise absolute error

$$E_{\text{exact}}^N(\varepsilon D^+U_\varepsilon) = \max_\varepsilon E_{\varepsilon,\text{exact}}^N(\varepsilon D^+U_\varepsilon).$$

In Table 3.4 these quantities are displayed for the numerical results obtained when the monotone Method 3.2 is applied to problem (2.4). In this table we observe that the errors $E_{\varepsilon,\text{exact}}^N(\varepsilon D^+U_\varepsilon)$ stabilize as ε decreases and that, for all $N \ge 32$ and all $\varepsilon \le 2^{-3}$, the errors $E_{\varepsilon,\text{exact}}^N(\varepsilon D^+U_\varepsilon)$ have stabilized. We note also that the uniform errors $E_{\text{exact}}^N(\varepsilon D^+U_\varepsilon)$ decrease monotonically as N increases.

Table 3.4 *Exact maximum pointwise errors $E^N_{\varepsilon,\text{exact}}(\varepsilon D^+ U_\varepsilon)$, $E^N_{\text{exact}}(\varepsilon D^+ U_\varepsilon)$ generated by Method 3.2 applied to problem (2.4) for various values of ε and N.*

ε	Number of intervals N						
	8	16	32	64	128	256	512
2^{-1}	0.153130	0.083441	0.043730	0.022412	0.011349	0.005711	0.002865
2^{-2}	0.158703	0.111998	0.073243	0.043730	0.022412	0.011349	0.005711
2^{-3}	0.158703	0.111998	0.073243	0.045399	0.027031	0.015655	0.008875
2^{-4}	0.158703	0.111998	0.073243	0.045399	0.027031	0.015655	0.008875
2^{-5}	0.186704	0.111998	0.073243	0.045399	0.027031	0.015655	0.008875
2^{-6}	0.216490	0.111998	0.073243	0.045399	0.027031	0.015655	0.008875
2^{-7}	0.232737	0.111998	0.073243	0.045399	0.027031	0.015655	0.008875
2^{-8}	0.241236	0.117451	0.073243	0.045399	0.027031	0.015655	0.008875
$\cdot$	$\cdot$	$\cdot$	$\cdot$	$\cdot$	$\cdot$	$\cdot$	$\cdot$
$\cdot$	$\cdot$	$\cdot$	$\cdot$	$\cdot$	$\cdot$	$\cdot$	$\cdot$
2^{-18}	0.249991	0.124992	0.073243	0.045399	0.027031	0.015655	0.008875
$E^N_{\text{exact}}(\varepsilon D^+ U_\varepsilon)$	0.249991	0.124992	0.073243	0.045399	0.027031	0.015655	0.008875

We recall the example at the end of §2.5, of a finite difference scheme on a uniform mesh that is exact at the mesh points when applied to a problem from Problem Class 2.1, but for which

$$\lim_{N \to \infty} \varepsilon |D^+ u_\varepsilon(0) - u'_\varepsilon(0)| = 2e^{-1} \neq 0.$$

In other words, on a uniform mesh, the scaled forward discrete derivative of the exact solution u_ε does not converge ε–uniformly to the exact scaled first derivative as the mesh is refined. We now demonstrate that the use of the piecewise–uniform mesh Ω_ε^N defined in (3.3c) overcomes this problem. We note first that for any i, $0 \leq i \leq N-1$

$$D^+ u_\varepsilon(x_i) - u'_\varepsilon(x) = \frac{1}{x_{i+1} - x_i} \int_{s=x_i}^{x_{i+1}} \int_{t=x_i}^{s} u''_\varepsilon(t)\, dt\, ds - \int_{t=x_i}^{x} u''_\varepsilon(t)\, dt \, .$$

It follows from Lemma 3.2 and the definition of σ that

$$\varepsilon |D^+ u_\varepsilon(0) - u'_\varepsilon(0)| = \frac{\varepsilon}{x_1} \left| \int_{s=0}^{x_1} \int_{t=0}^{s} u''_\varepsilon(t)\, dt\, ds \right| \leq C \frac{\sigma}{\varepsilon N} \leq C N^{-1} \ln N$$

and so at the point $x = 0$ we have convergence as the mesh is refined. We now show that, for any mesh point $x_i \in \Omega_\varepsilon^N \cup \{0\}$, the same bound is valid for $|D^+ u_\varepsilon(x_i) - u'_\varepsilon(x_i)|$, which is a special case of the following stronger result.

Lemma 3.13 *For each $x_i \in \Omega_\varepsilon^N \cup \{0\}$ and all $x \in \overline{\Omega}_i$ we have*

$$|\varepsilon(D^+ u_\varepsilon(x_i) - u'_\varepsilon(x))| \leq C N^{-1} \ln N$$
$$|D^+ v_\varepsilon(x_i) - v'_\varepsilon(x)| \leq C N^{-1}$$
$$|\varepsilon(D^+ w_\varepsilon(x_i) - w'_\varepsilon(x))| \leq C N^{-1} \ln N$$

where $u_\varepsilon = v_\varepsilon + w_\varepsilon$ is the solution of any problem from Problem Class 2.1 and C is a constant independent of N and ε.

Proof Since we have

$$|\varepsilon(D^+ u_\varepsilon(x_i) - u'_\varepsilon(x))| \leq |\varepsilon(D^+ v_\varepsilon(x_i) - v'_\varepsilon(x))| + |\varepsilon(D^+ w_\varepsilon(x_i) - w'_\varepsilon(x))|$$

it suffices to bound each term on the right–hand side separately. Any function $\phi \in C^2(\Omega_i)$ satisfies the identity

$$D^+\phi(x_i) - \phi'(x) = \frac{1}{x_{i+1} - x_i} \int_{s=x_i}^{x_{i+1}} \int_{t=x_i}^{s} \phi''(t) \, dt \, ds - \int_{t=x_i}^{x} \phi''(t) \, dt \quad (3.34)$$

from which it follows that

$$|D^+\phi(x_i) - \phi'(x)| \leq \frac{3}{2}(x_{i+1} - x_i)\|\phi''\|_{\overline{\Omega}_i}. \quad (3.35)$$

Applying this to v_ε and using Lemma 3.3 we see that

$$|D^+ v_\varepsilon(x_i) - v'_\varepsilon(x)| \leq C\frac{3}{2}(x_{i+1} - x_i)\|v''_\varepsilon\|_{\overline{\Omega}_i} \leq CN^{-1} \quad (3.36)$$

which gives the required bound in the first term. For the second term we have

$$|\varepsilon(D^+ w_\varepsilon(x_i) - w'_\varepsilon(x))| \leq C\frac{3}{2}(x_{i+1} - x_i)\|\varepsilon w''_\varepsilon\|_{\overline{\Omega}_i} \leq C\varepsilon^{-1}(x_{i+1} - x_i). \quad (3.37)$$

We now observe that in the case $\sigma = 1/2$ we have $x_{i+1} - x_i = N^{-1}$ and $\varepsilon^{-1} \leq C \ln N$. Moreover, in the case $\sigma = \frac{\varepsilon}{\alpha} \ln N$ and $x_i \in [0, \sigma)$ we have $x_{i+1} - x_i = 2\sigma N^{-1}$. In both of these cases the required result follows from (3.37).

It remains therefore to consider the case $\sigma = \frac{\varepsilon}{\alpha} \ln N$ and $x_i \in [\sigma, 1)$. From the differential equation for w_ε and integration by parts we obtain

$$\int_{t=x_i}^{s} \varepsilon w''_\varepsilon(t) \, dt = -\int_{t=x_i}^{s} a(t) w'_\varepsilon(t) \, dt = \int_{t=x_i}^{s} a'(t) w_\varepsilon(t) \, dt - a(t) w_\varepsilon(t)\Big|_{t=x_i}^{s}.$$

Using Lemma 3.3 we conclude that for all $s \in \overline{\Omega}_i$

$$\left|\int_{t=x_i}^{s} \varepsilon w''_\varepsilon(t) \, dt\right| \leq C\|w_\varepsilon\|_{\overline{\Omega}_i} \leq Ce^{-\alpha\sigma/\varepsilon} = CN^{-1}.$$

The required bound then follows by using this result and (3.34) for $\varepsilon w_\varepsilon$. □

It can be shown that the above estimate is also valid for solutions of problems from Problem Class 2.2. This estimate and the numerical results in Table 3.4, in conjunction with the triangle inequality, suggest that the scaled discrete derivative $\varepsilon D^+ U_\varepsilon$ derived from Method 3.2 converges to the scaled derivative $\varepsilon u'_\varepsilon$ for any problem from Problem Class 2.2.

In the remainder of this section, we show mathematically that this is indeed the case for Method 3.1 applied to Problem Class 2.1. The mathematical proofs are not especially revealing in themselves and so the reader may skip them without any loss of new ideas.

Lemma 3.14 *At each mesh point* $x_i \in \Omega_\varepsilon^N \cup \{0\}$

$$\left|\varepsilon D^+ \left(V_\varepsilon(x_i) - v_\varepsilon(x_i)\right)\right| \leq CN^{-1},$$

where v_ε *is the solution of (3.9a) and* V_ε *is the solution of (3.14)*

Proof For convenience we introduce the notation

$$e_i = \left(V_\varepsilon - v_\varepsilon\right)(x_i) \quad \text{and} \quad \tau_i = L_\varepsilon^N e_i.$$

We want to prove that for all $i, 0 \leq i \leq N-1$, $|\varepsilon D^+ e_i| \leq CN^{-1}$. Since $e_N = 0$ we have from Lemma 3.4

$$\left|\varepsilon D^+ e_{N-1}\right| = \frac{|\varepsilon e_{N-1}|}{1 - x_{N-1}} \leq C\varepsilon N^{-1}. \tag{3.38}$$

To prove the result for $0 \leq i \leq N-2$, we rewrite the relation $\tau_i = L_\varepsilon^N e_i$ in the form

$$\varepsilon D^+ e_j - \varepsilon D^+ e_{j-1} + \frac{1}{2}(x_{j+1} - x_{j-1})a(x_j)D^+ e_j = \frac{1}{2}(x_{j+1} - x_{j-1})\tau_j.$$

Summing this and rearranging we obtain

$$\left|\varepsilon D^+ e_i\right| \leq \left|\varepsilon D^+ e_{N-1}\right| + \frac{1}{2}\sum_{j=i+1}^{N-1}(x_{j+1} - x_{j-1})|\tau_j|$$

$$+ \frac{1}{2}\left|\sum_{j=i+1}^{N-1}(x_{j+1} - x_{j-1})a(x_j)D^+ e_j\right|.$$

We now bound each term separately. We have already bounded the first. From the proof of Lemma 3.4 we know that $|\tau_j| \leq CN^{-1}$ and so the second is also bounded by CN^{-1}. To bound the last term we observe that

$$(x_{j+1} - x_{j-1})a(x_j)D^+ e_j = \left(\frac{x_{j+1} - x_{j-1}}{x_{j+1} - x_j}a(x_j)e_{j+1} - \frac{x_j - x_{j-2}}{x_j - x_{j-1}}a(x_{j-1})e_j\right)$$

$$- \frac{x_{j+1} - x_{j-1}}{x_{j+1} - x_j}(a(x_j) - a(x_{j-1}))e_j - \left(\frac{x_{j+1} - x_{j-1}}{x_{j+1} - x_j} - \frac{x_j - x_{j-2}}{x_j - x_{j-1}}\right)a(x_{j-1})e_j.$$

We now sum both sides of this expression. We observe that the terms in the first bracket on the right–hand side telescope and that the last bracket on the right–hand side is non–zero only for $j = N/2$ and $j = N/2 + 1$. It follows that for $i < N/2$

$$\sum_{j=i+1}^{N-1} (x_{j+1} - x_{j-1})a(x_j)D^+e_j$$

$$= \left(\frac{x_N - x_{N-2}}{x_N - x_{N-1}}a(x_{N-1})e_N - \frac{x_{i+1} - x_{i-1}}{x_{i+1} - x_i}a(x_i)e_{i+1}\right)$$

$$- \sum_{j=i+1}^{N-1} \frac{x_{j+1} - x_{j-1}}{x_{j+1} - x_j}(a(x_j) - a(x_{j-1}))e_j$$

$$+ (1 - \frac{h}{H})\left(a(x_{N/2-1})e_{N/2} - a(x_{N/2})e_{N/2+1}\right)$$

and for $i \geq N/2$ some or all of the last term on the right–hand side will be zero. From Lemma 3.4 we know that $|e_j| \leq CN^{-1}$. We also have $\frac{x_{j+1}-x_{j-1}}{x_{j+1}-x_j} \leq 2$ and $|a(x_j) - a(x_{j-1})| \leq \|a'\|(x_j - x_{j-1})$. It follows that for all $0 \leq i \leq N - 2$

$$\left| \sum_{j=i+1}^{N-1} (x_{j+1} - x_{j-1})a(x_j)D^+e_j \right| \leq CN^{-1}$$

which completes the proof. $\square$

Lemma 3.15 Let w_ε be the solution of (3.9b) and W_ε the solution of (3.15). Then, when $\sigma = 1/2$ we have for all $x_i \in \overline{\Omega}_\varepsilon^N$

$$|\varepsilon(W_\varepsilon - w_\varepsilon)(x_i)| \leq CN^{-1}\ln N(1 - x_i) \qquad (3.39)$$

and when $\sigma = \frac{\varepsilon}{\alpha}\ln N$ we have for all $x_i \in [\sigma, 1)$

$$|W_\varepsilon(x_i)| \leq C(1 - x_i)N^{-1} \qquad (3.40)$$

and

$$|\varepsilon D^+W_\varepsilon(x_i)| \leq CN^{-1}. \qquad (3.41)$$

Proof When $\sigma = 1/2$ we have from the proof of Lemma 3.5

$$|L_\varepsilon^N(W_\varepsilon - w_\varepsilon)(x_i)| \leq C\varepsilon^{-2}N^{-1}e^{-\alpha x_{i-1}/\varepsilon}$$

We introduce the two mesh functions

$$\Psi_i^\pm = \frac{C}{\alpha}\varepsilon^{-2}N^{-1}(1 - x_i) \pm (W_\varepsilon - w_\varepsilon)(x_i)$$

and we observe that $\Psi_0^\pm > 0$, $\Psi_N^\pm = 0$ and

$$L_\varepsilon^N \Psi_i^\pm \leq -C\varepsilon^{-2}N^{-1} \pm C\varepsilon^{-2}N^{-1}\left(1 - e^{-\alpha x_{i-1}/\varepsilon}\right) \leq 0.$$

The discrete minimum principle for L_ε^N then gives $\Psi_i^\pm \geq 0$ and so

$$|(W_\varepsilon - w_\varepsilon)(x_i)| \leq \frac{C}{\alpha}\varepsilon^{-2}N^{-1}(1 - x_i)$$

or

$$|\varepsilon(W_\varepsilon - w_\varepsilon)(x_i)| \leq C(1 - x_i)N^{-1}\ln N$$

which is (3.39).

To prove (3.40) we first consider on $(\sigma, 1)$ the two mesh functions

$$\Psi_i^\pm = |W_\varepsilon(\sigma)|\frac{1 - x_i}{1 - \sigma} \pm W_\varepsilon(x_i).$$

Then $\Psi_{N/2}^\pm \geq 0, \Psi_N^\pm = 0$ and $L_\varepsilon^N \Psi_i^\pm < 0$ and so, by the discrete minimum principle for L_ε^N on $(\sigma, 1)$ we get $\Psi_i^\pm \geq 0$. Hence, using the proof of Lemma 3.5,

$$|W_\varepsilon(x_i)| \leq |W_\varepsilon(\sigma)|\frac{1 - x_i}{1 - \sigma} < CN^{-1}(1 - x_i)$$

as required.

To prove (3.41) for $i = N - 1$ we use (3.40) and obtain

$$|\varepsilon D^+ W_\varepsilon(x_{N-1})| = \frac{|\varepsilon W_\varepsilon(x_{N-1})|}{1 - x_{N-1}} \leq CN^{-1}$$

as required. For $N/2 \leq i \leq N - 2$ we write $L_\varepsilon^N W_\varepsilon(x_j) = 0$ in the form

$$\varepsilon D^+\left(W_\varepsilon(x_j) - W_\varepsilon(x_{j-1})\right) + \frac{1}{2}a(x_j)(x_{j+1} - x_{j-1})D^+ W_\varepsilon(x_j) = 0.$$

Summing and rearranging gives

$$\varepsilon D^+ W_\varepsilon(x_i) = \varepsilon D^+ W_\varepsilon(x_{N-1}) + \frac{1}{2}\sum_{j=i+1}^{N-1}(x_{j+1} - x_{j-1})a(x_j)D^+ W_\varepsilon(x_j)$$

and so for $x_i \in [\sigma, 1)$

$$\varepsilon\, D^+ W_\varepsilon(x_i) = \varepsilon D^+ W_\varepsilon(x_{N-1}) + \sum_{j=i+1}^{N-1}a(x_j)(W_\varepsilon(x_{i+1}) - W_\varepsilon(x_j))$$

$$= \varepsilon D^+ W_\varepsilon(x_{N-1})$$

$$+ \sum_{j=i+1}^{N-1}\left(a(x_j)W_\varepsilon(x_{j+1}) - a(x_{j-1})W_\varepsilon(x_j) - (a(x_j) - a(x_{j-1}))W_\varepsilon(x_j)\right)$$

$$= \varepsilon D^+ W_\varepsilon(x_{N-1}) - a(x_i)W_\varepsilon(x_{i+1}) - \sum_{j=i+1}^{N-1}(a(x_j) - a(x_{j-1}))W_\varepsilon(x_j)$$

Using the result for $i = N - 1$ and (3.24) in the proof of Lemma 3.5 we conclude that

$$|\varepsilon D^+ W_\varepsilon(x_i)| \le CN^{-1}$$

as required. $\qquad\square$

Lemma 3.16 *Let w_ε be the solution of (3.9b) and W_ε the solution of (3.15). Then, for all $x_i \in \Omega_\varepsilon^N \cup \{0\}$, we have*

$$|\varepsilon D^+ (W_\varepsilon - w_\varepsilon)(x_i))| \le CN^{-1} \ln N$$

where C is a constant independent of N and ε.

Proof We establish the result for the cases $\sigma = 1/2$ and $\sigma = (\varepsilon/\alpha) \ln N$ separately. First, suppose that $\sigma = (\varepsilon/\alpha) \ln N$. For all $x_i \in [\sigma, 1)$ we have from the triangle inequality

$$|\varepsilon D^+ (W_\varepsilon - w_\varepsilon)(x_i))| \le |\varepsilon (D^+ W_\varepsilon - w_\varepsilon')(x_i))| + |\varepsilon (D^+ w_\varepsilon - w_\varepsilon')(x_i))|.$$

We bound each term on the right–hand side separately. By Lemma 3.13

$$|\varepsilon (D^+ w_\varepsilon - w_\varepsilon')(x_i))| \le CN^{-1} \ln N.$$

To bound the first term on the right–hand side we have from the triangle inequality, Lemma 3.15 and (3.10)

$$\begin{aligned}
|\varepsilon (D^+ W_\varepsilon - w_\varepsilon')(x_i)| &= |\varepsilon D^+ W_\varepsilon(x_i)| + |\varepsilon w_\varepsilon'(x_i)| \\
&\le CN^{-1} + Ce^{-\alpha x_i/\varepsilon} \\
&\le CN^{-1}
\end{aligned}$$

since $x_i \in [\sigma, 1)$. For $x_i = \sigma$, we write $L_\varepsilon^N W_\varepsilon(\sigma) = 0$ in the form

$$\varepsilon D^+ W_\varepsilon(x_{N/2-1}) = \left(1 + \frac{h}{2\varepsilon} a(\sigma)\right) \varepsilon D^+ W_\varepsilon(\sigma) + \frac{1}{2} a(\sigma)\left(W_\varepsilon(x_{N/2+1}) - W_\varepsilon(\sigma)\right)$$

Using the estimate obtained at the point σ and the proof of Lemma 3.5 we obtain

$$|\varepsilon D^+ W_\varepsilon(x_{N/2-1})| \le (1 + CN^{-1} \ln N)CN^{-1} + CN^{-1} \le CN^{-1}$$

as required. We now consider $x_i \in [0, \sigma)$. For convenience we introduce the notation

$$\hat{e}_i = (W_\varepsilon - w_\varepsilon)(x_i) \quad \text{and} \quad \hat{\tau}_i = L_\varepsilon^N \hat{e}_i.$$

By Lemma 3.5 and its proof we have

$$|\hat{e}_i| \leq CN^{-1}\ln N \quad \text{and} \quad |\hat{\tau}_i| \leq C\sigma\varepsilon^{-2}N^{-1}e^{-\alpha x_{i-1}/\varepsilon}. \qquad (3.42)$$

We write the equation $\hat{\tau}_i = L_\varepsilon^N \hat{e}_i$ in the form

$$\varepsilon D^+(\hat{e}_j - \hat{e}_{j-1}) + \frac{1}{2}a(x_j)(x_{j+1} - x_{j-1})D^+\hat{e}_j = \frac{1}{2}(x_{j+1} - x_{j-1})\hat{\tau}_j. \qquad (3.43)$$

Summing and rearranging gives

$$\varepsilon D^+\hat{e}_i = \varepsilon D^+\hat{e}_{N/2-1} + \sum_{j=i+1}^{N/2-1}\left[a(x_j)(\hat{e}_{j+1} - \hat{e}_j) - h\hat{\tau}_j\right]$$

$$= \varepsilon D^+\hat{e}_{N/2-1} + \sum_{j=i+1}^{N/2-1}\left[a(x_j)\hat{e}_{j+1} - a(x_{j-1})\hat{e}_j - (a(x_j) - a(x_{j-1}))\hat{e}_j - h\hat{\tau}_j\right]$$

$$= \varepsilon D^+\hat{e}_{N/2-1} + a(x_{N/2-1})\hat{e}_{N/2} - a(x_{i-1})\hat{e}_i -$$
$$\sum_{j=i+1}^{N/2-1}\left[(a(x_j) - a(x_{j-1}))\hat{e}_j - h\hat{\tau}_j\right]$$

Hence, using the result at the point $x_{N/2-1}$ and (3.42), we have

$$|\varepsilon D^+\hat{e}_i| \leq CN^{-1}\ln N + Ch\sigma\varepsilon^{-2}N^{-1}\sum_{j=i+1}^{N/2-1}e^{-(j-1)\alpha h/\varepsilon}$$

$$\leq CN^{-1}\left(\ln N + \frac{\sigma}{\varepsilon}\frac{\alpha h/\varepsilon}{(1 - e^{-\frac{\alpha h}{\varepsilon}})}\right).$$

But $y = \alpha h/\varepsilon = 2N^{-1}\ln N$ and $B(y) = \frac{y}{1-e^{-y}}$ are bounded and it follows that $|\varepsilon D^+\hat{e}_i| \leq CN^{-1}\ln N$ as required.

It remains now to consider the case $\sigma = 0.5$. Then $\varepsilon^{-1} \leq C\ln N$ and the mesh is uniform. For $i = N - 1$ we have

$$\varepsilon D^+\hat{e}_{N-1} = -\frac{\varepsilon\hat{e}_{N-1}}{1 - x_{N-1}}$$

and so, by Lemma 3.15,

$$|\varepsilon D^+\hat{e}_{N-1}| \leq CN^{-1}\ln N \qquad (3.44)$$

as required. For $0 \leq i \leq N - 2$ we sum and rearrange (3.43) to obtain

$$\varepsilon D^+\hat{e}_i = \varepsilon D^+\hat{e}_{N-1} + a(x_{N-1})\hat{e}_N - a(x_i)\hat{e}_{i+1} - \sum_{j=i+1}^{N-1}\left[(a(x_j) - a(x_{j-1}))\hat{e}_j + h\hat{\tau}_j\right]$$

and so, using (3.42) and (3.44) we obtain

$$|\varepsilon D^+ \hat{e}_i| \leq CN^{-1} \ln N + C h \sigma \varepsilon^{-2} N^{-1} \sum_{j=i+1}^{N-1} e^{-(j-1)\alpha h / \varepsilon}$$

$$\leq CN^{-1} \ln N$$

as required, which completes the proof.   □

Theorem 3.17 *Let u_ε be the solution of a problem from Problem Class 2.1 or 2.2 and U_ε the corresponding numerical solution generated by Method 3.1 or 3.2 respectively. Then, for all $N \geq 4$ and each i, $0 \leq i \leq N - 1$, we have*

$$\sup_{0 < \varepsilon \leq 1} \|\varepsilon (D^+ U_\varepsilon(x_i) - u'_\varepsilon)\|_{\overline{\Omega}_i} \leq CN^{-1} \ln N$$

where C is a constant independent of N and ε.

Proof From the triangle inequality we have

$$|\varepsilon (D^+ U_\varepsilon(x_i) - u'_\varepsilon(x))| \leq |\varepsilon D^+ (U_\varepsilon - u_\varepsilon)(x_i)| + |\varepsilon (D^+ u_\varepsilon(x_i) - u'_\varepsilon(x))|$$

In the case of Problem Class 2.1 and Method 3.1 we see from Lemma 3.13 that

$$|\varepsilon (D^+ u_\varepsilon(x_i) - u'_\varepsilon(x))| \leq CN^{-1} \ln N.$$

To bound the first term on the right–hand side we write

$$|\varepsilon D^+ (U_\varepsilon - u_\varepsilon)(x_i)| \leq |\varepsilon D^+ (V_\varepsilon - v_\varepsilon)(x_i)| + |\varepsilon D^+ (W_\varepsilon - w_\varepsilon)(x_i)|$$

$$\leq CN^{-1} \ln N$$

where in this case the bound for each term on the right–hand side is given respectively by Lemma 3.14 and 3.16.

In the case of Problem Class 2.2 and Method 3.2 we obtain the same bounds using lemmas analogous to Lemmas 3.13, 3.14 and 3.16   □

Since $\varepsilon \overline{U}_\varepsilon$ is a linear function in the open interval $\Omega_i = (x_i, x_{i+1})$ for each i, $0 \leq i \leq N - 1$, we have $\varepsilon \overline{U}'_\varepsilon(x) = \varepsilon D^+ U_\varepsilon(x_i)$ for all $x \in \Omega_i$. It then follows, from Theorem 3.17, that $\varepsilon \overline{U}'_\varepsilon$ is an ε–uniform approximation to $\varepsilon u'_\varepsilon(x)$ for each $x \in (x_i, x_{i+1})$. We now show that this approximation can be extended in a natural way to the entire domain $\overline{\Omega}$. We define the piecewise constant function $\overline{D}^+ U_\varepsilon$ on $[0, 1)$ by

$$\varepsilon \overline{D}^+ U_\varepsilon(x) = \varepsilon D^+ U_\varepsilon(x_i), \quad \text{for} \quad x \in [x_i, x_{i+1}), \ i = 0, 1, ..., N - 1$$

and at the point $x = 1$ by

$$\varepsilon \overline{D}^+ U_\varepsilon(1) = \varepsilon D^+ U_\varepsilon(x_{N-1}).$$

Then, from the above theorem, $\overline{D}^+ U_\varepsilon$ is an ε–uniform global approximation to $\varepsilon u_\varepsilon'$ in the sense that

$$\sup_{0<\varepsilon\leq 1} \|\varepsilon \overline{D}^+ U_\varepsilon - \varepsilon u_\varepsilon'\|_{\overline{\Omega}} \leq CN^{-1} \ln N.$$

This enables us to give an alternative definition of a robust layer–resolving method, which is even stronger than that given in §1.4. We give the definition in the case of one–dimensional problems; it can be extended to multi–dimensional problems in an obvious way.

Definition 3.18 *Consider a class of mathematical problems (P_ε) parameterized by a singular perturbation parameter ε, where ε satisfies $0 < \varepsilon \leq 1$. Assume that each problem in (P_ε) has a unique solution denoted by u_ε, and that each u_ε is approximated by a sequence of numerical solutions $\{(U_\varepsilon, \overline{\Omega}_\varepsilon^N)\}_{N=1}^\infty$ obtained from a monotone numerical method (P_ε^N), where U_ε is defined on the mesh $\overline{\Omega}_\varepsilon^N$ and N is a discretization parameter. Let $\overline{U}_\varepsilon$ denote the piecewise linear interpolant over $\overline{\Omega}$ of the discrete solution U_ε on $\overline{\Omega}_\varepsilon^N$. Let $\overline{D}U_\varepsilon$ denote a piecewise constant function generated from the discrete derivative of U_ε, and assume that for some $q \geq 0$, $\varepsilon^q \|u_\varepsilon'\|_{\overline{\Omega}} \leq C'$. Then (P_ε^N) is said to be a C^1 robust layer–resolving method if the numerical solutions of (P_ε^N) are computable with an ε–uniform amount of computational work and they converge C^1 ε–uniformly to the exact solution of (P_ε), in the sense that there exist a positive integer N_0, and positive numbers C and p, where N_0, C and p are all independent of N and ε, such that for all $N \geq N_0$*

$$\sup_{0<\varepsilon\leq 1} \varepsilon^q \|\overline{D}U_\varepsilon - u_\varepsilon'\|_{\overline{\Omega}} + \|\overline{U}_\varepsilon - u_\varepsilon\|_{\overline{\Omega}} \leq CN^{-p}. \tag{3.45}$$

Here p is called the ε–uniform order of convergence and C is called the ε–uniform error constant of the numerical method (P_ε^N).

From Theorems 3.11 and 3.14 we see that Methods 3.1 and 3.2 applied to Problem Classes 2.1 and 2.2, respectively, are C^1 robust layer–resolving.

3.6 Piecewise–uniform meshes with alternative transition parameters

From the previous sections we know that robust layer–resolving numerical approximations are generated by a numerical method comprising a standard monotone finite difference operator on an appropriate piecewise–uniform fitted mesh of the form (3.2). It is natural to ask if we can improve or simplify the choice of the piecewise–uniform fitted mesh. In particular, we want to know if we can choose a simpler transition parameter σ depending on only ε rather than both ε and N.

In what follows we show that this is impossible by examining the specific problem (2.3) and the numerical method composed of the standard upwind finite difference operator

$$\varepsilon\delta^2 U_\varepsilon(x_i) + 2D^+ U_\varepsilon(x_i) = 0, \quad \text{for all} \quad x_i \in \Omega_A^N$$
$$U_\varepsilon(0) = 1, \quad U_\varepsilon(1) = 0$$

on a piecewise–uniform mesh of the form (3.2)

$$\overline{\Omega}_A^N = \{x_i | x_i = 2i\sigma_A/N, \ i \le N/2; \ x_i = x_{i-1} + 2(1 - \sigma_A)/Ni > N/2\},$$

where, instead of (3.2b), the transition parameter is now defined by

$$\sigma_A = \frac{\varepsilon}{2},$$

which depends on ε but not on N. The resulting numerical solutions can be written in the form

$$U_\varepsilon(x_i) = \begin{cases} c(\lambda^{-i} - 1) + 1, & i \le N/2 \\ 2c\lambda^{-N/2}\Lambda^{1-N/2}(\Lambda^{N-i} - 1)(\lambda + \Lambda)^{-1}, & N/2 \le i \end{cases}$$

where

$$c = \frac{\lambda^{N/2}(\lambda + \Lambda)}{(\lambda^{N/2} - 1)(\lambda + \Lambda) + 2\Lambda(1 - \Lambda^{-N/2})}$$

$$\lambda = 1 + 2/N, \quad \Lambda = 1 + 2\rho(1 - \varepsilon), \quad \rho = 2/(\varepsilon N).$$

Holding ρ fixed, we obtain the following limits

$$\lim_{N\to\infty} \lambda = 1, \quad \lim_{N\to\infty} \lambda^{N/2} = e^1, \quad \lim_{N\to\infty} \Lambda = 1 + 2\rho, \quad \lim_{N\to\infty} \Lambda^{-N/2} = 0.$$

Thus, at the transition point $\sigma_A = \varepsilon/2$, we have

$$\lim_{N\to\infty} U_\varepsilon(\sigma_A) = \frac{(1 + \rho)(e^{-1} - 1)}{1 + \rho + \rho e^{-1}} + 1.$$

Note also that

$$\lim_{\varepsilon\to 0} u_\varepsilon(\sigma_A) = e^{-1},$$

where u_ε is the exact solution of problem (2.3). It follows that, for fixed ρ, the error at the point σ_A satisfies

$$\lim_{N\to\infty} (U_\varepsilon - u_\varepsilon)(\sigma_A) = \frac{\rho e^{-1}(1 - e^{-1})}{1 + \rho + \rho e^{-1}} \ne 0$$

which shows that this finite difference method is not ε–uniform for Problem Class

2.1. Moreover, numerical results in Hegarty et al. (1995) demonstrate some further inadequacies of numerical solutions obtained from this choice of the transition parameter.

We now examine another possible choice for the transition parameter, which again depends on ε but not on N. This choice is motivated by the observation that the analytical width of the boundary layer for convection–diffusion problems in Problem Class 2.1 is $O(\varepsilon \ln 1/\varepsilon)$, whereas the transition parameter in the ε-uniform fitted mesh Method 3.1 is $O(\varepsilon \ln N)$, when ε is small. This suggests that a transition parameter depending on $\varepsilon \ln 1/\varepsilon$ may be satisfactory; indeed some authors advocate it.

To examine the numerical solutions obtained with such a transition parameter, we take the specific problem (2.4) from Problem Class 2.2, because the result is more decisive for problems in this class than in Problem Class 2.1. We construct a numerical method composed of the standard upwind finite difference equation

$$\varepsilon \delta^2 U_\varepsilon(x_i) + 2D^+ U_\varepsilon(x_i) = 0 , \quad \text{for all} \quad x_i \in \Omega_B^N$$
$$\varepsilon D^+ U_\varepsilon(0) = -2 , \quad U_\varepsilon(1) = 0$$

and the piecewise–uniform mesh

$$\overline{\Omega}_B^N = \{x_i | x_i = 2i\sigma_B/N, \ i \le N/2; \ x_i = x_{i-1} + 2(1-\sigma_B)/N, i > N/2\},$$

where the transition parameter is defined by

$$\sigma_B = \frac{\varepsilon}{2} \ln(1/\varepsilon).$$

The corresponding numerical solutions can be written in the form

$$U_\varepsilon(x_i) = \begin{cases} \lambda(\lambda^{-i} - \lambda^{-N/2}) + c(1 - \Lambda^{-N/2}), & 0 \le i \le N/2 \\ c(\Lambda^{N/2-i} - \Lambda^{-N/2}), & N/2 \le i \le N \end{cases}$$

where

$$c = \frac{2\lambda^{1-N/2}}{1 + \lambda\Lambda^{-1}}, \quad \lambda = 1 + \frac{4\sigma_B}{N\varepsilon}, \quad \Lambda = 1 + \frac{4(1-\sigma_B)}{N\varepsilon}.$$

Since, for each fixed N,

$$\lim_{\varepsilon \to 0} \lambda^{-1} = \lim_{\varepsilon \to 0} \Lambda^{-1} = \lim_{\varepsilon \to 0} \frac{\lambda}{\Lambda} = 0$$

it follows that

$$\lim_{\varepsilon \to 0} U_\varepsilon(0) \to \infty.$$

Observing that the exact solution is bounded at $x = 0$, we see that the error at the point $x = 0$ becomes unbounded as $\varepsilon \to 0$. More precisely, it is not hard to show that

$$|U_\varepsilon(0) - u_\varepsilon(0)| = O(\ln(1/\varepsilon)/N) \quad \text{as } \varepsilon \to 0, \text{with } N \text{ fixed.}$$

It follows that this finite difference method is not ε-uniform for Problem Class 2.2. Again, the numerical results in Hegarty et al. (1997) illustrate additional unsatisfactory features of the numerical approximations obtained with this choice of the transition parameter.

The above two examples of alternative transition parameters indicate that there may not be much leeway in the choice of a satisfactory transition parameter. We now state without proof some theoretical results concerning this question. We consider a numerical method composed of the upwind finite difference operator (2.8a) and a general piecewise–uniform mesh

$$\overline{\Omega}^N_{pw} = \{x_i = ih, i \leq N_1; x_{i+N_1} = \sigma + iH, i > N_1\} \tag{3.46}$$

where the transition parameter σ satisfies $0 < \sigma < 1$, and the mesh steps h, H are defined by

$$h = \sigma/N_1 \text{ and } H = (1 - \sigma)/(N - N_1).$$

Then, necessary conditions for this method to be ε-uniform for Problem Class 2.1 are given by the following theorem.

Theorem 3.19 *The numerical method composed of the upwind finite difference operator (2.8a) and the piecewise–uniform mesh (3.46) is not ε-uniform for Problem Class 2.1 for any values of h, H, if σ is a function of only ε. If it is an ε-uniform method then σ is a function of ε and N.*

Furthermore, for numerical methods with transition parameters σ of the form

$$\sigma = \sigma(\varepsilon, N) = \min\{m, \tau(\varepsilon, N)\}, \text{ any } m \in (0, 1) \tag{3.47}$$

we have the following result.

Theorem 3.20 *If the numerical method composed of the upwind finite difference operator (2.8a) on the piecewise–uniform mesh (3.46) with the transition parameter σ of the form (3.47) is ε-uniform for Problem Class 2.1, then the quantity $\tau(\varepsilon, N)$ in (3.47) satisfies*

$$\tau(\varepsilon, N) = \varepsilon\zeta(N), \text{ where } \frac{\zeta(N)}{N} \to 0 \text{ and } \zeta(N) \to \infty \text{ as } N \to \infty.$$

The original statement of the above theoretical results for more general problems than those of Problem Class 2.1 may be found in Shishkin (1992b, p. 207–8). The nature of the problems studied in the present book suggest the choice $\zeta(N) = c \ln N$, where c is a suitable constant independent of N and ε. It is clear that this choice fulfills the necessary conditions of the above two theorems.

To conclude this section we study the effect of varying the location of the transition point. To do so we introduce the following one–parameter family of numerical methods related to Method 3.1.

Method 3.3. Upwind finite difference operator on one–parameter family of piecewise–uniform fitted meshes for Problem Class 2.1.

$$L_\varepsilon^N U_\varepsilon = \varepsilon \delta^2 U_\varepsilon + a(x_i) D^+ U_\varepsilon = f(x_i), \quad x_i \in \Omega_\varepsilon^N, \qquad (3.48\text{a})$$
$$U_\varepsilon(0) = u_\varepsilon(0), \quad U_\varepsilon(1) = u_\varepsilon(1), \qquad (3.48\text{b})$$

where

$$\overline{\Omega}_\varepsilon^N = \{x_i | x_i = 2i\sigma/N, \quad i \leq N/2; \ x_i = x_{i-1} + 2(1 - \sigma_c)/N, \quad N/2 < i\}, \qquad (3.48\text{c})$$

and

$$\sigma_c = \min\{\frac{1}{2}, c\varepsilon \ln N\} \qquad (3.48\text{d})$$

where c is a positive parameter.

Here the location of the transition parameter σ varies as the parameter c is changed. We note that Method 3.1 is a special case of Method 3.3, corresponding to $c = \frac{1}{\alpha}$.

We now solve problem (2.3) using Method 3.3 for various values of c and N. The exact ε–uniform maximum pointwise error E_{exact}^N in the numerical solutions for various values of c and N is given in Table 3.5. For each fixed value of N, the

Table 3.5 *Exact ε–uniform maximum pointwise error E_{exact}^N generated by Method 3.3 applied to problem (2.3) for various values of c and N.*

	Number of intervals N					
c	8	16	32	64	128	256
0.25	0.214640	0.185616	0.146707	0.110331	0.080835	0.058441
0.50	0.190655	0.114633	0.065004	0.035908	**0.019681**	**0.010727**
0.75	0.167483	**0.098872**	**0.059310**	**0.035210**	0.020615	0.011838
1.00	**0.164549**	0.108328	0.069125	0.043656	0.026335	0.015402
2.00	0.207160	0.169612	0.115354	0.079352	0.049345	0.029778
3.00	0.201195	0.199818	0.162223	0.105467	0.069818	0.043285
4.00	0.197897	0.202588	0.189139	0.136710	0.090167	0.055999
5.00	0.197897	0.197998	0.201222	0.162222	0.103986	0.067049

figure in bold is the smallest error occurring for the 8 values of c considered here. We infer from the entries in this table that, for a given N and $c \in [0.25, 0.5]$, the error is sensitive to variations of c. However, we will see later in §4.4 that, relative to other numerical methods, this sensitivity is not excessively large.

CHAPTER 4

The limitations of non–monotone numerical methods

4.1 Non–physical behaviour of numerical solutions

One of the main goals of this chapter is to show that for problems with boundary layers robust numerical solutions cannot be guaranteed if the numerical method is ε–uniform; it is necessary also for it to be monotone. Various advantages and disadvantages of non–monotone methods are discussed in the present chapter for ordinary differential equations and in chapter 9 for partial differential equations. For problems in Problem Class 2.1 the upwind finite difference Method 3.1 is monotone and ε–uniform. In this chapter we examine, among other things, whether or not non–monotone methods on fitted piecewise–uniform meshes can lead to better methods. What better means in this context is ambiguous. It could mean, for example, that over a certain range of values of N the method has a smaller ε–uniform maximum pointwise error, or a higher ε–uniform order of convergence, or more physically–relevant solutions, or some combination of all of these properties. But it turns out that these properties are mutually conflicting. In what follows we show, in particular, that the benefits of non–monotone methods can be enjoyed only at the cost of introducing non–physical numerical oscillations or physically meaningless negative values into the numerical solutions.

It is easy to see why oscillations in numerical solutions are often inconsistent with the behaviour of the exact solution representing some real physical phenomenon. Consider, for example, the one–dimensional convection–diffusion problem (2.3). The derivatives of its exact solution satisfy the equation

$$\varepsilon u_\varepsilon''(x) = -2u_\varepsilon'(x) \quad \text{for all} \ \ x \in \Omega = (0,1).$$

This implies that, at each point $x \in \Omega$, the first and second derivatives are either both zero or of opposite sign. Therefore the exact solution cannot oscillate, because such behaviour requires the existence of points in the domain Ω at which the first and second derivatives are non–zero and have the same sign.

Furthermore, negative values are usually regarded as unacceptable in numerical approximations of physically non–negative quantities such as energy, or

momentum or mass. A simple example of a problem having a non–negative solution is the time–dependent convection–diffusion problem

$$\varepsilon \frac{\partial^2 c_\varepsilon}{\partial x^2} + 2 \frac{\partial c_\varepsilon}{\partial x} = \frac{\partial c_\varepsilon}{\partial t}, \quad (x,t) \in \Omega \times (0,T],$$
$$c_\varepsilon(0,t) = 1, \quad c_\varepsilon(1,t) = 0, \quad c_\varepsilon(x,0) = 1 - x^2.$$

This can be regarded as a model for the evolution in time of the concentration
$c_\varepsilon(x,t)$ of a substance dissolving in a fluid that is moving with velocity 2 within
the confines of the one–dimensional container Ω. The initial distribution of the
substance in the container is $1 - x^2$. The term $\varepsilon c_\varepsilon''$ is the diffusion term and $2c_\varepsilon'$
is the convection term. In the case of small ε, convection dominates and little
diffusion takes place. In the long term, the concentration of the substance evolves
so that it is highest near the boundary point $x = 0$ and decreases monotonically
with increasing distance from this point. It is clear that negative concentrations
make no physical sense and that it is undesirable that they appear in numerical
approximations of the concentration.

   Another serious disadvantage of numerical solutions with non–physical numerical oscillations is that difficulties arise in the theoretical error analysis of the
corresponding numerical methods. In what follows we show that, from a mathematical perspective, these non–physical numerical oscillations are a symptom
of the lack of monotonicity of the underlying numerical method. We recall from
chapter 2 that, if a numerical method satisfies a discrete maximum or minimum
principle, then it is a monotone method. For such methods classical arguments
based on stability and consistency can often be applied to establish ε–uniform
error estimates, provided that we take care to identify explicitly the dependence
of the estimates on the small parameter ε. But, without a discrete maximum or
minimum principle, proofs of ε–uniform convergence in the maximum norm are
difficult. Some progress can be made in one dimension using a discrete Green's
function (see Andreev and Kopteva (1996)), but a general theory for singularly
perturbed partial differential equations does not exist today. On the other hand,
extensive theoretical ε–uniform error bounds have been obtained in recent years
for monotone numerical methods using upwind finite difference operators on
piecewise–uniform fitted meshes for a wide class of singularly perturbed partial
differential equations.

4.2   A non–monotone method

In this section we construct a non–monotone method for Problem Class 2.1. From
Table 3.1 it is clear that the order of ε–uniform convergence of the monotone
Method 3.1 is first order. We can attempt to increase this order of convergence
by using a finite difference operator with a second order discretization error, for

example the centred difference operator (2.5a-c), and the piecewise–uniform fitted mesh Ω_ε^N (3.3d-f). The resulting non–monotone method applied to Problem Class 2.1 is:

Method 4.1. Centred finite difference operator on piecewise–uniform fitted mesh for Problem Class 2.1.

$$\varepsilon\delta^2 U_\varepsilon + a(x_i)D^0 U_\varepsilon = f(x_i), \quad x_i \in \Omega_\varepsilon^N, \tag{4.1a}$$
$$U_\varepsilon(0) = u_\varepsilon(0), \quad U_\varepsilon(1) = u_\varepsilon(1), \tag{4.1b}$$

where

$$\overline{\Omega}_\varepsilon^N = \{x_i | x_i = 2i\sigma/N, \ i \le N/2; \ x_i = x_{i-1} + 2(1-\sigma)/N, \ N/2 < i\} \tag{4.1c}$$

and

$$\sigma = \min\{\frac{1}{2}, \frac{\varepsilon}{\alpha}\ln N\}. \tag{4.1d}$$

It turns out that non–physical oscillations arise in the numerical solutions generated by this method, when it is applied to typical problems from Problem Class 2.1. A graph of the numerical solution obtained by applying Method 4.1 with $N = 8$ to problem (2.3) with $\varepsilon = 0.01$ is shown in Fig. 4.1. We see that the numerical solution has bounded non–physical oscillations in the coarse mesh region, but none in the fine mesh region. This is in contrast to the situation shown

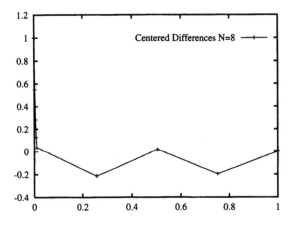

FIG. 4.1. Numerical solution of problem (2.3) for $\varepsilon = 0.01$ generated by Method 4.1 with $N = 8$.

in Fig. 2.1 where Method 2.1 is used to solve the same problem. In that figure we see that the numerical oscillations occur in the whole domain and that they may be large and unbounded. Thus, the use of the piecewise–uniform fitted mesh in Method 4.1, instead of the uniform mesh in Method 2.1, replaces unbounded by bounded non–physical numerical oscillations in the numerical solutions.

To gain greater insight into the underlying phenomena, we need to be more precise about what we mean by an oscillating mesh function. To do so we introduce the two mesh functions F and G on the uniform mesh $\overline{\Omega}_u^N = \{x_i|\ x_i = i/N\}_{i=0}^N$ defined by

$$F(x_i) = (-0.5)^i \quad \text{and} \quad G(x_i) = 1 - x_i - i(-2N)^{-i}.$$

It is not hard to see that at a typical interior mesh point x_i, $D^+F(x_i)$ and $D^-F(x_i)$ are of opposite sign and that $F(x_i)$ oscillates between positive and negative values as x_i moves across the mesh. On the other hand $D^+G(x_i) < 0$ for all $1 \leq i < N$, and hence G is a monotonically decreasing mesh function. Nevertheless, $\delta^2 G(x_i)$ and $\delta^2 G(x_{i+1})$ are of opposite sign, which shows that $G(x_i)$ oscillates about the line $g(x_i) = 1 - x_i$. Motivated by these two examples, we now formulate the following definitions concerning oscillating and non–oscillating mesh functions on an arbitrary mesh.

Definition 4.1 *A mesh function Z oscillates at an interior mesh point x_i if*

$$\delta^2 Z(x_i)\delta^2 Z(x_{i+1}) < 0.$$

Definition 4.2 *A mesh function Z oscillates strongly at an interior mesh point x_i if it oscillates at x_i and if, in addition,*

$$D^+ Z(x_i)D^- Z(x_i) < 0.$$

It is easy to see that the function F oscillates strongly and G oscillates at each interior mesh point x_i. Finally, we introduce the concept of an oscillation–free mesh function.

Definition 4.3 *A mesh function Z is oscillation–free if, at all interior mesh points x_i,*

$$\delta^2 Z(x_i)\delta^2 Z(x_{i+1}) \geq 0.$$

We return now to Method 4.1 and examine more closely its application to problem (2.3). In this case its solution can be written in the closed form

$$U_\varepsilon(x_i) = \begin{cases} c(\lambda_1^i (\lambda_1\lambda_2)^{-N/2} - 1), & \text{if } i \leq N/2 \\ c(\lambda_2^{i-N} - 1), & \text{if } i \geq N/2 \end{cases},$$

where

$$c = \frac{1}{(\lambda_1 \lambda_2)^{-N/2} - 1}, \quad \lambda_1 = \frac{1 - \frac{2\sigma}{N\varepsilon}}{1 + \frac{2\sigma}{N\varepsilon}} \quad \text{and} \quad \lambda_2 = \frac{1 - \frac{2(1-\sigma)}{N\varepsilon}}{1 + \frac{2(1-\sigma)}{N\varepsilon}}. \tag{4.2}$$

From (4.1f) we see that $\sigma = (\varepsilon/2)\ln N$, for all sufficiently small ε, and so

$$\lambda_1 = \frac{1 - \frac{\ln N}{N}}{1 + \frac{\ln N}{N}}.$$

Hence, for all $N > 1$ and all sufficiently small ε, we have

$$0 < \lambda_1 < 1, \quad \text{and} \quad |\lambda_2| < 1.$$

It follows that

$$1 - (\lambda_1 \lambda_2)^{N/2} > 0$$

and so on the fine mesh $\{x_i : 0 < i < N/2\}$ we have

$$D^+ U_\varepsilon(x_i) = \frac{\lambda_1^i(\lambda_1 - 1)}{(x_{i+1} - x_i)(1 - (\lambda_1 \lambda_2)^{N/2})} < 0,$$

and

$$\delta^2 U_\varepsilon(x_i) = \frac{\lambda_1^{i-1}(\lambda_1 - 1)^2}{(x_{i+1} - x_i)^2(1 - (\lambda_1 \lambda_2)^{N/2})} > 0.$$

Thus U_ε is oscillation–free on the fine mesh, which is also apparent from the graph in Fig. 3.1.

Correspondingly, on the coarse mesh $\{x_i : N/2 \le i < N\}$, we have

$$D^+ U_\varepsilon(x_i) = \frac{c\lambda_2^{i-N}(\lambda_2 - 1)}{(x_{i+1} - x_i)},$$

and from (4.2), for each fixed N, we have

$$\lim_{\varepsilon \to 0} \lambda_2^{i-N} = (-1)^{i-N}.$$

This shows that, for all sufficiently small ε, U_ε oscillates strongly on the coarse mesh. Also $|\lambda_2|^{i-N} > |\lambda_2|^{j-N}$ for all $j > i$ and hence, for $j > i \ge N/2$,

$$|D^+ U_\varepsilon(x_j)| < |D^+ U_\varepsilon(x_i)|$$

which shows that the numerical oscillations decrease in magnitude as x_i moves away from the fine mesh.

To see that Method 4.1 is non–monotone, fix N and assume that $N/2$ is even. Let ε be sufficiently small so that $\lambda_2 < 0$. Because $|\lambda_2| < 1$ we have $|\lambda_2|^{i-N} > 1$ for all $i < N$. Therefore, for $i - N$ odd, we have $\lambda_2^{i-N} - 1 < 0$ and, for $i - N$ even, we have $\lambda_2^{i-N} - 1 > 0$. From this it follows that $U_\varepsilon(x_i)U_\varepsilon(x_{i+1}) < 0$ for all $i \ge N/2$. From the discussion in §2.3, this implies that Method 4.1 is not monotone.

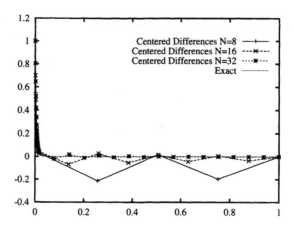

FIG. 4.2. Numerical solution of problem (2.3) for $\varepsilon = 0.01$ generated by Method
4.1 with $N = 8, 16, 32$

Graphs of the numerical solutions of problem (2.3) using Method 4.1 are
shown in Figs. 4.2 and 4.3 for a fixed value of ε and various values of N. We
see that, although there are small non–physical numerical oscillations outside
the boundary layer, as N increases their amplitudes decrease and the numerical
solutions converge to the exact solution. Comparing the graphs in Fig. 4.2 with
those in Fig. 2.1, we conclude that the use of a fitted piecewise–uniform mesh
has a dramatic impact on, but does not eliminate, the non–physical numerical
oscillations generated by the non–monotone Method 4.1. In the next section we

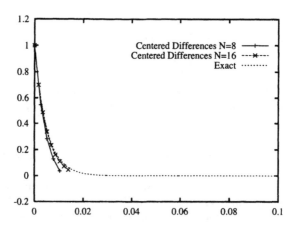

FIG. 4.3. Boundary layer in the numerical solution of problem (2.3) for $\varepsilon = 0.01$
generated by Method 4.1 with $N = 8$ and $N = 16$.

examine whether or not its use reduces the ε–uniform maximum pointwise errors corresponding to this non–monotone method.

4.3 Accuracy and order of convergence

In this section we examine the exact maximum pointwise errors $E^N_{\varepsilon,\text{exact}}$ and E^N_{exact} that arise when the non–monotone method Method 4.1 is used to compute numerical solutions of the convection–diffusion problem (2.3). This enables us to determine the ε–uniform accuracy of the method and its ε–uniform order of convergence.

Values of $E^N_{\varepsilon,\text{exact}}$ and E^N_{exact} for various values of ε and N are given in Table 4.1. Compare the results in Table 4.1 with the corresponding results in Table 3.1. We see that, when $N \geq 32$, the non–monotone centred finite difference operator (2.5a-c) leads to significantly smaller exact ε–uniform maximum pointwise errors than the monotone upwind finite difference operator, using the same piecewise–uniform fitted mesh in both cases. We conclude that, for this simple problem, the non–monotone Method 4.1 gives, for approximately the same amount of computational effort, significantly more accurate (in the sense that the exact ε–uniform monotone pointwise errors are smaller) numerical solutions than the monotone Method 3.1 on the same piecewise–uniform fitted mesh. However, we caution the reader that this is not the full story; there are hidden pitfalls and dangers in using non–monotone methods, which are discussed later in this chapter and also in chapter 9.

Since the non–monotone Method 4.1 gives more accurate numerical solutions to problem (2.3) than those given by Method 3.1, it is natural to ask if the method also has a higher ε–uniform order of convergence. Since the exact solution

Table 4.1 *Exact maximum pointwise errors $E^N_{\varepsilon,\text{exact}}$ and E^N_{exact} generated by Method 4.1 applied to problem (2.3) for various values of ε and N.*

ε	Number of intervals N						
	8	16	32	64	128	256	512
2^{-2}	0.010400	0.003684	0.001432	0.000478	0.000119	0.000030	0.000007
2^{-4}	0.083284	0.019392	0.003487	0.000518	0.000176	0.000058	0.000018
2^{-6}	0.189976	0.055199	0.018243	0.005274	0.000990	0.000123	0.000018
2^{-8}	0.246220	0.097380	0.031410	0.012061	0.004656	0.001357	0.000259
2^{-10}	0.263946	0.119850	0.049481	0.016718	0.006878	0.003033	0.001170
2^{-12}	0.268661	0.126882	0.059405	0.024947	0.008617	0.003667	0.001723
2^{-14}	0.269858	0.128746	0.062512	0.029587	0.012520	0.004373	0.001893
2^{-16}	0.270159	0.129219	0.063335	0.031041	0.014759	0.006270	0.002202
2^{-18}	0.270234	0.129338	0.063544	0.031427	0.015462	0.007369	0.003137
2^{-20}	0.270253	0.129367	0.063596	0.031525	0.015648	0.007714	0.003681
2^{-22}	0.270257	0.129375	0.063609	0.031549	0.015695	0.007805	0.003852
2^{-24}	0.270259	0.129377	0.063613	0.031555	0.015707	0.007829	0.003897
E^N_{exact}	0.270259	0.129377	0.063613	0.031555	0.015707	0.007829	0.003897

of problem (2.3) is known, we can find the exact order of convergence $p_{\varepsilon,\text{exact}}^N$, defined by

$$p_{\varepsilon,\text{exact}}^N = \log_2 \frac{E_{\varepsilon,\text{exact}}^N}{E_{\varepsilon,\text{exact}}^{2N}} \qquad (4.3)$$

and the corresponding exact ε–uniform order of convergence, defined by

$$p_{\text{exact}}^N = \log_2 \frac{E_{\text{exact}}^N}{E_{\text{exact}}^{2N}}. \qquad (4.4)$$

The values of these quantities, obtained when Method 4.1 is applied to problem (2.3), are given in Table 4.2. It is apparent from these values that the method is ε–uniform, and also that its ε–uniform order of convergence is 1.0 rather than the hoped for higher order 2.0. The well–known fact that Method 4.1 has order of convergence two for non–singularly perturbed problems is confirmed by the numerical results in the first row of Table 4.2. This illustrates how different the behaviour of numerical methods can be for problems with boundary layers compared to that for problems with smooth solutions.

As a further comparison of monotone and non–monotone methods, using the same piecewise–uniform fitted mesh, we plot in Fig. 4.4 the maximum pointwise errors E_{exact}^N, as a function of N, for the monotone Method 3.1 and the non–monotone Method 4.1 applied to problem (2.3). It is clear from these graphs that, for N sufficiently large, in this case for $N \geq 32$, the non–monotone method has smaller maximum pointwise errors than the monotone method. It can also be seen that the exact ε–uniform order of convergence in both cases is 1.0.

Table 4.2 *Orders of convergence $p_{\varepsilon,\text{exact}}^N$ and p_{exact}^N generated by Method 4.1 applied to problem (2.3) for various values of ε and N.*

ε	Number of intervals N					
	8	16	32	64	128	256
2^{-2}	1.50	1.36	1.58	2.00	2.00	2.00
2^{-4}	2.10	2.48	2.75	1.56	1.62	1.66
2^{-6}	1.78	1.60	1.79	2.41	3.00	2.76
2^{-8}	1.34	1.63	1.38	1.37	1.78	2.39
2^{-10}	1.14	1.28	1.57	1.28	1.18	1.37
2^{-12}	1.08	1.09	1.25	1.53	1.23	1.09
2^{-14}	1.07	1.04	1.08	1.24	1.52	1.21
2^{-16}	1.06	1.03	1.03	1.07	1.24	1.51
2^{-18}	1.06	1.03	1.02	1.02	1.07	1.23
2^{-20}	1.06	1.02	1.01	1.01	1.02	1.07
2^{-22}	1.06	1.02	1.01	1.01	1.01	1.02
2^{-24}	1.06	1.02	1.01	1.01	1.00	1.01
p_{exact}^N	1.06	1.02	1.01	1.01	1.00	1.01

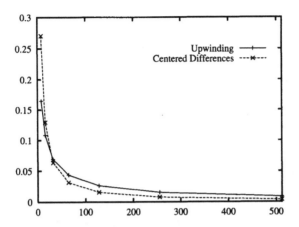

FIG. 4.4. Exact maximum pointwise error E^N_{exact} as a function of N generated by Methods 3.1 and 4.1 applied to problem (2.3).

4.4 Tuning non–monotone methods

In this section we tune the non–monotone method by varying the location of the transition point to obtain an 'optimal' piecewise–uniform fitted mesh. We change its location by introducing an extra tuning parameter c into the definition of the transition parameter σ associated with the piecewise–uniform fitted mesh (3.3c-e). In particular, we attempt to increase the ε–uniform order of convergence of the non–monotone Method 4.1 by a suitable choice of the parameter c. To be precise we introduce the same one–parameter family of transition parameters

$$\sigma_c = \min\{\frac{1}{2}, c\varepsilon \ln N\} \tag{4.5}$$

as in §3.6, where the value of the parameter c determines the location of the transition point. Note that the transition parameter σ in (3.3e) is the special case corresponding to $c = 1/\alpha$. The one–parameter family of non–monotone numerical methods is then defined by Method 4.2.

Method 4.2. Centred finite difference operator on one–parameter family of piecewise–uniform fitted meshes for Problem Class 2.1.

$$\varepsilon\delta^2 U_\varepsilon + a(x_i)D^0 U_\varepsilon = f(x_i), \quad x_i \in \Omega^N_\varepsilon, \tag{4.6a}$$

$$U_\varepsilon(0) = u_\varepsilon(0), \quad U_\varepsilon(1) = u_\varepsilon(1), \tag{4.6b}$$

$$\overline{\Omega}^N_\varepsilon = \{x_i | x_i = 2i\sigma_c/N, \ i \leq N/2; \ x_i = x_{i-1} + 2(1-\sigma_c)/N, \ N/2 < i\} \tag{4.6c}$$

and

$$\sigma_c = \min\{\frac{1}{2}, c\varepsilon \ln N\}. \qquad (4.6d)$$

where c is a positive parameter.

Graphs of the numerical solutions produced by applying Method 4.2 to problem (2.3), for values of c varying from $c = 0.25$ to $c = 5.0$, are shown in Fig. 4.5 and Fig. 4.6. We see that as c increases the amplitudes of the oscillations outside the boundary layer decrease, while if c becomes too large then oscillations arise in the boundary layer region; note the difference in scale of the horizontal axes of these graphs.

To clarify the situation, we observe that the numerical solution of problem (2.3) given by Method 4.2 can be expressed in closed form. On the fine mesh the solution is

$$U_\varepsilon(x_i) = \frac{\lambda_1^i - (\lambda_1\lambda_2)^{N/2}}{1 - (\lambda_1\lambda_2)^{N/2}}, \quad 0 \le i \le N/2,$$

where

$$\lambda_1 = \frac{1 - \frac{2c\ln N}{N}}{1 + \frac{2c\ln N}{N}} \quad \text{and} \quad \lambda_2 = \frac{1 - \frac{2(1-\sigma_c)}{N\varepsilon}}{1 + \frac{2(1-\sigma_c)}{N\varepsilon}}.$$

From this expression it is clear that $|\lambda_1| < 1$, $|\lambda_2| < 1$ and that U_ε is oscillation-free on the fine mesh, when $\lambda_1 > 0$; that is when $\frac{2c\ln N}{N} < 1$. The numerical

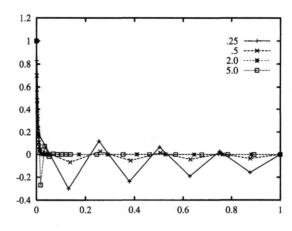

FIG. 4.5. Comparison of the numerical solutions of problem (2.3) for $\varepsilon = 0.01$ generated by Method 4.2 with $N = 16$ for values of c between 0.25 and 5.0

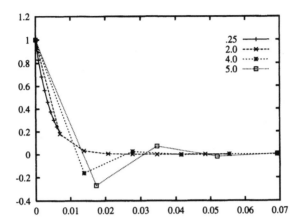

FIG. 4.6. *Comparison of the boundary layer in the numerical solutions generated by Method 4.2 applied to problem (2.3) for $\varepsilon = 0.01$ and for values of c between 0.25 and 5.0 and $N = 16$.*

solutions for $N = 16$ and various values of c are shown in Fig. 4.5. We see that visible oscillations occur with this number of nodes only if c is sufficiently large, in this case only if

$$c > \frac{16}{2\ln 16} \approx 2.9.$$

We also see from these expressions that, for any fixed choice of c, U_ε is oscillation free on the fine mesh if N is large enough to guarantee that $\lambda_1 > 0$, which is the case if N satisfies $N/\ln N > 2c$. On the other hand, repeating the argument from §3.2, we see that U_ε oscillates strongly in the coarse mesh region for all sufficiently small ε, since for fixed N, irrespective of the choice of the tuning parameter c, we have $\lim_{\varepsilon \to 0} \lambda_2^{i-N} = (-1)^{i-N}$. In fact, taking $\varepsilon \leq \frac{1}{2N}$, ensures that $\lambda_2 < 0$ for all c.

In Table 4.3 and Fig. 4.7 the exact ε–uniform maximum pointwise errors are shown for Method 4.2 applied to problem (2.3) with values of c varying in the range of values $c = 0.25$ to $c = 5.0$. For each fixed value of N, the figure in bold is the smallest error observed over this range of values of c. We see that, roughly speaking, the value of c yielding the smallest exact ε–uniform maximum pointwise errors for this particular problem, is about $c = 1.0$. This suggests that for Problem Class 2.1 the corresponding 'optimal' choice for c is approximately $c = 2/\alpha$. However, the errors in Table 4.3 indicate that, for a given N, the exact ε–uniform maximum pointwise error is sensitive to changes in the value of the parameter c. This is in sharp contrast to the results in Table 3.5, which arise when the monotone Method 3.3 is used to solve problem (2.3). We see that, for

Table 4.3 *Exact ε–uniform maximum pointwise error E_{exact}^N generated by Method 4.2 applied to problem (2.3) for various values of c and N.*

c	\multicolumn{6}{c}{Number of intervals N}
	8	16	32	64	128	256
0.25	1.083890	0.663576	0.428585	0.285464	0.193826	0.133244
0.50	0.270259	0.129377	0.063613	0.031555	0.015707	0.007829
0.75	0.077116	0.028749	0.010601	0.003835	0.001368	0.000483
1.00	**0.044524**	**0.016888**	**0.006333**	**0.002235**	**0.000743**	**0.000239**
2.00	0.144474	0.068769	0.024986	0.008527	0.002836	0.000923
3.00	0.265606	0.144474	0.060378	0.019545	0.006467	0.002085
4.00	0.386856	0.224381	0.105215	0.037643	0.011480	0.003727
5.00	0.508438	0.299452	0.154491	0.060378	0.018284	0.005823

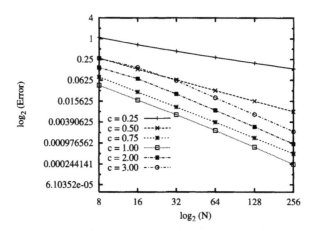

FIG. 4.7. Graph of $\log_2(E_{exact}^N)$ versus $\log_2(N)$ generated by Method 4.2 applied to problem (2.3) for various values of c.

a given N, the error is much less sensitive to the choice of c than is the present case.

To determine the ε–uniform order of convergence of Method 4.2, when it is applied to problem (2.3), we give, in Table 4.4, the exact ε–uniform order of convergence p_{exact}^N for various values of c and N. The figure in bold in each column is the highest order of convergence observed over the same range of c as in Table 4.3. We conclude that for this particular problem the 'optimal' choice of c, corresponding to the highest ε–uniform order of convergence, is again about $c = 1.0$. We also see that the ε–uniform orders of convergence approach 2.0 asymptotically as $N \to \infty$. We are thus tempted to propose this member of the family of non–monotone Methods 4.2 as the 'optimally tuned' method for problem (2.3), and we now describe it formally as follows.

Table 4.4 *Exact ε–uniform order of convergence p^N_{exact} generated by Method 4.2 applied to problem (2.3) for various values of c and N.*

c	Number of intervals N								
	8	16	32	64	128	256	512	1024	2048	4096
0.25	0.71	0.63	0.59	0.56	0.54	0.53	0.53	0.55	0.64	0.89
0.50	1.06	1.02	1.01	1.01	1.00	1.00	1.01	1.03	1.13	1.38
0.75	**1.42**	1.44	1.47	1.48	1.49	1.50	1.51	1.53	1.62	**1.88**
1.00	1.40	1.41	1.50	1.59	1.64	**1.68**	**1.72**	**1.73**	**1.75**	1.77
2.00	1.07	**1.46**	1.55	1.59	1.62	1.66	1.70	1.72	1.75	1.77
3.00	0.88	1.26	**1.63**	1.60	1.63	1.67	1.70	1.73	1.75	1.77
4.00	0.79	1.09	1.48	1.71	1.62	1.67	1.70	1.73	1.75	1.77
5.00	0.76	0.95	1.36	**1.72**	**1.65**	1.67	1.70	1.73	1.75	1.77

Method 4.3. Centred finite difference operator on 'optimally tuned' piecewise–uniform fitted mesh for Problem Class 2.1.

$$\varepsilon \delta^2 U_\varepsilon + a(x_i) D^0 U_\varepsilon = f(x_i), \quad x_i \in \Omega^N_\varepsilon, \tag{4.7a}$$

$$U_\varepsilon(0) = u_\varepsilon(0), \quad U_\varepsilon(1) = u_\varepsilon(1), \tag{4.7b}$$

$$\overline{\Omega}^N_\varepsilon = \{x_i | x_i = 2i\sigma_{opt}/N, \ i \le N/2; \ x_i = x_{i-1} + 2(1 - \sigma_{opt})/N, \ N/2 < i\} \tag{4.7c}$$

with the transition parameter

$$\sigma_{opt} = \min\{\frac{1}{2}, \frac{2\varepsilon}{\alpha} \ln N\}. \tag{4.7d}$$

Here 'optimally tuned' is used in the sense that the ε–uniform maximum pointwise errors, generated when this method is used to solve any problem from Problem Class 2.1, are approximately minimized by this choice of the tuning parameter c. We only conjecture that Method 3.3 is optimally tuned for Problem Class 2.1; it is for this reason that the words 'optimally tuned' are placed in quotation marks.

In Fig. 4.8 we compare graphs of the exact and numerical solutions of problem (2.3) with ε = 0.01 obtained using the 'optimally tuned' non–monotone Method 4.3, with N = 16 and N = 32. In this case no oscillations are visible to the eye, although we know, from the above analysis of the explicit expression for the numerical solutions, that small oscillations are, in fact, present. Then, for the same method and problem, in Table 4.5 we give the exact maximum pointwise errors for various values of ε and N. Comparing the numerical results in this table with those in Table 3.1, we see that the exact maximum pointwise errors, obtained using the 'optimally tuned' non–monotone Method 4.3, are significantly

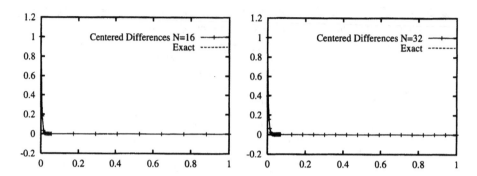

FIG. 4.8. Graphs of the exact solution and of numerical solutions of problem
(2.3) for $\varepsilon = 0.01$, generated by Method 4.3, with $N = 16$ and $N = 32$.

Table 4.5 *Exact maximum pointwise errors $E^N_{\varepsilon,\text{exact}}$ and E^N_{exact} generated by Method 4.3 applied to problem (2.3) for various values of ε and N.*

	Number of intervals N						
ε	8	16	32	64	128	256	512
2^{-2}	0.034436	0.007848	0.001921	0.000478	0.000119	0.000030	0.000007
2^{-4}	0.038284	0.014749	0.005823	0.002085	0.000705	0.000230	0.000073
2^{-6}	0.041728	0.014942	0.005823	0.002085	0.000705	0.000230	0.000073
2^{-8}	0.043703	0.015951	0.005890	0.002085	0.000705	0.000230	0.000073
2^{-10}	0.044311	0.016610	0.006132	0.002105	0.000705	0.000230	0.000073
2^{-12}	0.044471	0.016816	0.006274	0.002176	0.000711	0.000230	0.000073
2^{-14}	0.044511	0.016870	0.006319	0.002218	0.000729	0.000231	0.000073
2^{-16}	0.044522	0.016884	0.006330	0.002231	0.000740	0.000236	0.000073
2^{-18}	0.044524	0.016888	0.006333	0.002235	0.000743	0.000239	0.000074
E^N_{exact}	0.044524	0.016888	0.006333	0.002235	0.000743	0.000239	0.000074

smaller than those corresponding to either the monotone Method 3.1 or the non–monotone Method 4.1. The improvement is such that the non–monotone Method 4.3, with $N = 64$ produces smaller errors than those of the monotone Method 3.1 with $N = 512$. Also, with $N = 32$, it produces smaller errors than those of the non–monotone Method 4.1 with $N = 512$. In Table 4.6 we give the computed ε–uniform orders of convergence for the 'optimally tuned' non–monotone Method 4.3 applied to problem (2.3). We see that these orders are all greater than 1.4 and that they increase as the number of mesh points N increases. This indicates that Method 4.3 is an ε–uniform method of order not less than 1.4 at least for this problem.

We have thus achieved our original aim of constructing an ε–uniform method of higher order, at least for the particular problem (2.3). We see that the numerical results given by this method for problem (2.3) are dramatically better than those given by the monotone Method 3.1. This has been achieved, however,

Table 4.6 *Orders of convergence $p^N_{\varepsilon,\text{exact}}$ and p^N_{exact} generated by Method 4.3 applied to problem (2.3) for various values of ε and N.*

ε	Number of intervals N					
	8	16	32	64	128	256
2^{-2}	2.13	2.03	2.01	2.00	2.00	2.00
2^{-4}	1.38	1.34	1.48	1.56	1.62	1.66
2^{-6}	1.48	1.36	1.48	1.56	1.62	1.66
2^{-8}	1.45	1.44	1.50	1.56	1.62	1.66
2^{-10}	1.42	1.44	1.54	1.58	1.62	1.66
2^{-12}	1.40	1.42	1.53	1.61	1.63	1.66
2^{-14}	1.40	1.42	1.51	1.61	1.66	1.67
2^{-16}	1.40	1.42	1.50	1.59	1.65	1.69
2^{-18}	1.40	1.41	1.50	1.59	1.64	1.68
p^N_{exact}	1.40	1.41	1.50	1.59	1.64	1.68

at the cost of introducing small non–physical oscillations, sometimes invisible to the naked eye, into the numerical solutions and of losing the discrete maximum principle, since the method is non–monotone. The cost seems small, but we need to be careful. The excellent numerical solutions given by this method are only for the particular problem (2.3), and they are obtained only by careful tuning of the artificial parameter c. Furthermore, the behaviour of the apparently 'optimally tuned' method is sensitive to the choice of this parameter, and it is not at all clear how the parameter should be chosen for more complicated problems. In this sense this 'better' method is not robust.

4.5   Neumann boundary conditions

In this section we examine the behaviour of non–monotone numerical methods, when they are used to solve problems with a Neumann boundary condition. In particular, we examine the numerical results obtained when a non–monotone method analogous to Method 4.3 is used to solve problem (2.4). Since there is no general consensus in the literature on the most appropriate finite difference approximation to a Neumann boundary condition, we use the approximation generated automatically when the standard finite element method with a basis of piecewise linear polynomials is applied to the problem. This seems a sensible approach, because Neumann boundary conditions are natural boundary conditions for the finite element method, and their discretization is generated by the method itself. Indeed we can obtain the complete finite difference approximation using this finite element approach in the following way.

The finite element method generates a discrete solution of the form

$$U_\varepsilon(x) = \sum_{i=0}^{N} U_i \phi_i(x).$$

where $\{\phi_i\}_{i=0}^N$ is the basis of piecewise–linear polynomials defined by $\phi_i(x_j) = \delta_{i,j}$ and the nodal values $\{U_i\}_{i=0}^N$ are determined by the system of equations

$$-(\varepsilon U_\varepsilon', \phi_i') + 2(U_\varepsilon', \phi_i) = -2\phi_0(0), \quad i = 0, 1, ...N - 1, \qquad (4.8a)$$
$$U_\varepsilon(1) = u_\varepsilon(1) \qquad (4.8b)$$

with the scalar product $(v, w) = \int_0^1 vw \, dx$ for any functions v and w. The nodes are chosen to be the mesh points of the piecewise–uniform fitted mesh with the 'optimal' transition parameter σ_{opt} given by (4.7d). This is equivalent to the following finite difference method

Method 4.4.   Centred finite difference operator on 'optimally tuned' piecewise–uniform fitted mesh for Problem Class 2.2.

$$\varepsilon \delta^2 U_\varepsilon + a(x_i) D^0 U_\varepsilon = f(x_i), \quad x_i \in \Omega_\varepsilon^N, \qquad (4.9a)$$
$$\left(U_\varepsilon(x_1) - U_\varepsilon(x_0)\right)\left(\frac{\varepsilon}{x_1 - x_0} + 1\right) = \varepsilon u_\varepsilon'(0), \quad U_\varepsilon(1) = u_\varepsilon(1), \qquad (4.9b)$$

$$\overline{\Omega}_\varepsilon^N = \{x_i | x_i = 2i\sigma_{opt}/N, \ i \leq N/2; \ x_i = x_{i-1} + 2(1 - \sigma_{opt})/N, \ N/2 < i\} \qquad (4.9c)$$

and

$$\sigma_{opt} = \min\{\frac{1}{2}, \frac{2\varepsilon}{\alpha} \ln N\}. \qquad (4.9d)$$

Note that the approximation in (4.9b) of the Neumann boundary condition at the point $x = 0$ is obtained from (4.8a) with $i = 0$.

We now apply the 'optimally tuned' non–monotone Method 4.4 to problem (2.4). In Table 4.7 we present the resulting exact maximum pointwise errors $E_{\varepsilon,\text{exact}}^N$ and E_{exact}^N for various values of ε and N. Again, these errors are dramatically smaller than those in Table 3.2 obtained when the monotone Method 3.2 is applied to the same problem. For example, we see from the results in these tables that the non–monotone Method 4.4 with $N = 8$ has a smaller maximum pointwise error than the monotone Method 3.2 with $N = 256$. However, as in the previous section, we caution the reader that these excellent results for this and other one–dimensional linear problems should not persuade us that the method is a useful numerical method for finding satisfactory approximations to the solution of a wide class of problems. Here the results are achieved by the fine tuning of a parameter for a specific problem and the resulting good behaviour for this problem is not necessarily robust with respect to changes in the data of the problem.

Table 4.7 *Exact maximum pointwise errors $E^N_{\varepsilon,\text{exact}}$ and E^N_{exact} generated by Method 4.4 applied to problem (2.4) for various values of ε and N.*

	Number of intervals N						
ε	8	16	32	64	128	256	512
2^{-2}	0.034363	0.007826	0.001915	0.000476	0.000119	0.000030	0.000007
2^{-4}	0.038578	0.014749	0.005823	0.002085	0.000705	0.000230	0.000073
2^{-6}	0.043579	0.015124	0.005823	0.002085	0.000705	0.000230	0.000073
2^{-8}	0.046424	0.016603	0.005937	0.002085	0.000705	0.000230	0.000073
2^{-10}	0.047296	0.017461	0.006352	0.002116	0.000705	0.000230	0.000073
2^{-12}	0.047525	0.017728	0.006593	0.002226	0.000713	0.000230	0.000073
2^{-14}	0.047584	0.017799	0.006669	0.002290	0.000742	0.000232	0.000073
2^{-16}	0.047598	0.017817	0.006689	0.002310	0.000758	0.000239	0.000073
2^{-18}	0.047602	0.017822	0.006694	0.002316	0.000763	0.000244	0.000075
2^{-20}	0.047603	0.017823	0.006696	0.002317	0.000765	0.000245	0.000076
E^N_{exact}	0.047603	0.017823	0.006696	0.002317	0.000765	0.000245	0.000076

Also, the explicit solution of Method 4.4 applied to problem (2.4) can be written in the form

$$U^N_\varepsilon(x_i) = \begin{cases} \lambda_1^i - (\lambda_1\lambda_2)^{N/2}, & \text{if } i \leq N/2 \\ (\lambda_1\lambda_2)^{N/2}(\lambda_2^{i-N} - 1), & \text{if } i \geq N/2 \end{cases}$$

where

$$\lambda_1 = \frac{1 - \frac{a\sigma}{N\varepsilon}}{1 + \frac{a\sigma}{N\varepsilon}} \quad \text{and} \quad \lambda_2 = \frac{1 - \frac{a(1-\sigma)}{N\varepsilon}}{1 + \frac{a(1-\sigma)}{N\varepsilon}}$$

from which it is clear that, for all sufficiently small values of ε, λ_2 is negative and so oscillations appear in U^N_ε for all $i \geq N/2$. In contrast we recall from §2.1 that, for all values of ε, the exact solution u_ε of problem (2.4) is monotonically decreasing, and so the oscillations in the numerical solutions given by Method 4.4 are non–physical.

4.6 Approximation of scaled derivatives

The exact maximum pointwise errors $E^N_{\varepsilon,\text{exact}}$ and E^N_{exact} in approximating the scaled derivative $\varepsilon u'_\varepsilon$ of the solution of problem (2.4) by the appropriate scaled discrete derivative obtained from the numerical solutions given by Method 3.2 are displayed In Table 3.4. Analogous results corresponding to the 'optimally tuned' non–monotone Method 4.4 are shown in Table 4.8. At first sight both methods appear to be ε–uniform. However, the approximations of the solution of problem (2.4), generated by the non–monotone Method 4.4 are more accurate than the approximations of its scaled first derivative, given by the scaled discrete derivative of the numerical solution. Indeed, we see that the exact maximum pointwise relative errors in the approximation of the scaled first derivative by

Table 4.8 *Exact maximum pointwise errors* $E_{\varepsilon,\text{exact}}^N(\varepsilon D^+ U_\varepsilon)$, $E_{\text{exact}}^N(\varepsilon D^+ U_\varepsilon)$ *generated by Method 4.4 applied to problem (2.4) for various values of ε and N.*

	Number of intervals N						
ε	8	16	32	64	128	256	512
2^{-1}	0.400000	0.222222	0.117647	0.060606	0.030769	0.015504	0.007782
2^{-2}	0.666667	0.400000	0.222222	0.117647	0.060606	0.030769	0.015504
2^{-3}	0.684090	0.514749	0.356086	0.222222	0.117647	0.060606	0.030769
2^{-4}	0.684090	0.514749	0.356086	0.230034	0.140941	0.083046	0.047578
$\cdot$	$\cdot$	$\cdot$	$\cdot$	$\cdot$	$\cdot$	$\cdot$	$\cdot$
2^{-20}	0.684090	0.514749	0.356086	0.230034	0.140941	0.083046	0.047578
$E_{\text{exact}}^N(\varepsilon D^+ U_\varepsilon)$	0.684090	0.514749	0.356086	0.230034	0.140941	0.083046	0.047578

Table 4.9 *Exact maximum pointwise errors* $E_{\varepsilon,\text{exact}}^N(\varepsilon D^0 U_\varepsilon)$, $E_{\text{exact}}^N(\varepsilon D^0 U_\varepsilon)$ *generated by Method 4.4 applied to problem (2.4) for various values of ε and N.*

	Number of intervals N						
ε	8	16	32	64	128	256	512
2^{-1}	0.066939	0.022645	0.006632	0.001798	0.000468	0.000120	0.000030
2^{-2}	0.153130	0.066939	0.022645	0.006632	0.001798	0.000468	0.000120
2^{-3}	0.158703	0.102986	0.054387	0.022645	0.006632	0.001798	0.000468
2^{-4}	0.158703	0.102986	0.054387	0.024179	0.009431	0.003349	0.001113
$\cdot$	$\cdot$	$\cdot$	$\cdot$	$\cdot$	$\cdot$	$\cdot$	$\cdot$
2^{-20}	0.158703	0.102986	0.054387	0.024179	0.009431	0.003349	0.001113
$E_{\text{exact}}^N(\varepsilon D^0 U_\varepsilon)$	0.158703	0.102986	0.054387	0.024179	0.009431	0.003349	0.001113

the scaled discrete derivative is significantly smaller for the monotone method than for the non–monotone method.

Of course it is possible to retune the non–monotone method to give good results for this particular problem. In Table 4.9 the exact maximum pointwise errors $E_{\varepsilon,\text{exact}}^N$ and E_{exact}^N in approximating the scaled derivative $\varepsilon u_\varepsilon'$ of the solution of problem (2.4) are displayed for the 'optimally tuned' non–monotone Method 4.4 where the scaled derivative $\varepsilon u_\varepsilon'$ is approximated by the scaled centred discrete derivative $\varepsilon D^0 U_\varepsilon(x_i)$ at all internal mesh points x_i, $i = 1, \ldots N-1$ and by the scaled discrete derivative $\varepsilon \left(D^+ U_\varepsilon(0) + \frac{x_1}{2\varepsilon}(a(0)D^+ U_\varepsilon(0) - f(0)) \right)$ at the point $x = 0$. The non–monotone Method 4.4 then produces better approximations to the scaled derivative than those obtained with Method 3.1. We stress that this improved performance is achieved only when the method is tuned to this specific problem, which shows that the method is not robust.

4.7   Further considerations

Results in this chapter show that non–monotone numerical methods, consisting of centred finite difference operators on piecewise–uniform fitted meshes, can lead to smaller ε–uniform maximum pointwise errors than monotone numerical methods based on upwind finite difference operators on the same piecewise–uniform fitted meshes, at least for some singularly perturbed linear ordinary differential equations similar to problems (2.3) and (2.4). The question as to whether or not this is also the case for a wide class of singularly perturbed partial differential equations is deferred to chapter 9. It is important to observe that this increase in accuracy is at a cost. First, the increased accuracy of the method is sensitive to the particular choice of the tuning parameter c in the definition of the fitted mesh, which means that this good performance is not robust. Also, small non–physical oscillations between positive and negative values occur in the numerical solutions, which implies that the discrete problem does not have a discrete maximum or minimum principle analogous to that of the continuous problem. Finally, rigorous theoretical analysis of the convergence of non–monotone methods on piecewise–uniform fitted meshes is still in its infancy and consequently, at present, there is little theoretical support for the use of these methods.

Another entirely different approach, to eliminating or controlling non–physical oscillations and to improving the accuracy, is to apply appropriate post–processing techniques to the computed solution based on, for example, high quality interpolation of the numerical solutions from the mesh points to the whole domain. However, it is not clear that appropriate techniques can be found for a sufficiently wide class of singularly perturbed partial differential equations to make this approach of practical interest.

Because of the hidden dangers and deceptive benefits of ε–uniform non–monotone methods, which are illustrated above for ordinary differential equations and in chapter 9 for partial differential equations, the emphasis in this book is on the development of ε–uniform methods that are monotone.

CHAPTER 5

Convection–diffusion problems in a moving medium

5.1 Motivation

We motivate the study of the problems considered in this chapter by describing some physical situations, which involve the flow of a substance through a moving medium. In such situations the flow of the substance is a superposition of the primary flow of the medium and the secondary flow of the substance through the medium. Our first example is the transport of a pollutant spilling onto the surface of a river, which is illustrated in Fig. 5.1. The pollutant sinks through the flowing water and is deposited on the river bed. The velocity of any particle of the pollutant can be decomposed into two components; one due to the movement of the water and the other (its intrinsic velocity) due to gravity. The environmental

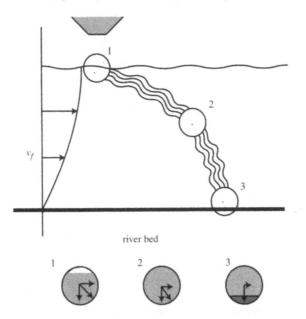

FIG. 5.1. Transport of a pollutant in a moving medium.

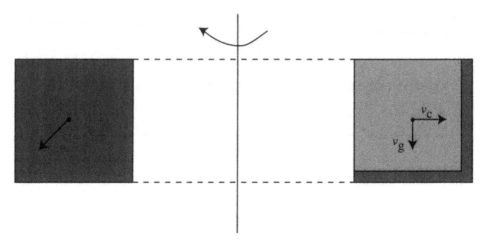

FIG. 5.2. Velocity components due to centrifugal and gravitational forces in a centrifuge.

scientist may be interested in the long term distribution of the pollutant over a given cross–section of the river. Our second example involves the operation of a centrifuge, which is a rotating vessel used typically to separate materials from a solvent. As the centrifuge rotates the heavier materials in the solvent are thrown onto the side walls of the centrifuge by the centrifugal force. Here a secondary radial flow is imposed on the primary rotational flow, as shown in Fig. 5.2. Many similar examples arise in sedimentation processes, such as the flow of a substance through a porous medium, for example oil in a reservoir, and the dispersal of a pollutant downwind of a tall chimney stack.

Physical situations, of the kind described above, are often modelled mathematically by the linear steady–state convection–diffusion equation

$$-\varepsilon \Delta c + \mathbf{v} \cdot \nabla c = f \tag{5.1}$$

in a rectangular domain Ω, where c is the concentration of a substance in a medium moving with velocity $\mathbf{v}$. The first term on the left hand side represents the diffusion of the substance in the medium, while the second term represents its convection by the medium. When ε is small, changes in the concentration are convection–dominated and boundary layers of different kinds can appear in the concentration at different parts of the boundary Γ of Ω. For example, at points of Γ where the velocity field $\mathbf{v}$ is not parallel to an edge, that is where $\mathbf{v} \cdot \mathbf{n} \neq 0$, $\mathbf{n}$ denoting the unit outward normal to Γ, a regular boundary layer typically arises. On the other hand, at points of Γ where $\mathbf{v}$ is parallel to an edge, that is where $\mathbf{v} \cdot \mathbf{n} = 0$, we find a parabolic boundary layer. In the vicinity of a corner of the domain, further layers can exist; these are called corner boundary layers.

In this chapter, we consider only regular and corner boundary layers. Parabolic boundary layers are considered in subsequent chapters.

In mathematical models of such problems the usual convention is that the flow is from left to right. This results in the sign pattern in (5.1). However, for notational reasons, it is sometimes more convenient to have a different sign pattern. We can obtain this new sign pattern by changing the independent spatial variables from (x, y) to $(\xi, \eta) = (1 - x, 1 - y)$, which transforms (5.1) into the mathematically equivalent equation

$$\varepsilon \tilde{\Delta}\tilde{c} + \tilde{\mathbf{v}} \cdot \tilde{\nabla}\tilde{c} = \tilde{f} \tag{5.2}$$

where, for any function h, we write $\tilde{h}(\xi, \eta) = h(x, y)$ and $\tilde{\Delta}$, $\tilde{\nabla}$ in (5.2) involve derivatives with respect to ξ and η.

With the exception of the final two chapters, we assume henceforth that the domain Ω is the unit square $\Omega = (0, 1) \times (0, 1)$ with boundary Γ. We denote its four edges by

$$\Gamma_{\mathrm{B}} = \{(x, 0) | \ 0 < x < 1\}, \quad \Gamma_{\mathrm{R}} = \{(1, y) | \ 0 < y < 1\},$$

$$\Gamma_{\mathrm{T}} = \{(x, 1) | \ 0 < x < 1\}, \quad \Gamma_{\mathrm{L}} = \{(0, y) | \ 0 < y < 1\}$$

and its four corners by

$$C_{\mathrm{BL}} = (0, 0), \quad C_{\mathrm{BR}} = (1, 0), \quad C_{\mathrm{TR}} = (1, 1), \quad C_{\mathrm{TL}} = (0, 1).$$

Thus $\Gamma = \Gamma_{\mathrm{B}} \cup \Gamma_{\mathrm{R}} \cup \Gamma_{\mathrm{T}} \cup \Gamma_{\mathrm{L}} \cup \{C_{\mathrm{BL}}, C_{\mathrm{BR}}, C_{\mathrm{TR}}, C_{\mathrm{TL}}\}$.

For notational convenience we use the unit square as the problem domain. It is important to observe that all of the techniques, described here for problems on the unit square, are equally valid for problems on any rectangular domain. It is only in the final two chapters, that we need to distinguish between the unit square and other rectangular domains. Note, however, that if Ω is a more complex polygonal domain, for example, additional technical difficulties arise in the construction of a suitable mesh, and the mathematical derivation of ε–uniform error estimates requires ε–explicit bounds for mixed derivatives of the solution. Appropriate constructions for convex polygonal domains are developed in Shishkin (1992b).

5.2   Convection–diffusion problems

In this chapter, we consider two–dimensional convection-diffusion problems of the form

$$\varepsilon \Delta u_\varepsilon + \mathbf{a}(x, y) \cdot \nabla u_\varepsilon = f(x, y), \quad (x, y) \in \Omega \tag{5.3a}$$

$$u_\varepsilon = g, \quad (x, y) \in \Gamma \tag{5.3b}$$

where $0 < \varepsilon \le 1$. The problem is singularly perturbed when $\varepsilon << 1$, in which case boundary layers of various kinds are usually present in the solution.

The fact that the singular perturbation problem (5.3) involves a partial differential equation is of crucial importance. Although some fundamental boundary layer phenomena arise in problems with ordinary differential equations, many additional technical issues arise with partial differential equations, such as the smoothness of the solution, the compatibility of the data and the geometry of the domain. In this book we place little emphasis on the details of the mathematical technicalities, because we want to concentrate on distinguishing between good and bad numerical methods by examining how well they perform on concrete problems. Nevertheless we now introduce some technical terminology and notation which helps us to explain some of these technical issues. Readers may safely skip the remainder of this section without any serious loss of understanding.

The appropriate functions are the Hölder continuous functions $C^{k,\gamma}(\Omega)$ defined as follows: for each integer $k \geq 0$ and $\gamma \in (0,1]$, a function $w \in C^{k,\gamma}(\Omega)$, if $w \in C^k(\Omega)$ and if, for any two points $(x,y),(x',y') \in \Omega$, all of the derivatives $D^\beta w, |\beta| \leq k$ satisfy the following Hölder condition of order γ

$$|D^\beta w(x,y) - D^\beta w(x',y')| \leq M((x - x')^2 + (y - y')^2)^{\gamma/2}$$

for some constant M, with $D^\beta = \frac{\partial^{\beta_1 + \beta_2}}{\partial x^{\beta_1} \partial y^{\beta_2}}$, $\beta = (\beta_1, \beta_2)$, and $|\beta| = \beta_1 + \beta_2$ where β_1, β_2 are non–negative integers.

In this chapter we assume that the data of (5.3) satisfy the following conditions

$$\mathbf{a} \cdot \mathbf{n} \neq 0 \quad \text{on} \quad \bar{\Omega} \qquad (5.3\text{c})$$

$$a_1, a_2 \in C^{0,1}(\bar{\Omega}), \ f \in C^{0,\gamma}(\bar{\Omega}), \ \gamma > 0 \ , \quad g \in C^0(\Gamma) \qquad (5.3\text{d})$$

where $\mathbf{n}$ denotes the unit outward normal on Γ. Condition (5.3c) ensures that no parabolic boundary layers arise in the solution, while (5.3d) imposes conditions on the smoothness of the data . Under these assumptions on the data, both regular and corner boundary layers can arise in the solution of the problem. Without (5.3c) parabolic boundary layers are in general also present in the solution. Without the smoothness condition (5.3d) singularities other than boundary layers occur in the solution because, for example, a discontinuity in the boundary data, at a point on the boundary, propagates into the interior of the domain and forms an interior layer. Numerical methods, appropriate for the resolution of interior layers, are not dealt with in this book. Some such methods are described in Shishkin (1988b, 1993, 1994, 1996, 1997) and Farrell et al. (1996). It is interesting to observe that condition (5.3d) allows continuous but non–differentiable data of the form

$$f(x,y) = \sqrt{1 - x}, \quad a_1(x,y) = 1 + |x - 0.5|.$$

5.3 Location of regular and corner boundary layers

In this section we determine the location of the various kinds of boundary layers that can occur in the solution of problem (5.3). The reduced problem corresponding to (5.3) is obtained by formally putting $\varepsilon = 0$. This leads to the first order hyperbolic equation

$$\mathbf{a} \cdot \nabla v_0 = f$$

and we have to impose appropriate boundary conditions to obtain a well defined problem. It is clear that not all of the boundary conditions (5.3b) can be satisfied by the solution of this first order equation. To find the appropriate boundary conditions we introduce the characteristic curves $\mathbf{c}(s) = (x(s), y(s))$ where $\frac{dx}{ds} = a_1$ and $\frac{dy}{ds} = a_2$. In terms of the new independent variable s, the reduced equation can be written as the first order ordinary differential equation

$$\frac{dv_0}{ds} = f$$

which shows that, if v_0 is specified at any point on a characteristic, then its value at any other point of this characteristic is fully determined. This means that on these characteristic curves, which are illustrated in Fig. 5.3, the boundary layer structure is similar to that for one–dimensional problems and the boundary layer

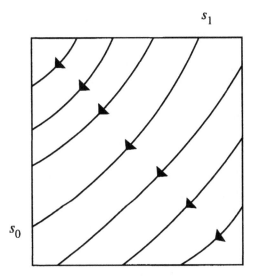

FIG. 5.3. Characteristic curves of the reduced problem corresponding to problem (5.3).

is called a regular boundary layer. Each characteristic cuts the boundary at two points $\{s_0, s_1\}$, and the set of all points at which characteristics begin is called the inflow boundary, while that at which characteristics end is called the outflow boundary.

When the velocity profile $\mathbf{a}$ is independent of x and y, and the inhomogeneous term f is identically zero, the characteristics associated with (5.3) are straight lines. It is then convenient to introduce the new independent variables

$$\xi = a_2 x - a_1 y, \quad \eta = a_1 x + a_2 y$$

and the change of variables $(x, y) \to (\xi, \eta)$ transforms (5.3a) into

$$\varepsilon \tilde{\Delta} \tilde{u}_\varepsilon + \frac{\partial \tilde{u}_\varepsilon}{\partial \eta} = 0$$

where $\tilde{u}_\varepsilon(\xi, \eta) = u_\varepsilon(x, y)$. It is easy to see that for any constants A and B the function $u(\xi, \eta) \equiv (A\xi + B)S(\eta)$ is a solution of this transformed differential equation, provided that S is a solution of the problem

$$\varepsilon \frac{d^2 S}{d\eta^2} + \frac{dS}{d\eta} = 0 \quad \text{on } (s_0, s_1)$$
$$S(s_0) = g(s_0), \qquad S(s_1) = g(s_1)$$

where s_0 and s_1 are points on the outflow and inflow boundaries respectively. The boundary layer in the solution of this one–dimensional problem occurs in a neighbourhood of the outflow boundary $\eta = s_0$. The boundary layer function is an exponential function in the variable η, which shows that it is essentially one–dimensional in nature.

Because regular boundary layers are essentially one–dimensional in nature, the solutions of the original and the reduced problems differ significantly only at points near the outflow boundary, and the difference near each outflow point is influenced by the value of the exact solution only at that boundary point and not by its values at other boundary points. Note that corner boundary layers arise, if there is a marked difference between the solutions of the full and the reduced problems near a corner of the domain.

It is clear from the above discussion that the location of a regular boundary layer in the solution of problem (5.3) depends on the sign of the components of $\mathbf{a}$, which is illustrated in Fig. 5.4. To be specific we assume in the remainder of this chapter that these components are strictly positive throughout the domain. Thus, for all $(x, y) \in \bar{\Omega}$, we assume that the velocity profile $\mathbf{a}$ fulfills the condition

$$\mathbf{a}(x, y) = (a_1(x, y), a_2(x, y)) \geq \alpha = (\alpha_1, \alpha_2) > (0, 0).$$

With this pattern of signs regular boundary layers of width $O(\varepsilon)$ typically appear in the solution of problem (5.3) in a neighbourhood of the edges Γ_L and Γ_B, as shown in Fig. 5.5.

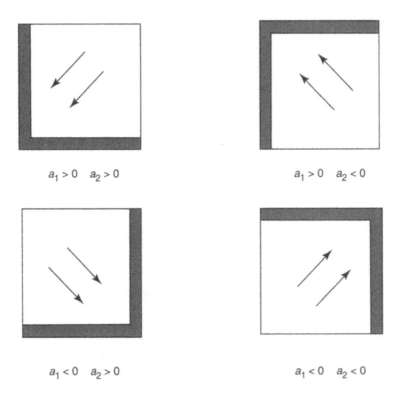

FIG. 5.4. Locations of the boundary layers in problem (5.3) for various velocity profiles $\mathbf{a} = (a_1, a_2)$.

We now introduce the following class of convection–diffusion problems, in which we impose weak continuity conditions in order to obtain a wide class.

Problem Class 5.1. Linear convection diffusion in two dimensions with regular and corner boundary layers.

$$\varepsilon \Delta u_\varepsilon + \mathbf{a}(x,y) \cdot \nabla u_\varepsilon = f(x,y), \quad (x,y) \in \Omega \quad \text{(5.4a)}$$
$$u_\varepsilon = g, \quad (x,y) \in \Gamma \quad \text{(5.4b)}$$
$$\mathbf{a}(x,y) \geq \alpha > 0, \quad (x,y) \in \bar{\Omega} \quad \text{(5.4c)}$$
$$a_1, a_2 \in C^{0,1}(\bar{\Omega}), \ f \in C^{0,\gamma}(\bar{\Omega}), \ \gamma > 0 , \quad g \in C^0(\Gamma). \quad \text{(5.4d)}$$

We can consider an even wider class of problems than those in Problem Class 5.1 by requiring that, instead of (5.4d), the data satisfy the weaker Hölder continuity conditions, $a_1, a_2, f \in C^{0,\gamma}(\bar{\Omega})$, $g \in C^\gamma$ for some $\gamma \in (0,1)$. Then

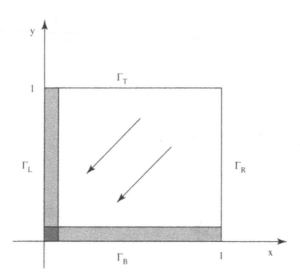

FIG. 5.5. Location of boundary layers for a typical problem in Problem Class 5.1.

the solution u_ε of the corresponding problem is Hölder continuous in the sense that $u_\varepsilon \in C^{0,\gamma}(\overline{\Omega}) \cap C^2(\Omega)$ (Ladyzhenskaya and Ural'tseva (1968)). If we require a smoother solution, such as $u_\varepsilon \in C^{2,\gamma}(\overline{\Omega}) \cap C^{4,\gamma}(\Omega)$, this can be obtained by imposing extra compatibility conditions on the data (Han and Kellogg (1990)). For example, in the case of the subclass of Problem Class 5.1 corresponding to homogeneous boundary condition $g = 0$, the extra compatibility conditions

$$f(0,0) = f(0,1) = f(1,0) = f(1,1) = 0$$

and the stronger smoothness conditions

$$a_1, a_2, f \in C^{0,\gamma}(\overline{\Omega}) \cap C^{2,\gamma}(\Omega) \quad \text{for some } \gamma \in (0,1)$$

are sufficient to guarantee that $u_\varepsilon \in C^{2,\gamma}(\overline{\Omega}) \cap C^{4,\gamma}(\Omega)$ (Volkov (1965)).

Before developing numerical methods for problems in Problem Class 5.1, we examine in the next section some asymptotic properties of their solutions.

5.4 Asymptotic nature of boundary layers

In order to gain more insight into the analytic nature of the boundary layers in the solutions of problems from Problem Class 5.1, we derive an asymptotic approximation u_{asy} to the solution u_ε of a typical problem. A function u_{asy} is an asymptotic expansion of u_ε if

$$u_\varepsilon(x,y) = u_{\text{asy}}(x,y) + R(x,y)$$

where, for some $r > 0$

$$\|R\|_{\overline{\Omega}} \leq C\varepsilon^r.$$

If u_{asy} is to contain useful information about the boundary layers, it is essential that the remainder term R is small at each point of the domain. The discussion in this section is based on well known results about matched asymptotic expansions, which may be found in, for example, O'Malley (1991), Eckhaus (1973, 1979) and Il'in (1992).

To find u_{asy} we begin by observing that the solution of problem (5.4) can be decomposed into the sum of two functions

$$u_\varepsilon = v_\varepsilon + w_\varepsilon$$

where the smooth component v_ε and the singular component w_ε are defined in the following way. The smooth component v_ε is the solution of the problem

$$\varepsilon \Delta v_\varepsilon + \mathbf{a}(x, y) \cdot \nabla v_\varepsilon = f(x, y), \quad (x, y) \in \Omega \qquad (5.5a)$$

$$v_\varepsilon(1, y) = g(1, y) \text{ on } \Gamma_{\text{R}}, \quad v_\varepsilon(x, 1) = g(x, 1) \text{ on } \Gamma_{\text{T}}, \qquad (5.5b)$$

$$v_\varepsilon(0, y) = h(0, y) \text{ on } \Gamma_{\text{L}}, \quad v_\varepsilon(x, 0) = h(x, 0) \text{ on } \Gamma_{\text{B}}, \qquad (5.5c)$$

where the boundary values $h(0, y)$ and $h(x, 0)$ are chosen so that v_ε has no boundary layers. More precisely, v_ε satisfies the original differential equation (5.3a) and the inflow boundary conditions. On the outflow boundary new boundary conditions are specified, which guarantee that the first derivatives of v_ε are bounded independently of ε at all points of $\overline{\Omega}$. Consequently, the singular component w_ε satisfies the corresponding homogeneous differential equation, vanishes on the inflow boundary and satisfies appropriate (in general non–zero) boundary conditions on the outflow boundary. That is, w_ε is the solution of the problem

$$\varepsilon \Delta w_\varepsilon + \mathbf{a} \cdot \nabla w_\varepsilon = 0, \quad (x, y) \in \Omega \qquad (5.6a)$$

$$w_\varepsilon(1, y) = 0 \text{ on } \Gamma_{\text{R}}, \quad w_\varepsilon(x, 1) = 0 \text{ on } \Gamma_{\text{T}}, \qquad (5.6b)$$

$$w_\varepsilon(0, y) = g(0, y) - v_\varepsilon(0, y) \text{ on } \Gamma_{\text{L}}, \qquad (5.6c)$$

$$w_\varepsilon(x, 0) = g(x, 0) - v_\varepsilon(x, 0) \text{ on } \Gamma_{\text{B}}. \qquad (5.6d)$$

It is shown in Shishkin (1992b) that the magnitude of the singular component w_ε and its derivatives are significant only near the outflow boundary.

The above decomposition of the solution motivates the following method of constructing the asymptotic expansion u_{asy}. The leading term is taken to be the solution of the reduced problem

$$\mathbf{a}(x, y) \cdot \nabla v_0 = f(x, y), \quad (x, y) \in \Omega \qquad (5.7a)$$

$$v_0(1, y) = g(1, y) \text{ on } \Gamma_{\text{R}}, \quad v_0(x, 1) = g(x, 1) \text{ on } \Gamma_{\text{T}}. \qquad (5.7b)$$

Then, for ε small, the expectation is that v_0 and u_ε are almost equal at points some distance from the outflow edges Γ_{L} and Γ_{B}. At points near these edges, it is necessary to add correction terms $w_{0,L}, w_{0,B}$ with the property that the resulting sum $v_0 + w_{0,L} + w_{0,B}$ is close to u_ε. We also require that the correction terms $w_{0,L}, w_{0,B}$ are small at points some distance from the edges Γ_L and Γ_B, because v_0 is already an approximation to u_ε. The correction terms $w_{0,L}, w_{0,B}$ describe the asymptotic nature of the boundary layers in neighbourhoods of the corresponding edges Γ_L and Γ_B, and so they are called boundary layer functions. To find an expression for the boundary layer function $w_{0,L}$, corresponding to the regular boundary layer on Γ_L, we introduce the stretched variable

$$X = \frac{x}{\varepsilon}.$$

This variable is $O(1)$ only in an ε–neighbourhood of Γ_L. The change of variables

$$(x, y) \to (X, y)$$

and multiplication by ε, transforms (5.6a) into the equation

$$\frac{\partial^2 \tilde{w}}{\partial X^2} + \varepsilon^2 \frac{\partial^2 \tilde{w}}{\partial y^2} + \tilde{a}_1(X, y) \frac{\partial \tilde{w}}{\partial X} + \varepsilon \tilde{a}_2(X, y) \frac{\partial \tilde{w}}{\partial y} = 0$$

where $\tilde{w}(X, y) = w(x, y)$. Neglecting the $0(\varepsilon)$ terms then leads to the equation

$$\frac{\partial^2 \tilde{w}_{0,L}(X, y)}{\partial X^2} + \tilde{a}_1(X, y) \frac{\partial \tilde{w}_{0,L}(X, y)}{\partial X} = 0 \qquad (5.8a)$$

where $\tilde{w}_{0,L}$ is the boundary layer function associated with the edge Γ_L. Treating y as a parameter, (5.8) is a second order ordinary differential equation in X. To obtain a well defined problem for $\tilde{w}_{0,L}$, we need to specify boundary conditions for $X = 0$ and $X \to \infty$. The appropriate outflow boundary condition at $X = 0$ is

$$\tilde{w}_{0,L}(0, y) = g(0, y) - v_0(0, y) \qquad (5.8b)$$

which gives $v_0(0, y) + w_{0,L}(0, y) = u_\varepsilon(0, y)$. The inflow boundary condition for $X \to \infty$ is chosen so that, at some distance from the outflow edge Γ_L, the boundary layer function $w_{0,L}$ is of negligible magnitude. The appropriate condition is

$$\tilde{w}_{0,L}(\infty, y) = 0. \qquad (5.8c)$$

Using these boundary conditions it is easy to check that the function

$$\tilde{w}_{0,L}(X, y) = (g(0, y) - v_0(0, y))e^{-a_1 X}, \quad X = x/\varepsilon$$

is the solution of (5.8) in the special case when a_1 is independent of x and y. In an analogous fashion, we get

$$\tilde{w}_{0,B}(x, Y) = (g(x, 0) - v_0(x, 0))e^{-a_2 Y}, \quad Y = y/\varepsilon$$

in the case when a_2 is independent of x and y.

At the outflow corner C_{BL} we observe that

$$(u_\varepsilon - v_0 - w_{0,L} - w_{0,B})(0,0) = v_0(0,0) - g(0,0).$$

Since this is not, in general, zero it follows that we need to determine one more correction term. In the case when a_1 and a_2 are independent of x and y the appropriate term is

$$\tilde{w}_{0,BL}(X,Y) = (v_0(0,0) - g(0,0))e^{-a_1 X}e^{-a_2 Y}, \quad X = x/\varepsilon, Y = y/\varepsilon$$

which is interpreted as the corner boundary layer function corresponding to the corner C_{BL}. Note that this is a particularly simple corner boundary layer, since it is the product of two exponentials. In other situations, for example when a_1 and a_2 depend on x and y, corner boundary layer functions can be more complicated.

When the data are sufficiently smooth (smoother than the conditions in (5.4d)) and stronger compatibility conditions are imposed, then it can be shown (see e.g., Emel'ianov (1973), Hegarty (1986), Linss and Stynes (1998)) that

$$\|u_\varepsilon - u_{\mathrm{asy}}\|_{\overline{\Omega}} \leq C\varepsilon \qquad (5.9a)$$

where the required asymptotic expansion u_{asy} is defined by

$$u_{\mathrm{asy}}(x,y) = v_0(x,y) + w_{0,L}(x,y) + w_{0,B}(x,y) + w_{0,BL}(x,y). \qquad (5.9b)$$

From this expression we conclude that the asymptotic behaviour of the regular and corner boundary layers, for this problem, is described by pure exponential functions.

It is important to realise that an asymptotic expansion u_{asy} of the kind discussed in this section is not sufficient for the design and analysis of ε–uniform numerical methods for all problems in Problem Class 5.1. This is because this traditional asymptotic analysis requires that the boundary data are smooth and also that there is a high degree of compatibility. This is too restrictive, because engineers often have applied problems that do not satisfy these smoothness conditions. Note also that asymptotic analysis gives us an asymptotic expansion of the solution, but an asymptotic expansion for the derivatives of the solution usually cannot be derived from it. Moreover, even if we have an asymptotic expansion for the derivatives, it does not follow that this representation of the derivatives suffices for the construction of an ε–uniform method and the analysis of the ε–uniform convergence in the maximum norm of its numerical solutions. Therefore, a special technique (Shishkin (1992b)) is required to obtain estimates of the appropriate form for proving the convergence of the numerical solutions and their scaled discrete derivatives to the exact solution and its scaled derivatives. These estimates need to be valid under minimal restrictions on the smoothness of the boundary data and minimal compatibility conditions. In fact, it is possible to construct an ε–uniform method for problem (5.3) which requires only the smoothness conditions (5.3d) on the data and no compatibility conditions. This is discussed more fully in the next section. It should be noted that the ε–uniform order of convergence of the numerical solutions depends on the smoothness of the boundary data and the compatibility conditions.

5.5   Monotone parameter–uniform methods

We now construct a monotone numerical method which is ε–uniform for problems in Problem Class 5.1. Mesh refinement is required near the outflow boundaries Γ_B, Γ_L, since regular boundary layers of width $O(\varepsilon)$ appear in the solution near these edges, and a corner boundary layer occurs near the corner C_{BL} of these edges. For the construction of our monotone ε–uniform method we use the standard upwind finite difference operator and, because the domain is rectangular, a tensor product of two one–dimensional piecewise–uniform fitted meshes, as shown in Fig. 5.6. The appropriate piecewise–uniform fitted meshes are specified in the following definition.

Method 5.1.   Upwind finite difference operator on piecewise–uniform fitted mesh for Problem Class 5.1.

$$[\varepsilon(\delta_x^2 + \delta_y^2) + a_1(x_i, y_j)D_x^+ + a_2(x_i, y_j)D_y^+]U_\varepsilon = f(x_i, y_j) \quad (x_i, y_j) \in \Omega_\varepsilon^N \tag{5.10a}$$

$$U_\varepsilon = g, \quad (x_i, y_j) \in \Gamma^N \tag{5.10b}$$

where

$$\overline{\Omega}_\varepsilon^N = \overline{\Omega}_{\sigma_1}^N \times \overline{\Omega}_{\sigma_2}^N, \quad \Gamma^N = \Gamma \cap \overline{\Omega}_\varepsilon^N \tag{5.10c}$$

$$\overline{\Omega}_{\sigma_1}^N = \{x_i : 0 \le i \le N\} \quad \text{and} \quad \overline{\Omega}_{\sigma_2}^N = \{y_j : 0 \le j \le N\} \tag{5.10d}$$

with

$$x_i = \begin{cases} 2i\sigma_1/N, & \text{for } 0 \le i \le N/2 \\ \sigma_1 + 2(i - (N/2))(1 - \sigma_1)/N, & \text{for } N/2 \le i \le N \end{cases} \tag{5.10e}$$

$$y_j = \begin{cases} 2j\sigma_2/N, & \text{for } 0 \le j \le N/2 \\ \sigma_2 + 2(j - (N/2))(1 - \sigma_2)/N, & \text{for } N/2 \le j \le N \end{cases} \tag{5.10f}$$

and

$$\sigma_1 = \min\{1/2, c_1\varepsilon \ln N\}, \quad c_1 \ge \alpha_1^{-1} \tag{5.10g}$$

$$\sigma_2 = \min\{1/2, c_2\varepsilon \ln N\}, \quad c_2 \ge \alpha_2^{-1} \tag{5.10h}$$

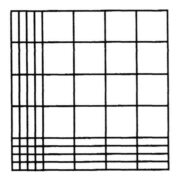

FIG. 5.6. Piecewise–uniform fitted mesh $\overline{\Omega}_\varepsilon^N$ with $N = 8$ for Problem Class 5.1.

We remark that in (5.10) any values for the constants c_i, $i = 1, 2$ satisfying the conditions $c_i \geq \alpha_i^{-1}$, may be used. Here too, as we saw in chapter 3 for the one–dimensional case, it can be shown by numerical experiments that the maximum pointwise errors in the resulting numerical solutions are not particularly sensitive to the values of these constants. For all of the numerical computations in this chapter we take $c_i = 1$, $i = 1, 2$, since $\alpha_i \geq 1$ in our examples. It should be noted that, in the published theoretical work on ε–uniform error estimates for numerical methods using these fitted meshes, the strict inequalities $c_i > \alpha_i^{-1}$ for $i = 1, 2$ are usually assumed.

It is proved in Shishkin (1992b) that, with sufficient smoothness of the data (smoother than (5.4d)) and sufficiently strong compatibility conditions, a numerical method for Problem Class 5.1, and in particular Method 5.1, consisting of any monotone finite difference operator on the piecewise–uniform fitted mesh Ω_ε^N (5.10d-h), satisfies, for all $N \geq N_0$, the following ε–uniform error estimate of order essentially one

$$\sup_{0 < \varepsilon \leq 1} \|\bar{U}_\varepsilon^N - u_\varepsilon\|_{\overline{\Omega}} \leq CN^{-1}(\ln N)^2 \qquad (5.11)$$

where the constants N_0 and C are independent of ε and N. In Shishkin (1992b) the smoothness and compatibility assumptions are implicit. We now state them explicitly in the special case of homogeneous boundary conditions $g = 0$: the smoothness conditions are

$$a_1, a_2, f \in C^{5,\gamma}, \quad \text{for some } \gamma \in (0, 1)$$

and the compatibility conditions are

$$D^\beta f(0,0) = D^\beta f(0,1) = D^\beta f(1,0) = 0, \quad \text{for all } |\beta| \leq 1$$

$$D^\beta f(1,1) = 0, \quad \text{for all } |\beta| \leq 5$$

where $\beta = (\beta_1, \beta_2)$, β_1, β_2 are non–negative integers and $|\beta| = \beta_1 + \beta_2$.

Returning to the case of general boundary conditions, if we weaken the assumptions on the smoothness of the data to condition (5.4d) and if we require no compatibility, then it is proved in Shishkin (2000) that, for all $N \geq N_0$, we have the following ε–uniform error estimate for Method 5.1 applied to Problem Class 5.1

$$\sup_{0<\varepsilon\leq 1} \|\bar{U}_\varepsilon - u_\varepsilon\|_{\overline{\Omega}} \leq \mu(N), \tag{5.12}$$

where μ is independent of ε and $\mu(N) \to 0$ as $N \to \infty$. Furthermore, if we restrict the boundary data to $g \in C^\gamma(\Gamma), \gamma > 0$, then for all $N \geq N_0$ and some $p > 0$ independent of N and ε, we have the following global ε–uniform error estimate of order p

$$\sup_{0<\varepsilon\leq 1} \|\bar{U}_\varepsilon - u_\varepsilon\|_{\overline{\Omega}} \leq CN^{-p}. \tag{5.13}$$

We emphasize that for numerical methods for Problem Class 5.1 standard proofs of ε–uniform error estimates of the form (5.13) depend on strong smoothness assumptions on the exact solution u_ε, namely $u_\varepsilon \in C^{2,\gamma}(\overline{\Omega}) \cap C^{4,\gamma}(\Omega)$. On the other hand, the ε–uniform error estimate given in Shishkin (2000) is valid under the original assumptions (5.4d), which means that no extra regularity or compatibility conditions are required, and that the reduced solution is not even in $C^{1,\gamma}(\Omega)$. To illustrate this we consider the following constant coefficient reduced problem

$$\frac{\partial v_0}{\partial x} + \frac{\partial v_0}{\partial y} = 1, \quad (x,y) \in \Omega$$
$$v_0 = x, \quad (x,y) \in \Gamma_T, \qquad v_0 = y, \quad (x,y) \in \Gamma_R$$

which has the solution

$$v_0(x,y) = \begin{cases} y, & \text{if } y \leq x , \\ x, & \text{if } y \geq x. \end{cases}$$

It is not hard to verify that there is a discontinuity in each of the first derivatives of v_0 at each point of the line $y = x$.

5.6   Computed errors and computed orders of convergence

Each of the one–dimensional problems considered in the previous chapters has an exact solution which can be expressed in a simple explicit closed form. Therefore the exact error and exact order of convergence can be computed for our numerical approximations to these solutions. In the remainder of the book, more complicated problems are considered, which usually do not have a known solution available in a simple closed form. Therefore we find the computed error, as

opposed to the exact error, by replacing the exact solution in the expression for
the error by the numerical solution calculated using an appropriate monotone
fitted mesh method on the finest available mesh. Furthermore, the computed
order of convergence is calculated using the two–mesh method.

All of the problems in the next five chapters are solved on a sequence of
piecewise–uniform fitted meshes Ω_ε^N, for $N = 8, 16, 32, 64, 128, 256, 512$, where
the superscript N indicates the number of mesh elements used in each coordinate
direction. The computed error e_ε^N at a typical point (x_i, y_j) on each of the coarser
meshes is defined by

$$e_\varepsilon^N(x_i, y_j) = |U_\varepsilon^N(x_i, y_j) - \bar{U}_\varepsilon^{512}(x_i, y_j)|,$$

where the overline denotes piecewise bilinear interpolation from the finest fitted
mesh Ω_ε^{512} to the whole domain $\bar{\Omega}$. Correspondingly, for each ε and N, the
maximum pointwise error is approximated by the computed maximum pointwise
error

$$E_\varepsilon^N = \|e_\varepsilon^N\|_{\Omega_\varepsilon^N}$$

and for each N the ε–uniform maximum pointwise error is approximated by the
computed ε–uniform maximum pointwise error

$$E^N = \max_\varepsilon E_\varepsilon^N.$$

The order of convergence of the numerical solutions is estimated using the two–
mesh method. See chapter 8 for a fuller description than that given here. We
define the two–mesh differences as follows

$$D_\varepsilon^N = \|U_\varepsilon^N - \bar{U}_\varepsilon^{2N}\|_{\Omega_\varepsilon^N}, \quad \text{and} \quad D^N = \max_\varepsilon D_\varepsilon^N. \tag{5.14}$$

The computed order of convergence p_ε^N and the computed ε–uniform order of
convergence p^N are then defined by

$$p_\varepsilon^N = \log_2 \frac{D_\varepsilon^N}{D_\varepsilon^{2N}} \tag{5.15}$$

and

$$p^N = \log_2 \frac{D^N}{D^{2N}}. \tag{5.16}$$

It should be noted that the computed orders of convergence, which are based
on the numerical solution generated with the finest available mesh as opposed to
the two–mesh method, must be used with care. To illustrate the fact that they
can be misleading, let us assume that the ε–uniform error for a particular problem

is given exactly by $E^N = N^{-1}$. Then, on the basis of the two–mesh method, the entries in the final column of the table of computed orders of convergence are

$$\log_2\left(\frac{E^N - E^{2N}}{E^{2N} - E^{4N}}\right) = \log_2\left(\frac{N^{-1} - (2N)^{-1}}{(2N)^{-1} - (4N)^{-1}}\right) = 1$$

which is the correct value. On the other hand, if the computed orders of convergence are based on the solutions on the finest available mesh, then the entries in the final column of the table of computed orders of convergence are

$$\log_2\left(\frac{E^N - E^{4N}}{E^{2N} - E^{4N}}\right) = \log_2\left(\frac{N^{-1} - (4N)^{-1}}{(2N)^{-1} - (4N)^{-1}}\right) = \log_2 3$$

which is an incorrect overestimate of the actual order of convergence. It is for this reason that we use the two–mesh method to estimate the orders of convergence.

5.7   Numerical results

In this section we use Method 5.1 to compute numerical solutions for the following convection–diffusion problem from Problem Class 5.1.

$$\varepsilon\Delta u_\varepsilon + (2 + x^2 y)\frac{\partial u_\varepsilon}{\partial x} + (1 + xy)\frac{\partial u_\varepsilon}{\partial y} = x^2 + y^3 + \cos(x + 2y) \quad (5.17a)$$

$$u_\varepsilon(x, 0) = 0, \quad u_\varepsilon(x, 1) = \begin{cases} 4x(1 - x), & x < 1/2 \\ 1, & x \geq 1/2 \end{cases} \quad (5.17b)$$

$$u_\varepsilon(0, y) = 0, \quad u_\varepsilon(1, y) = \begin{cases} 8y(1 - 2y), & y < 1/4 \\ 1, & y \geq 1/4 \end{cases} \quad (5.17c)$$

Table 5.1 *Computed maximum pointwise errors E_ε^N and E^N generated by Method 5.1 applied to problem (5.17) for various values of ε and N.*

ε	Number of intervals N					
	8	16	32	64	128	256
1	0.132D-01	0.706D-02	0.379D-02	0.187D-02	0.854D-03	0.343D-03
2^{-2}	0.999D-01	0.582D-01	0.310D-01	0.153D-01	0.671D-02	0.227D-02
2^{-4}	0.187D+00	0.114D+00	0.653D-01	0.351D-01	0.169D-01	0.619D-02
2^{-6}	0.309D+00	0.194D+00	0.112D+00	0.603D-01	0.285D-01	0.102D-01
2^{-8}	0.362D+00	0.248D+00	0.150D+00	0.820D-01	0.387D-01	0.136D-01
2^{-10}	0.378D+00	0.275D+00	0.172D+00	0.964D-01	0.466D-01	0.168D-01
2^{-12}	0.382D+00	0.282D+00	0.178D+00	0.100D+00	0.489D-01	0.178D-01
2^{-14}	0.383D+00	0.283D+00	0.179D+00	0.101D+00	0.495D-01	0.181D-01
2^{-16}	0.383D+00	0.284D+00	0.180D+00	0.101D+00	0.496D-01	0.181D-01
$\cdot$	$\cdot$	$\cdot$	$\cdot$	$\cdot$	$\cdot$	$\cdot$
$\cdot$	$\cdot$	$\cdot$	$\cdot$	$\cdot$	$\cdot$	$\cdot$
$\cdot$	$\cdot$	$\cdot$	$\cdot$	$\cdot$	$\cdot$	$\cdot$
2^{-34}	0.383D+00	0.284D+00	0.180D+00	0.101D+00	0.497D-01	0.181D-01
E^N	0.383D+00	0.284D+00	0.180D+00	0.101D+00	0.497D-01	0.181D-01

Table 5.2 *Computed orders of convergence p_ε^N and p^N generated by Method 5.1 applied to problem (5.17) for various values of ε and N.*

ε	Number of intervals N				
	8	16	32	64	128
1	1.54	0.77	0.90	0.96	0.62
2^{-2}	0.62	0.78	0.89	0.95	0.97
2^{-4}	0.74	0.66	0.77	0.71	0.80
2^{-6}	0.59	0.66	0.71	0.80	0.84
2^{-8}	0.46	0.53	0.67	0.79	0.88
2^{-10}	0.42	0.46	0.61	0.74	0.82
2^{-12}	0.41	0.45	0.60	0.73	0.81
.	.	.	.	.	.
.	.	.	.	.	.
.	.	.	.	.	.
.	.	.	.	.	.
2^{-34}	0.41	0.44	0.59	0.72	0.80
p^N	0.41	0.44	0.59	0.72	0.80

Note that condition (5.4c) is fulfilled by this problem, but that the boundary data are not differentiable at the points $(0.5, 1)$ and $(1, 0.25)$, and so $g \notin C^1(\Gamma)$.

The values of the computed errors E_ε^N and E^N for Method 5.1 applied to problem (5.17) for various values of ε and N are presented in Table 5.1. Examining each row of the table in turn we see that the computed maximum pointwise error E_ε^N decreases as the number of mesh elements N increases, irrespective of the value of ε, and that the same is true of the maximum value in each column of the table. These numerical results for Method 5.1 applied to problem (5.17) validate the known theoretical ε–uniform estimates.

The values of the computed orders of convergence p_ε^N and p^N obtained when Method 5.1 is applied to problem (5.17) are given in Table 5.2 for various values of ε and N. It is important to observe that the values of p^N in the last row of this table increase with increasing N.

The ε–uniform Method 5.1 for Problem Class 5.1 can be used to generate accurate graphical representations of the exact solution of any specific problem at each point of the domain Ω and for any required value of ε. Therefore Method 5.1 provides us with a flexible experimental tool to study visually the changes as ε varies in the solution, of problems in Problem Class 5.1 for which the exact solution cannot be written in a form from which a graph can be easily generated. An example of its use for problem (5.17) for various values of ε with $N = 64$ is shown in Fig. 5.7.

5.8   Neumann boundary conditions

In this section we consider the following class of singularly perturbed convection-diffusion problems with Dirichlet boundary conditions on the inflow edges and Neumann boundary conditions on the outflow edges of the square domain Ω.

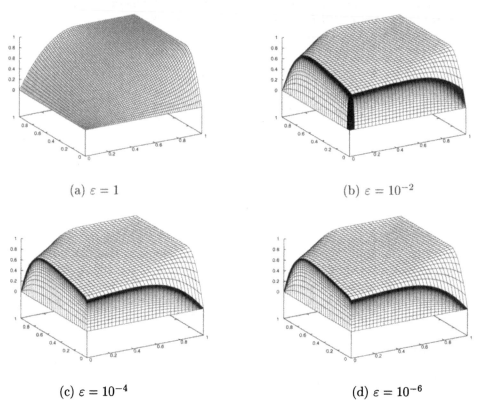

(a) $\varepsilon = 1$ (b) $\varepsilon = 10^{-2}$

(c) $\varepsilon = 10^{-4}$ (d) $\varepsilon = 10^{-6}$

FIG. 5.7. Numerical solution generated by Method 5.1 applied to problem (5.17) with $N = 64$.

Problem Class 5.2. Linear convection diffusion in two dimensions with Dirichlet and Neumann boundary conditions.

$$\varepsilon \Delta u_\varepsilon + \mathbf{a}(x,y) \cdot \nabla u_\varepsilon = f(x,y), \quad (x,y) \in \Omega \tag{5.18a}$$

$$u_\varepsilon = g, \quad (x,y) \in \Gamma_R \cup \Gamma_T, \tag{5.18b}$$

$$\frac{\partial u_\varepsilon}{\partial x}(0,y) = 0, \quad \varepsilon \frac{\partial u_\varepsilon}{\partial y} = h, \tag{5.18c}$$

$$\mathbf{a}(x,y) \geq (\alpha_1, \alpha_2) > \mathbf{0} \quad (x,y) \in \overline{\Omega} \tag{5.18d}$$

$$a_1, a_2 \in C^{0,1}(\bar{\Omega}), \quad f \in C^{0,\gamma}(\bar{\Omega}), \ \gamma > 0, \tag{5.18e}$$

$$g \in C^0(\Gamma_R \cup \Gamma_T), \quad h \in C^0(\Gamma_B) \tag{5.18f}$$

Problems from this problem class typically have a regular boundary layer on the edge Γ_B and a corner boundary layer at C_{BL}. Also, analogously to the one–dimensional case discussed in §2.6, a weak regular boundary layer can appear in the solution on the edge Γ_L. For such two–dimensional problems we say that the regular boundary layer is weak at a point if, at this point, all of its first derivatives are bounded independently of ε, but some of its higher derivatives are not bounded independently of ε.

We now present numerical results for the following specific problem from Problem Class 5.2.

$$\varepsilon \Delta u_\varepsilon + \frac{\partial u_\varepsilon}{\partial x} + 2\frac{\partial u_\varepsilon}{\partial y} = x^2(1-x)^2(1-y)^2 y^2, \quad (x,y) \in \Omega \qquad (5.19a)$$

$$u_\varepsilon = 1, \quad (x,y) \in \Gamma_R \cup \Gamma_T \qquad (5.19b)$$

$$\frac{\partial u_\varepsilon}{\partial x} = 0, \quad (x,y) \in \Gamma_L, \quad \varepsilon\frac{\partial u_\varepsilon}{\partial y} = -1, \quad (x,y) \in \Gamma_B \qquad (5.19c)$$

It is easy to see that the solution of this problem has a regular boundary layer on the edge Γ_B and a weak regular boundary layer on the edge Γ_L, because the Neumann boundary condition is homogeneous. It also has a corner boundary layer at the corner C_{BL}.

We introduce the following monotone method composed of an upwind finite difference operator on a piecewise–uniform fitted mesh for the numerical solution of problems from Problem Class 5.2.

Method 5.2.   Upwind finite difference operator on piecewise–uniform fitted mesh for Problem Class 5.2.

$$[\varepsilon(\delta_x^2 + \delta_y^2) + a_1(x_i, y_j)D_x^+ + a_2(x_i, y_j)D_y^+]U_\varepsilon = f(x_i, y_j) \quad (x_i, y_j) \in \Omega_\varepsilon^N \qquad (5.20a)$$

$$U_\varepsilon(1, y_j) = g(1, y_j), \quad U_\varepsilon(x_i, 1) = g(x_i, 1) \qquad (5.20b)$$

$$D_x^+ U_\varepsilon(x_0, y_j) = 0, \quad \varepsilon D_y^+ U_\varepsilon(x_i, y_0) = h(x_i, 0) \qquad (5.20c)$$

where

$$\overline{\Omega}_\varepsilon^N = \overline{\Omega}_u^N \times \overline{\Omega}_\sigma^N \qquad (5.20d)$$

where $\overline{\Omega}_u^N$ is a one-dimensional uniform mesh and

$$\overline{\Omega}_\sigma^N = \{y_j : 0 \le j \le N\} \qquad (5.20e)$$

Table 5.3 *Computed maximum pointwise errors E_ε^N and E^N generated by Method 5.2 applied to problem (5.19) for various values of ε and N.*

| | \multicolumn{6}{c}{Number of intervals N} | | | | | |
ε	8	16	32	64	128	256
1	0.123D+00	0.605D-01	0.293D-01	0.137D-01	0.586D-02	0.195D-02
2^{-2}	0.246D+00	0.121D+00	0.586D-01	0.273D-01	0.117D-01	0.391D-02
2^{-4}	0.492D+00	0.242D+00	0.117D+00	0.547D-01	0.234D-01	0.781D-02
2^{-6}	0.514D+00	0.331D+00	0.201D+00	0.109D+00	0.469D-01	0.156D-01
2^{-8}	0.527D+00	0.327D+00	0.193D+00	0.106D+00	0.515D-01	0.190D-01
2^{-10}	0.539D+00	0.331D+00	0.193D+00	0.106D+00	0.515D-01	0.190D-01
2^{-12}	0.546D+00	0.333D+00	0.194D+00	0.106D+00	0.515D-01	0.190D-01
2^{-14}	0.550D+00	0.334D+00	0.194D+00	0.106D+00	0.515D-01	0.190D-01
2^{-16}	0.552D+00	0.334D+00	0.194D+00	0.106D+00	0.515D-01	0.190D-01
.	.	.	.	.	.	.
.	.	.	.	.	.	.
2^{-28}	0.554D+00	0.335D+00	0.195D+00	0.106D+00	0.515D-01	0.190D-01
E^N	0.554D+00	0.335D+00	0.195D+00	0.106D+00	0.515D-01	0.190D-01

with

$$y_j = \begin{cases} 2j\sigma/N, & \text{for } 0 \leq j \leq N/2 \\ \sigma + 2(j - (N/2))(1 - \sigma)/N, & \text{for } N/2 \leq j \leq N \end{cases} \qquad (5.20\text{f})$$

and

$$\sigma = \min\{1/2, c\varepsilon \ln N\}, \quad c \geq \alpha_1^{-1}. \qquad (5.20\text{g})$$

We observe that in Method 5.2 the piecewise–uniform mesh is fitted only to the regular boundary layer on the edge Γ_B and not to the weak regular boundary layer on the edge Γ_L. We use Method 5.2 to solve problem (5.19) for various values

Table 5.4 *Computed orders of convergence p_ε^N and p^N generated by Method 5.2 applied to problem (5.19) for various values of ε and N.*

| | \multicolumn{5}{c}{Number of intervals N} | | | | |
ε	8	16	32	64	128
1	1.00	1.00	1.00	1.00	1.00
2^{-2}	1.00	1.00	1.00	1.00	1.00
2^{-4}	1.00	1.00	1.00	1.00	1.00
2^{-6}	0.49	0.51	0.55	1.00	1.00
2^{-8}	0.56	0.63	0.69	0.74	0.78
2^{-10}	0.61	0.64	0.69	0.74	0.78
.	.	.	.	.	.
.	.	.	.	.	.
2^{-28}	0.64	0.66	0.70	0.74	0.78
p^N	0.64	0.66	0.70	0.74	0.78

of ε and N. The computed maximum pointwise errors E_ε^N and E^N are given in Table 5.3. It is apparent from this table that the method is ε-uniform for this problem, which supports the view that it is not necessary to fit the mesh to the weak boundary layer on the edge Γ_L.

The computed orders of convergence p_ε^N and the computed ε-uniform orders of convergence p^N, for various values of ε and N, are given in Table 5.4. Again, it is important to observe that the values of p^N in the last row of Table 5.4 increase with increasing N.

5.9   Corner boundary layers

It should be observed that, near the corner C_{BL}, the analytic form of the exact solution of a problem from Problem Class 5.1 is often more complicated that implied by the regular expression (5.9b). For example, if the data of the problem are not sufficiently smooth for (5.9a) to hold, then the corner boundary layer function near C_{BL} is not in general the product of two one–dimensional exponentials (Han and Kellogg (1990)). It is a remarkable feature of the piecewise–uniform fitted mesh $\overline{\Omega}_\varepsilon^N$ defined in (5.10d-h) that, without any modification, it leads to numerical solutions which converge ε-uniformly even in this corner boundary layer. To verify this experimentally, in a case with a complicated corner layer, we consider the following convection–diffusion problem with data that are not smooth

$$\varepsilon \Delta u_\varepsilon + \frac{\partial u_\varepsilon}{\partial x} + 4\frac{\partial u_\varepsilon}{\partial y} = x^2 - y^2 \quad (x,y) \in \Omega \tag{5.21a}$$

$$u_\varepsilon(x,0) = x^{1/6} \quad u_\varepsilon(x,1) = 1 \tag{5.21b}$$

$$u_\varepsilon(0,y) = \sqrt{y} \quad u_\varepsilon(1,y) = 1. \tag{5.21c}$$

The exact solution u_ε of this problem is not differentiable at all points of $\overline{\Omega}$, because on the edges Γ_B and Γ_L respectively, we have

$$\frac{\partial u_\varepsilon}{\partial x}(x,0) = \frac{1}{6}x^{-5/6}, \quad \frac{\partial u_\varepsilon}{\partial y}(0,y) = \frac{1}{2}y^{-1/2}$$

which shows that these first derivatives are not well defined at the outflow corner C_{BL}. This means that the data of this problem satisfy only weak regularity and compatibility hypotheses. Nevertheless, the numerical results in Tables 5.5 and 5.6 indicate that Method 5.1 is ε-uniform for this problem. It is noteworthy that such a simple numerical method generates satisfactory numerical solutions even in cases where the underlying analytic behaviour is complicated. It underlines the crucial importance and effectiveness of using an appropriate piecewise–uniform fitted mesh. Surface plots of the numerical solution are given in Fig. 5.8 and Fig. 5.9, where we see the remarkable detail that can be achieved using numerical methods based on these simple piecewise–uniform fitted meshes.

Table 5.5 *Computed maximum pointwise errors E_ε^N and E^N generated by Method 5.1 applied to problem (5.21) for various values of ε and N.*

	Number of intervals N					
ε	8	16	32	64	128	256
1	0.674D-02	0.745D-02	0.566D-02	0.456D-02	0.356D-02	0.228D-02
2^{-2}	0.470D-01	0.267D-01	0.141D-01	0.700D-02	0.414D-02	0.249D-02
2^{-4}	0.919D-01	0.641D-01	0.412D-01	0.240D-01	0.121D-01	0.459D-02
2^{-6}	0.117D+00	0.919D-01	0.611D-01	0.355D-01	0.178D-01	0.666D-02
2^{-8}	0.134D+00	0.112D+00	0.776D-01	0.464D-01	0.233D-01	0.868D-02
2^{-10}	0.146D+00	0.125D+00	0.877D-01	0.535D-01	0.274D-01	0.103D-01
2^{-12}	0.157D+00	0.135D+00	0.942D-01	0.580D-01	0.300D-01	0.114D-01
2^{-14}	0.166D+00	0.143D+00	0.994D-01	0.612D-01	0.318D-01	0.120D-01
2^{-16}	0.174D+00	0.150D+00	0.104D+00	0.639D-01	0.333D-01	0.126D-01
2^{-18}	0.181D+00	0.156D+00	0.107D+00	0.661D-01	0.345D-01	0.130D-01
2^{-20}	0.186D+00	0.161D+00	0.110D+00	0.679D-01	0.355D-01	0.134D-01
2^{-22}	0.191D+00	0.164D+00	0.112D+00	0.693D-01	0.363D-01	0.137D-01
2^{-24}	0.194D+00	0.167D+00	0.114D+00	0.704D-01	0.369D-01	0.139D-01
2^{-26}	0.197D+00	0.170D+00	0.116D+00	0.713D-01	0.374D-01	0.141D-01
2^{-28}	0.199D+00	0.171D+00	0.117D+00	0.721D-01	0.378D-01	0.143D-01
2^{-30}	0.201D+00	0.173D+00	0.118D+00	0.726D-01	0.381D-01	0.144D-01
2^{-32}	0.202D+00	0.174D+00	0.118D+00	0.731D-01	0.384D-01	0.145D-01
2^{-34}	0.203D+00	0.175D+00	0.119D+00	0.735D-01	0.386D-01	0.145D-01
E^N	0.203D+00	0.175D+00	0.119D+00	0.735D-01	0.386D-01	0.145D-01

A second example of complicated corner layer behaviour is examined in the following convection–diffusion problem, which has the same differential equation as before but different boundary conditions

$$\varepsilon \Delta u_\varepsilon + \frac{\partial u_\varepsilon}{\partial x} + 4\frac{\partial u_\varepsilon}{\partial y} = x^2 - y^2, \quad (x,y) \in \Omega \tag{5.22a}$$

$$u_\varepsilon(x,0) = 1 \quad u_\varepsilon(x,1) = (1-x)^{1/6} \tag{5.22b}$$

$$u_\varepsilon(0,y) = 1 \quad u_\varepsilon(1,y) = \sqrt{1-y}. \tag{5.22c}$$

Here the lack of smoothness in the boundary data occurs at the inflow corner C_{TR}, because

$$\frac{\partial u_\varepsilon}{\partial x}(x,1) = -\frac{1}{6}(1-x)^{-5/6}, \quad \frac{\partial u_\varepsilon}{\partial y}(1,y) = -\frac{1}{2}(1-y)^{-1/2}$$

from which we see that these first derivatives are not well defined at the inflow corner C_{TR}. The numerical results in Tables 5.7 and 5.8 indicate that Method 5.1 is again ε–uniform for this problem. The computed orders of convergence in Table 5.8 are smaller than those in Table 5.6, which is not surprising because the incompatibility in the inflow boundary data has an influence throughout the domain. A graph of the numerical solution of problem (5.22) using Method 5.1 is displayed in Fig. 5.10. We see, along a line joining the points $(1,1)$ to $(0.75,0)$, that there is a considerable amount of smearing, which occurs because the mesh

Table 5.6 *Computed orders of convergence p_ε^N and p^N generated by Method 5.1 applied to problem (5.21) for various values of ε and N.*

	Number of intervals N				
ε	8	16	32	64	128
1	-0.25	0.16	0.20	0.22	0.21
2^{-2}	0.69	0.81	0.89	0.40	0.25
2^{-4}	0.27	0.46	0.53	0.61	0.72
2^{-6}	-0.31	0.29	0.51	0.63	0.75
2^{-8}	-0.57	0.08	0.41	0.62	0.76
2^{-10}	-0.60	0.01	0.34	0.58	0.73
2^{-12}	-0.66	-0.02	0.32	0.55	0.72
2^{-14}	-0.72	-0.03	0.32	0.55	0.72
2^{-16}	-0.78	-0.04	0.32	0.54	0.72
2^{-18}	-0.83	-0.04	0.32	0.54	0.72
2^{-20}	-0.86	-0.04	0.32	0.54	0.72
2^{-22}	-0.85	-0.04	0.32	0.53	0.72
⋮	⋮	⋮	⋮	⋮	⋮
2^{-34}	-0.81	-0.04	0.32	0.53	0.72
p^N	-0.60	-0.04	0.32	0.53	0.72

Table 5.7 *Computed maximum pointwise errors E_ε^N and E^N generated by Method 5.1 applied to problem (5.22) for various values of ε and N.*

	Number of intervals N					
ε	8	16	32	64	128	256
1	0.178D-01	0.108D-01	0.725D-02	0.521D-02	0.382D-02	0.237D-02
2^{-2}	0.719D-01	0.421D-01	0.233D-01	0.118D-01	0.529D-02	0.280D-02
2^{-4}	0.181D+00	0.112D+00	0.711D-01	0.391D-01	0.189D-01	0.690D-02
2^{-6}	0.307D+00	0.220D+00	0.142D+00	0.828D-01	0.413D-01	0.152D-01
2^{-8}	0.392D+00	0.294D+00	0.206D+00	0.136D+00	0.758D-01	0.305D-01
2^{-10}	0.433D+00	0.331D+00	0.239D+00	0.172D+00	0.104D+00	0.460D-01
2^{-12}	0.447D+00	0.344D+00	0.252D+00	0.187D+00	0.116D+00	0.537D-01
2^{-14}	0.452D+00	0.348D+00	0.256D+00	0.191D+00	0.120D+00	0.561D-01
2^{-16}	0.453D+00	0.349D+00	0.257D+00	0.192D+00	0.121D+00	0.568D-01
2^{-18}	0.453D+00	0.349D+00	0.257D+00	0.192D+00	0.121D+00	0.569D-01
⋮	⋮	⋮	⋮	⋮	⋮	⋮
2^{-34}	0.453D+00	0.349D+00	0.258D+00	0.192D+00	0.121D+00	0.570D-01
E^N	0.453D+00	0.349D+00	0.258D+00	0.192D+00	0.121D+00	0.570D-01

is not fitted to any internal layers that are present. However, the numerical results confirm experimentally that Method 5.1 is still an ε–uniform numerical method for this problem.

To emphasize how strongly the smoothness of the data influences the ε–uniform order of convergence, we now present numerical results for Method 5.1 applied to the following particular problem from Problem Class 5.1.

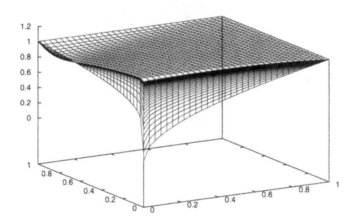

FIG. 5.8. *Surface plot of the numerical solution generated by Method 5.1 applied to problem (5.21) for $\varepsilon = 10^{-6}$ and $N = 64$.*

Table 5.8 *Computed orders of convergence p_ε^N and p^N generated by Method 5.1 applied to problem (5.22) for various values of ε and N.*

| | \multicolumn{6}{c}{Number of intervals N} | | | | | |
ε	8	16	32	64	128	256
1	0.64	0.51	0.40	0.32	0.26	0.43
2^{-2}	0.69	0.74	0.84	0.83	0.40	0.70
2^{-4}	0.60	0.55	0.68	0.72	0.80	0.67
2^{-6}	0.27	0.38	0.55	0.67	0.76	0.53
2^{-8}	0.18	0.26	0.30	0.43	0.57	0.35
2^{-10}	0.16	0.22	0.17	0.30	0.33	0.24
2^{-12}	0.16	0.21	0.13	0.24	0.22	0.19
2^{-14}	0.16	0.21	0.12	0.23	0.19	0.18
$\cdot$	$\cdot$	$\cdot$	$\cdot$	$\cdot$	$\cdot$	$\cdot$
$\cdot$	$\cdot$	$\cdot$	$\cdot$	$\cdot$	$\cdot$	$\cdot$
$\cdot$	$\cdot$	$\cdot$	$\cdot$	$\cdot$	$\cdot$	$\cdot$
2^{-34}	0.16	0.20	0.12	0.22	0.18	0.18
p^N	0.16	0.20	0.12	0.22	0.18	0.18

$$\varepsilon \Delta u_\varepsilon + 2\frac{\partial u_\varepsilon}{\partial x} + \frac{\partial u_\varepsilon}{\partial y} = x^2(1-x)^2 y^2(1-y)^2, \quad (x,y) \in \Omega \qquad (5.23\text{a})$$

$$u_\varepsilon(x,y) \equiv 0, \quad (x,y) \in \Gamma. \qquad (5.23\text{b})$$

The data in this problem are chosen such that the exact solution is much

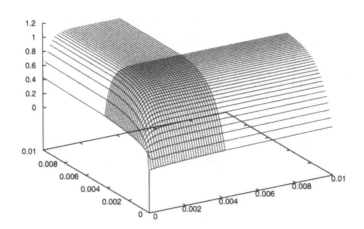

FIG. 5.9. *Surface plot of the numerical solution near the corner* $(0,0)$ *generated by Method 5.1 applied to problem* (5.21) *for* $\varepsilon = 10^{-6}$ *and* $N = 64$.

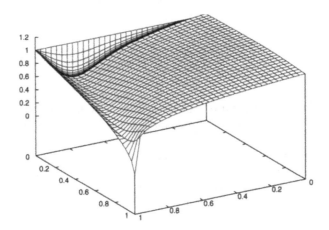

FIG. 5.10. *Numerical solution generated by Method 5.1 applied to problem* (5.22) *for* $\varepsilon = 10^{-6}$ *and* $N = 32$.

Table 5.9 *Computed maximum pointwise errors E_ε^N and E^N generated by Method 5.1 applied to problem (5.23) for various values of ε and N.*

	Number of intervals N					
ε	8	16	32	64	128	256
1	0.265D-02	0.156D-02	0.810D-03	0.393D-03	0.171D-03	0.576D-04
2^{-2}	0.239D-01	0.142D-01	0.744D-02	0.363D-02	0.159D-02	0.537D-03
2^{-4}	0.617D-01	0.384D-01	0.208D-01	0.106D-01	0.486D-02	0.171D-02
2^{-6}	0.925D-01	0.576D-01	0.336D-01	0.179D-01	0.836D-02	0.297D-02
2^{-8}	0.103D+00	0.658D-01	0.388D-01	0.207D-01	0.974D-02	0.346D-02
2^{-10}	0.105D+00	0.681D-01	0.402D-01	0.215D-01	0.101D-01	0.360D-02
2^{-12}	0.106D+00	0.686D-01	0.406D-01	0.217D-01	0.102D-01	0.363D-02
.	.	.	.	.	.	.
.	.	.	.	.	.	.
.	.	.	.	.	.	.
2^{-22}	0.106D+00	0.688D-01	0.407D-01	0.218D-01	0.103D-01	0.365D-02
E^N	0.106D+00	0.688D-01	0.407D-01	0.218D-01	0.103D-01	0.365D-02

Table 5.10 *Computed orders of convergence p_ε^N and p^N generated by Method 5.1 applied to problem (5.23) for various values of ε and N.*

	Number of intervals N				
ε	8	16	32	64	128
1	0.61	0.84	0.92	0.96	0.98
2^{-2}	0.64	0.82	0.91	0.95	0.98
2^{-4}	0.69	0.79	0.81	0.87	0.88
2^{-6}	0.52	0.63	0.74	0.81	0.86
2^{-8}	0.45	0.60	0.73	0.80	0.86
2^{-10}	0.43	0.59	0.73	0.80	0.86
.	.	.	.	.	.
.	.	.	.	.	.
.	.	.	.	.	.
2^{-22}	0.42	0.58	0.73	0.80	0.86
p^N	0.42	0.58	0.73	0.80	0.86

smoother than the exact solutions of the previous problems considered in this section. This is reflected in the significantly reduced pointwise errors in Table 5.9 and the increased orders of convergence in Table 5.10, compared with the corresponding values in Tables 5.7 and 5.8 respectively for problem (5.22). We observe that in Table 5.10 the orders of convergence tend to 1.0 as N increases, while in Table 5.8 they are approximately 0.16 for all values of N.

5.10   Computational work

It is obvious from the construction of a piecewise–uniform fitted mesh Ω_ε^N that there is an abrupt change in the mesh–size at the transition point when ε is small. In fact, for a fixed number N of mesh intervals in a given coordinate direction, the ratio of the coarse and fine mesh sizes is $\frac{1-\sigma}{\sigma}$, and since $\sigma \to 0$ as $\varepsilon \to 0$ we have

$$\frac{1-\sigma}{\sigma} \to \infty \text{ as } \varepsilon \to 0.$$

In other words the ratio of the mesh sizes increases without limit as the singular perturbation parameter decreases. It is known theoretically that Method 5.1 is ε–uniform and satisfies the error estimate (5.11) for problem (5.23) (Shishkin (1992b)). Therefore the unbounded behaviour of the ratio of the mesh–sizes has no adverse effect on the theoretical ε–uniform order of convergence of the numerical solutions. It is important to determine also whether or not the amount of computational work required to find the numerical solution is ε–uniform.

In the case of the one–dimensional problems of the previous chapters, the linear systems are solved by a direct method and it is clear that the amount of computational work required to determine the numerical solutions depends only on N. A new issue arises in the case of problems in two or more dimensions, because normally an iterative method is used and it is not immediately clear if the computational work is independent of ε. We first observe that the linear systems arising from the numerical methods used for the two–dimensional problems considered in this chapter are solved using the preconditioned Conjugate Gradient Squared (CGS) method of Sonneveld (1989) with incomplete block LU–factorisation as the preconditioner; that is, block LU–factorisation of the coefficient matrix with no fill–in allowed. The initial guess is always taken to be $\mathbf{u}_0$, where $\mathbf{u}_0$ is the solution of $\tilde{\mathbf{L}}\tilde{\mathbf{U}}\mathbf{u}_0 = \mathbf{f}$. Here $\tilde{\mathbf{L}}$ and $\tilde{\mathbf{U}}$ are the incomplete block factors. Thus $\mathbf{u}_0$ is obtained by forward and backward substitution in the usual way. For the particular case when Method 5.1 is applied to problem (5.23) we give in Table 5.11 the number of iterations required for various values of

Table 5.11 *Number of solver iterations required for convergence for Method 5.1 applied to problem (5.23) for various values of ε and N.*

ε	8	16	32	64	128	256	512
1	2	3	6	18	26	63	133
2^{-2}	2	3	6	12	29	61	127
2^{-4}	2	3	6	11	21	41	80
2^{-6}	2	3	6	11	21	42	146
2^{-8}	2	3	5	9	18	38	102
2^{-10}	2	3	5	9	16	28	70
2^{-12}	2	4	5	8	15	29	51
2^{-14}	2	3	5	9	15	29	55
2^{-16}	2	3	5	8	15	31	57
2^{-18}	2	3	5	8	15	31	61
2^{-20}	2	3	5	8	15	31	56
2^{-22}	2	3	5	8	15	31	56

Number of intervals N

N and ε. It is remarkable that the required number of iterations stabilizes and in fact diminishes as $\varepsilon \to 0$. Thus, at least for Method 5.1 applied to problem (5.23), we can infer that the amount of computational work required to compute a solution is bounded independently of ε. We note that other iterative solvers such as BiCGStab or GMRES could also be used with the same preconditioner without giving rise to significantly different iteration counts.

CHAPTER 6

Convection–diffusion problems with frictionless walls

6.1 The origin of parabolic boundary layers

In this chapter we consider convection–diffusion problems with solutions having parabolic boundary layers, and we discuss some of the associated physical phenomena. It is well known that, if a fluid flows over a surface or inside a tube, the component of the velocity tangential to the streamlines often has a parabolic profile in the direction orthogonal to the streamlines. If a boundary layer function in the solution of such a problem satisfies a partial differential equation of parabolic type, then it is called a parabolic boundary layer. The variable in the direction of the streamlines plays the role of the time variable. It is important to recognize that the word *parabolic* is employed here in two completely different senses: first, the velocity profile can be parabolic in shape for a particular problem, second the dominant part of the solution near some points of the boundary may satisfy a parabolic differential equation.

We remark that there are various kinds of parabolic boundary layers. However, in this chapter, we restrict the discussion to the kind that arises typically when we study the temperature in a fluid flowing parallel to a hot frictionless wall, as illustrated in Fig. 6.1. Because there is no friction on the wall the fluid slips easily along the wall, and no boundary layer is formed in the velocity profile of the fluid. However, there is a parabolic boundary layer in the temperature profile of the fluid near the hot wall, because it heats the fluid close to the wall. Parabolic boundary layers are different from regular boundary layers, because they are rich in complexity. Moreover, they are not as thin as regular boundary layers.

We see later that, in contrast to a regular boundary layer, in the case of a parabolic boundary layer the difference between the reduced solution and the solution itself, at any point on the boundary, depends on the boundary data at all points of the boundary in the boundary layer. Moreover, parabolic boundary layers cannot occur in the solutions of ordinary differential equations; they occur only in the solutions of partial differential equations.

The examples considered in this and the next chapter are all special cases of linear convection–diffusion problems of the form

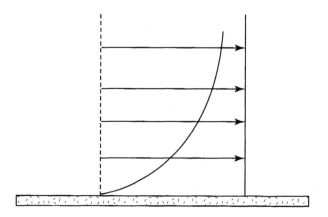

FIG. 6.1. Boundary layer in the temperature profile for uniform flow parallel to a hot frictionless wall.

$$\varepsilon \Delta u_\varepsilon + \mathbf{a}(x,y) \cdot \nabla u_\varepsilon = f(x,y), \quad (x,y) \in \Omega \qquad (6.1\text{a})$$

$$u_\varepsilon(x,y) = g(x,y), \quad (x,y) \in \Gamma \qquad (6.1\text{b})$$

$$\mathbf{a} = (a_1(x,y), a_2(x,y)) \geq (\alpha_1, \alpha_2) \geq 0, \quad (x,y) \in \overline{\Omega} \qquad (6.1\text{c})$$

where, as in the previous chapter, the domain Ω is the unit square. The reduced problem is $\mathbf{a} \cdot \nabla v_0 = f$ with the values of v_0 given on the inflow boundary. Its characteristics are the curves $\mathbf{c}(s) = (x(s), y(s))$ where $\frac{dx}{ds} = a_1$ and $\frac{dy}{ds} = a_2$. These curves are the streamlines of the flow. Thus, if either $a_1 = 0$ or $a_2 = 0$, parabolic boundary layers appear in its solution because the tangents to the characteristics of the reduced problem are parallel to the horizontal or vertical edges of the boundary Γ, that is $\mathbf{a} \cdot \mathbf{n} = 0$ on these edges.

In general, for problems of the form problem (6.1), two distinct kinds of parabolic boundary layers can be identified. The first kind are non–degenerate parabolic boundary layers, which occur at all points of the boundary Γ such that $\mathbf{a}(x,y) > \mathbf{0}$ in an open neighbourhood of this boundary point. Thus, non–degenerate parabolic boundary layers are described by singularly perturbed parabolic differential equations in which the coefficients multiplying the time–like derivatives are strictly greater than zero on the boundary adjacent to the boundary layer. Non–degenerate parabolic boundary layers are discussed in the present chapter. The second kind are the degenerate parabolic boundary layers, which occur at all points of the boundary Γ such that $|\mathbf{a}(x,y)| > 0$ in an open neighbourhood of this boundary layer and $\mathbf{a}(x,y) = \mathbf{0}$ at each point of the boundary in the boundary layer. Boundary layers of this kind are described by singularly perturbed parabolic differential equations in which the coefficients multiplying the time–like derivatives vanish on the boundary. Parabolic boundary layers of this kind are discussed in the next chapter.

In this chapter, we assume that the minimum values of the velocity components α_1, α_2 in (6.1c) satisfy the condition

$$\alpha_1^2 + \alpha_2^2 > 0 \qquad (6.2)$$

Given this condition the locations of the regular, parabolic and corner boundary layers, that can occur in the solution of problem (6.1), are illustrated in Fig. 6.2 for four typical cases. In this chapter we construct numerical methods for the following problem class, corresponding to one of these cases. The construction of numerical methods in the other cases is analogous.

Problem Class 6.1. Linear convection–diffusion in two–dimensions with non–degenerate parabolic layers.

$$\varepsilon \Delta u_\varepsilon + a_1(x,y)\frac{\partial u_\varepsilon}{\partial x} = f(x,y), \quad (x,y) \in \Omega \qquad (6.3a)$$

$$u_\varepsilon(x,y) = g(x,y), \quad (x,y) \in \Gamma \qquad (6.3b)$$

$$a_1(x,y) \geq \alpha_1 > 0, \quad (x,y) \in \overline{\Omega} \qquad (6.3c)$$

$$a_1, f \in C^{0,\gamma}(\overline{\Omega}) \ , \ g \in C^{\gamma}(\Gamma), \ \gamma > 0. \qquad (6.3d)$$

In the solution u_ε of a typical problem in this class, a regular boundary layer occurs on the outflow boundary Γ_L, non–degenerate parabolic boundary layers

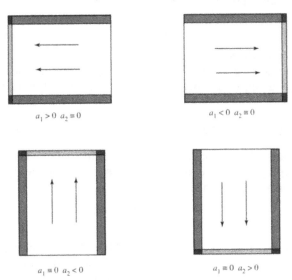

FIG. 6.2. Locations of the regular, parabolic and corner boundary layers in the solution of problem (6.1) in four cases.

occur on the edges Γ_B and Γ_T and a corner boundary layer occurs at each of the four corners.

6.2   Asymptotic nature

In this section we study the asymptotic nature of the exact solution u_ε of a problem from Problem Class 6.1. We show that the width of a non–degenerate parabolic boundary layer occurring in the solution is $0(\sqrt{\varepsilon})$ and that its profile is influenced by the boundary data at all points of the boundary associated with the boundary layer.

To be specific we consider the singularly perturbed convection–diffusion problem

$$\varepsilon \Delta u_\varepsilon + \frac{\partial u_\varepsilon}{\partial x} = 0, \quad (x,y) \in \Omega \tag{6.4a}$$

$$u_\varepsilon = g, \quad (x,y) \in \Gamma \tag{6.4b}$$

which is a problem from Problem Class 6.1.

The corresponding reduced problem is

$$\frac{\partial v_0}{\partial x} = 0, \quad (x,y) \in \Omega, \tag{6.5a}$$

$$v_0 = g, \quad (x,y) \in \Gamma_L \tag{6.5b}$$

which has the exact solution

$$v_0(x,y) \equiv g(1,y).$$

This is a function of the single variable y.

Proceeding as in the previous chapter, it is easy to see that the boundary layer function corresponding to the regular boundary layer on the outflow boundary Γ_L is

$$\tilde{w}_{0,L}(X,y) = (g(0,y) - g(1,y))e^{-X}, \quad \text{where} \quad X = x/\varepsilon. \tag{6.6}$$

We now consider the parabolic boundary layer in a neighbourhood of the bottom edge Γ_B. Proceeding in the same manner as we did for the regular boundary layer, we introduce the stretched variable

$$Y = \frac{y}{\varepsilon}$$

the values of which are $O(1)$ only in an ε–neighbourhood of the edge Γ_B. Making the same change of variables $(x,y) \to (x,Y)$ as for a regular boundary layer, and multiplying by ε, transforms the equation (6.4a) to

$$\varepsilon^2 \frac{\partial^2 \tilde{w}_\varepsilon}{\partial x^2} + \frac{\partial^2 \tilde{w}_\varepsilon}{\partial Y^2} + \varepsilon \frac{\partial \tilde{w}_\varepsilon}{\partial x} = 0.$$

Neglecting the $O(\varepsilon)$ terms in this equation then gives

$$\frac{\partial^2 \tilde{w}_{0,B}(x,Y)}{\partial Y^2} = 0$$

where the solution $\tilde{w}_{0,B}$ is the boundary layer function associated with the parabolic boundary layer on the edge Γ_B. Treating x as a parameter, this is a second order ordinary differential equation in Y for $\tilde{w}_{0,B}$, which has a solution of the form

$$\tilde{w}_{0,B}(x,Y) = A(x) + YB(x)$$

where A and B are determined from suitable matching conditions at the boundary points $Y = 0$ and $Y = \infty$. For this problem these conditions are

$$\tilde{w}_{0,B}(x,0) = g(x,0) - g(1,0), \quad \tilde{w}_{0,B}(x,\infty) = 0.$$

These imply that $A(x) = g(x,0) - g(1,0)$ and $B(x) \equiv 0$, and so $\tilde{w}_{0,B}(x,Y) = g(x,0) - g(1,0)$. But, in general, $g(x,0) \not\equiv g(1,0)$ and so this solution does not fulfill the second matching condition. This means that the use above of the same stretched variable as for regular boundary layers is not satisfactory in the case of parabolic boundary layers. Therefore we try a different stretched variable

$$\tilde{Y} = \frac{y}{\sqrt{\varepsilon}}$$

which corresponds to the change of variable $(x,y) \to (x,\tilde{Y})$. Combining this with multiplication by ε transforms the equation (6.4a) to

$$\varepsilon \frac{\partial^2 \hat{w}}{\partial x^2} + \frac{\partial^2 \hat{w}}{\partial \tilde{Y}^2} + \frac{\partial \hat{w}}{\partial x} = 0.$$

Neglecting the $O(\varepsilon)$ term in this equation then gives the parabolic partial differential equation

$$\frac{\partial^2 \hat{w}_{0,B}(x,\tilde{Y})}{\partial \tilde{Y}^2} + \frac{\partial \hat{w}_{0,B}(x,\tilde{Y})}{\partial x} = 0$$

whose solution $\hat{w}_{0,B}$ is the parabolic boundary layer function. We transform this parabolic problem into canonical form, by making the further change of variable

$$x_1 = 1 - x.$$

This leads to an equation of the same form as the standard heat equation

$$\frac{\partial^2 \hat{w}_{0,B}(x_1,\tilde{Y})}{\partial \tilde{Y}^2} - \frac{\partial \hat{w}_{0,B}(x_1,\tilde{Y})}{\partial x_1} = 0$$

where x_1 is the time–like variable. If we impose the following boundary conditions at the boundary points $\tilde{Y} = 0$ and $\tilde{Y} = \infty$

$$\hat{w}_{0,B}(x_1, 0) = g(x_1, 0) - g(1, 0), \quad \hat{w}_{0,B}(x_1, \infty) = 0$$

and the following initial condition at the boundary point $x_1 = 0$

$$\hat{w}_{0,B}(0, \tilde{Y}) = 0$$

we obtain a well-posed problem, which has the closed form solution

$$\hat{w}_{0,B}(x_1, \tilde{Y}) = \sqrt{\frac{2}{\pi}} \int_{\tilde{Y}/\sqrt{2x_1}}^{\infty} exp(-\frac{t^2}{2}) \Big(g(x_1 - \frac{\tilde{Y}^2}{2t^2}, 0) - g(1, 0)\Big) dt. \qquad (6.7)$$

Notice that in this case the parabolic boundary layer function $\tilde{w}_{0,B}$ is two–dimensional in nature, since it is a function of the two variables x_1 and $\tilde{Y}$. Moreover, its magnitude is governed by the expression

$$\frac{y}{\sqrt{\varepsilon x_1}}.$$

If $y = O(\sqrt{\varepsilon})$ and $0 < x \le b < 1$, where b is some constant independent of ε, we have

$$\frac{y}{\sqrt{\varepsilon x_1}} = O(1)$$

which shows that the width of the parabolic boundary layer is $0(\sqrt{\varepsilon})$ at all points of the edge Γ_B outside an open neighbourhood of the corner C_{BR}. We see also that the boundary data g at all points of the edge Γ_B are involved intrinsically in the integral representation (6.7) of the parabolic boundary layer function $\tilde{w}_{0,B}$. This is in contrast to the regular boundary layer function (6.6), where the boundary data influence only the amplitude of the layer.

So far we have shown that when the tangents to the characteristics of the reduced problem are parallel to the tangents to the boundary at some part of the boundary, then the boundary layer function associated with that part of the boundary satisfies a parabolic differential equation. We know also that the width of the associated parabolic boundary layer is $0(\sqrt{\varepsilon})$ rather than $0(\varepsilon)$, as is the case for a regular boundary layer.

We now highlight some other aspects of parabolic boundary layers by considering convection–diffusion problems with the same differential equation as in (6.4) but with different boundary conditions. First, we illustrate the change in the time–like direction of the width of a parabolic boundary layer by considering the problem

$$\varepsilon \Delta u_\varepsilon + \frac{\partial u_\varepsilon}{\partial x} = 0 \quad \text{on} \quad \Omega \tag{6.8a}$$

$$u_\varepsilon(x, y) = 16x^2(1 - x)^2, \quad (x, y) \in \Gamma_{\text{B}} \tag{6.8b}$$

$$u_\varepsilon(x, y) = 0, \quad (x, y) \in \Gamma \setminus \Gamma_{\text{B}}. \tag{6.8c}$$

Its solution has a parabolic boundary layer on the edge Γ_{B} and a corner boundary layer at the corner C_{BL}. A contour plot of its solution u_ε is given in Fig. 6.3. We see that the width of the boundary layer increases as we move along the boundary Γ_{B} away from the inflow edge Γ_{R}. It is in this sense that the parabolic boundary layer evolves as we move in the time–like direction horizontally from right to left away from its point of inception C_{BR}. Our second example demonstrates that all of the relevant boundary data influence the profile of the boundary layer. We consider the convection–diffusion problem

$$\varepsilon \Delta u_\varepsilon + \frac{\partial u_\varepsilon}{\partial x} = 0, (x, y) \in \Omega \tag{6.9a}$$

$$u_\varepsilon(x, y) = (6\sqrt{3}x(1 - x)(2x - 1))^3, \quad (x, y) \in \Gamma_{\text{B}} \tag{6.9b}$$

$$u_\varepsilon(x, y) = 0, \quad (x, y) \in \Gamma \setminus \Gamma_{\text{B}}. \tag{6.9c}$$

Its solution has a parabolic boundary layer on the edge Γ_{B} and a corner boundary layer at the corner C_{BL}. This differs from the previous example only in the form of the boundary condition on Γ_{B}. We have chosen this boundary condition so that to the left of the midpoint of Γ_{B} the boundary values are negative and to the right they are positive. From Fig. 6.4 we see that the solution is positive at some points of the left–hand subdomain $[0, 0.5] \times [0, 0.1]$. But the values of

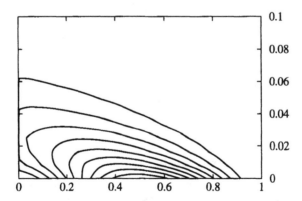

FIG. 6.3. Contour plot of the solution u_ε of problem (6.8) for $\varepsilon = 0.001$ near the edge Γ_{B} .

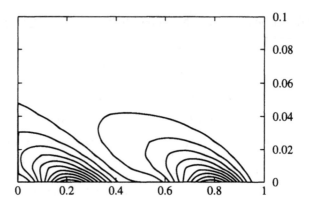

FIG. 6.4. Contour plot of the solution u_ε of problem (6.9) for $\varepsilon = 0.001$ near the edge Γ_B.

the boundary data on Γ_B to the left of the midpoint are negative, and so the positive values of the boundary data on Γ_B to the right of the midpoint must have an influence on the solution in the left–hand subdomain, otherwise the solution would take on negative values there.

Finally, we demonstrate the difference in the thickness of regular and parabolic boundary layers, by considering the boundary layers in the solution of the linear convection–diffusion problem

$$\varepsilon \Delta u_\varepsilon + \frac{\partial u_\varepsilon}{\partial x} = 0(x, y) \in \Omega \tag{6.10a}$$

$$u_\varepsilon(x, y) = (1 - x)^2, \quad (x, y) \in \Gamma_B \tag{6.10b}$$

$$u_\varepsilon(x, y) = y^2, (x, y) \in \Gamma_L \tag{6.10c}$$

$$u_\varepsilon(x, y) = 0, \quad (x, y) \in \Gamma_R \cup \Gamma_T. \tag{6.10d}$$

This solution has a parabolic boundary layer on the edge Γ_B, a regular boundary layer on the edge Γ_L and a corner boundary layer at the corner C_{BL}. A contour plot of u_ε for $\varepsilon = 0.05$ is given in Fig. 6.5. From this we see that the gradient of u_ε is considerably steeper in the region of the regular boundary layer near the edge Γ_L than in the region of the parabolic boundary layer on the edge Γ_B. This observation confirms that the parabolic boundary layer is wider than the regular one.

6.3   Inadequacy of uniform meshes

In this and the next section we construct various numerical methods on uniform and piecewise uniform meshes for solving singularly perturbed linear convection–diffusion problems the solutions of which have parabolic boundary layers. In this section we confine the discussion to uniform meshes, which suggests that the

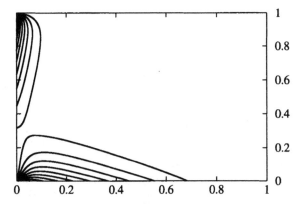

FIG. 6.5. Contour plot of the solution u_ε of problem (6.10) for $\varepsilon = 0.05$.

corresponding numerical solutions do not converge ε–uniformly. Therefore, this section is of more theoretical than practical interest. Our objective is to illustrate the fact that even for these thicker parabolic layers, standard numerical methods on uniform meshes are not adequate for their resolution. The standard method we use is the following monotone numerical method composed of an upwind finite difference operator on a uniform rectangular mesh.

Method 6.1.   Upwind finite difference operator on uniform rectangular mesh for Problem Class 6.1.

$$[\varepsilon(\delta_x^2 + \delta_y^2) + a_1(x_i, y_j)D_x^+]U_\varepsilon = f(x_i, y_j), \quad (x_i, y_j) \in \Omega_u^N \qquad (6.11a)$$

$$U_\varepsilon = g, \quad (x_i, y_j) \in \Gamma^N \qquad (6.11b)$$

where

$$\overline{\Omega}_u^N = \{(x_i, y_j) : x_i = i/N, \quad y_j = j/N \quad 0 \le i, \quad j \le N\}. \qquad (6.11c)$$

In Table 6.1 we present the computed maximum pointwise errors for this method applied to problem (6.8) for various values of ε and N. We see that, especially for small values of ε, the maximum pointwise errors are small. But for each fixed value of ε these errors first increase, with increasing N, to approximately 0.06 and thereafter they decrease monotonically. This means that the method is not ε–uniform, because for each value of N, there is a value of ε for which the error is no smaller than 0.06. This shows that the ε–uniform error does not vanish as $N \to \infty$. Instead there is a persistent, essentially constant, error along a diagonal of the error table.

Table 6.1 *Computed maximum pointwise error $E^N_{\varepsilon,\text{comp}}$ generated by Method 6.1 applied to problem (6.8) generated by Method 6.1 for various values of ε and N.*

ε	Number of intervals N					
	8	16	32	64	128	256
1	0.805D-02	0.357D-02	0.202D-02	0.103D-02	0.458D-03	0.156D-03
2^{-2}	0.308D-01	0.162D-01	0.814D-02	0.388D-02	0.168D-02	0.563D-03
2^{-4}	0.754D-01	0.426D-01	0.229D-01	0.116D-01	0.534D-02	0.204D-02
2^{-6}	**0.757D-01**	**0.679D-01**	0.595D-01	0.371D-01	0.206D-01	0.103D-01
2^{-8}	0.663D-01	0.656D-01	0.478D-01	0.552D-01	0.561D-01	0.365D-01
2^{-10}	0.304D-01	0.636D-01	0.551D-01	0.313D-01	0.343D-01	0.518D-01
2^{-12}	0.819D-02	0.307D-01	**0.617D-01**	0.487D-01	0.218D-01	0.186D-01
2^{-14}	0.207D-02	0.820D-02	0.307D-01	**0.607D-01**	0.451D-01	0.167D-01
2^{-16}	0.520D-03	0.207D-02	0.821D-02	0.307D-01	**0.602D-01**	0.433D-01
2^{-18}	0.130D-03	0.520D-03	0.207D-02	0.821D-02	0.307D-01	**0.600D-01**
2^{-20}	0.325D-04	0.130D-03	0.520D-03	0.208D-02	0.821D-02	0.307D-01
2^{-22}	0.814D-05	0.325D-04	0.130D-03	0.520D-03	0.208D-02	0.821D-02
2^{-24}	0.203D-05	0.814D-05	0.326D-04	0.130D-03	0.520D-03	0.208D-02
2^{-26}	0.509D-06	0.203D-05	0.814D-05	0.326D-04	0.130D-03	0.520D-03
2^{-28}	0.127D-06	0.509D-06	0.203D-05	0.814D-05	0.326D-04	0.130D-03
2^{-30}	0.318D-07	0.127D-06	0.509D-06	0.203D-05	0.814D-05	0.325D-04
2^{-32}	0.795D-08	0.318D-07	0.127D-06	0.509D-06	0.203D-05	0.814D-05
2^{-34}	0.199D-08	0.795D-08	0.318D-07	0.127D-06	0.509D-06	0.203D-05

The reader may wonder how we compute the errors in Table 6.1, because we do not have a simple closed expression for the exact solution of problem (6.8). Moreover, we cannot expect the interpolant of the solution $\bar{U}^{512}_\varepsilon$ on the finest available mesh to be ε–uniformly accurate, because Method 6.1 is not an ε–uniform method. To compute the errors we use the ε–uniform method for Problem Class 6.1 constructed in the next section, which is composed of an upwind finite difference operator on an appropriately fitted piecewise–uniform mesh. Using this ε–uniform method, we compute a numerical approximation on the finest available mesh ($N = 512$) to the solution u_ε of problem (6.8). The piecewise bilinear interpolant of this numerical solution is denoted by $\bar{u}_{\varepsilon,\text{comp}}$. This enables us to estimate the maximum pointwise error in any numerical method, ε–uniform or not, by

$$E^N_{\varepsilon,\text{comp}} = \|U^N_\varepsilon - \bar{u}_{\varepsilon,\text{comp}}\|_{\Omega^N}$$

rather than by

$$E^N_\varepsilon = \|U^N_\varepsilon - \bar{U}^{512}_\varepsilon\|_{\Omega^N}.$$

It is of interest to compare the errors in Table 6.1 with those in Table 2.1. In the latter table the errors are given for the numerical solution of a convection–diffusion problem in one dimension with regular boundary layers using a numerical method composed of an upwind finite difference operator on a uniform mesh.

Since the errors are smaller in Table 6.1 than those in Table 2.1 we are led to the conjecture that the deleterious consequences of using a uniform mesh in the presence of parabolic boundary layers may not be as serious as in the case of regular boundary layers. That this conjecture is false is seen later in §6.4. This is another instance where insight gained by solving standard problems, leads to incorrect conclusions when singular perturbation problems are involved.

We now give some theoretical hints as to why Method 6.1 cannot be expected to be ε–uniform for Problem Class 6.1. To do this we consider an even simpler problem than problem (6.8), namely the singularly perturbed heat equation

$$\varepsilon \frac{\partial^2 u_\varepsilon}{\partial x^2} - \frac{\partial u_\varepsilon}{\partial t} = 0, \quad (x,t) \in (0,1) \times (0,1] \tag{6.12a}$$

$$u_\varepsilon(x,0) = u_\varepsilon(1,t) = 0, \quad u_\varepsilon(0,t) = t^2. \tag{6.12b}$$

On the edge Γ_L, a parabolic boundary layer occurs in the solution u_ε for small values of ε. The classical numerical method with a finite difference operator, implicit in time and centred in space, on a uniform mesh Ω_u^N for problem (6.12) is

$$L_\varepsilon^N U_\varepsilon \equiv \varepsilon \delta_x^2 U_\varepsilon - D_t^- U_\varepsilon = 0, \quad (x_i, y_j) \in \Omega_u^N \tag{6.13a}$$

$$U_\varepsilon(x_i, 0) = U_\varepsilon(1, t_j) = 0, \quad U_\varepsilon(0, t_j) = t_j^2. \tag{6.13b}$$

A Taylor expansion of the integral representation of the exact solution u_ε, see Miller et al. (1996) (ch.14, p.135), then gives

$$L_\varepsilon^N \left(u_\varepsilon(x_i, t_j) - U_\varepsilon(x_i, t_j) \right) = \frac{h^2}{6\varepsilon} + O(t_j R^3 + h^2 \varepsilon^{-1} R + k + \varepsilon^2)$$

where $R = x_i / \sqrt{\varepsilon t_j}$. Taking R, h and k small and fixing t_j and the ratio h^2/ε (e.g., $i = 1, \quad t_j = 1$ and $\varepsilon = 1000 h^2$), we get

$$L_\varepsilon^N \left(u_\varepsilon(x_i, t_j) - U_\varepsilon(x_i, t_j) \right) \geq \frac{h^2}{12\varepsilon} = \frac{1}{12,000} > 0$$

and so

$$\lim_{h \to 0} L_\varepsilon^N (U_\varepsilon - u_\varepsilon)(x_i, t_j) \neq 0 \tag{6.14}$$

which shows that the numerical method (6.13) is not ε–uniformly consistent with problem (6.11). Since the finite difference operator L_ε^N in (6.13) satisfies a discrete minimum principle, it then follows from (6.14) that, with h^2/ε constant in the limiting process, we have

$$\lim_{h \to 0} \|U_\varepsilon - u_\varepsilon\| \neq 0$$

which proves that method (6.13) is not ε–uniform at the mesh points for problem (6.12).

We now examine the possibility or otherwise of constructing an ε–uniform method on a uniform mesh which uses a fitted finite difference operator instead of the simpler upwind finite difference operator in Method 6.1. For the singularly perturbed convection–diffusion equation

$$\varepsilon \Delta u_\varepsilon + \mathbf{a}(x,y) \cdot \nabla u_\varepsilon = f(x,y), \quad (x,y) \in \Omega \qquad (6.15)$$

a general fitted finite difference operator has the form

$$\varepsilon \gamma_1 \delta_x^2 U_\varepsilon + \varepsilon \gamma_2 \delta_y^2 U_\varepsilon + a_1 D_x^0 U_\varepsilon + a_2 D_y^0 U_\varepsilon = f, \quad (x_i, y_j) \in \Omega_u^N \qquad (6.16a)$$

where the fitting factors γ_1, γ_2 are chosen in an appropriate manner. For example, if the solution of the problem has only regular boundary layers, then an appropriate choice of these fitting factors is

$$\gamma_1 = \frac{\rho_1}{2} \coth \frac{\rho_1}{2} \quad \text{and} \quad \gamma_2 = \frac{\rho_2}{2} \coth \frac{\rho_2}{2} \qquad (6.16b)$$

where

$$\rho_1(x_i, y_j) = \frac{a_1(x_i, y_j)}{\varepsilon N} \quad \text{and} \quad \rho_2(x_i, y_j) = \frac{a_2(x_i, y_j)}{\varepsilon N}. \qquad (6.16c)$$

It is demonstrated both theoretically in Emel'ianov (1973), Hegarty (1986) and numerically in Hegarty (1986) that the resulting numerical method is ε–uniform for problems involving equations of the form (6.15) if only regular boundary layers occur in the solution u_ε. But if we now apply the same numerical method to problem (6.8), which has a parabolic boundary layer on the edge Γ_B, we obtain the errors shown in Table 6.2. Just as in Table 6.1 we see that there is a persistent, essentially constant, error of about 6% along a diagonal of Table 6.2, which indicates that this fitted operator method is not ε–uniform for this problem.

In principle, an ε–uniform fitted operator method on a uniform rectangular mesh could be specifically designed for a given problem of the form (6.8), but the same fitted operator method would not be ε–uniform for a problem involving the same differential equation and different boundary conditions. Indeed, it is shown in Shishkin (1989, 1995) that no fitted operator method on a uniform rectangular mesh exists that is ε–uniform for a class of problems whose solutions contain parabolic boundary layers. A more readable version of the proof, in a simpler case, is given in Miller et al. (1996).

This situation, where parabolic layers are present, is in sharp contrast to the situation for problems with only regular boundary layers. For the latter problems it is always possible to construct a numerical method consisting of a fitted finite difference operator on a uniform mesh, which generates numerical solutions that converge ε–uniformly at the mesh points. But for a method to

Table 6.2 *Computed maximum pointwise error $E_{\varepsilon,comp}^N$ generated by the fitted operator (6.16) on a uniform mesh applied to problem (6.8) for various values of ε and N.*

	Number of intervals N					
ε	8	16	32	64	128	256
1	0.679D-02	0.244D-02	0.690D-03	0.263D-03	0.183D-03	0.165D-03
2^{-2}	0.984D-02	0.288D-02	0.913D-03	0.596D-03	0.567D-03	0.566D-03
2^{-4}	0.270D-01	0.674D-02	0.134D-02	0.106D-02	0.128D-02	0.135D-02
2^{-6}	0.592D-01	0.262D-01	0.644D-02	0.810D-03	0.158D-02	0.183D-02
2^{-8}	**0.663D-01**	0.526D-01	0.232D-01	0.588D-02	0.886D-03	0.186D-02
2^{-10}	0.307D-01	**0.635D-01**	0.481D-01	0.194D-01	0.458D-02	0.765D-03
2^{-12}	0.820D-02	0.307D-01	**0.617D-01**	0.453D-01	0.171D-01	0.362D-02
2^{-14}	0.207D-02	0.820D-02	0.307D-01	**0.607D-01**	0.437D-01	0.157D-01
2^{-16}	0.520D-03	0.208D-02	0.821D-02	0.307D-01	**0.602D-01**	0.429D-01
2^{-18}	0.130D-03	0.520D-03	0.208D-02	0.821D-02	0.307D-01	**0.600D-01**
2^{-20}	0.325D-04	0.130D-03	0.520D-03	0.208D-02	0.821D-02	0.307D-01
2^{-22}	0.814D-05	0.325D-04	0.130D-03	0.520D-03	0.208D-02	0.821D-02
2^{-24}	0.203D-05	0.814D-05	0.326D-04	0.130D-03	0.520D-03	0.208D-02
2^{-26}	0.509D-06	0.203D-05	0.814D-05	0.326D-04	0.130D-03	0.520D-03
2^{-28}	0.127D-06	0.509D-06	0.203D-05	0.814D-05	0.326D-04	0.130D-03
2^{-30}	0.318D-07	0.127D-06	0.509D-06	0.203D-05	0.814D-05	0.325D-04
2^{-32}	0.795D-08	0.318D-07	0.127D-06	0.509D-06	0.203D-05	0.814D-05
2^{-34}	0.199D-08	0.795D-08	0.318D-07	0.127D-06	0.509D-06	0.203D-05

be ε–uniform we insist that the piecewise linear interpolants of its numerical solutions converge ε–uniformly at each point of the domain. As the examples in §2.2 and 2.4 show, standard finite difference operators on uniform meshes are of little use in the construction of ε–uniform pointwise–accurate numerical methods for one dimensional problems with regular boundary layers. On the other hand, in §2.5 we construct a fitted finite difference operator method on a uniform mesh which, with a suitable form of interpolation, is ε–uniform for Problem Class 2.1. The situation is quite different in two dimensions when parabolic boundary layers are present, because the results in Table 6.2 show that a method with a fitted finite difference operator on a uniform mesh is not ε–uniform even at the mesh points for problems from Problem Class 6.1.

6.4   Fitted meshes for parabolic boundary layers

The discussion in the previous section shows that if we want to construct an ε–uniform method for a singular perturbation problem with a parabolic boundary layer in its solution, it is necessary to use a non–uniform mesh. In this section we show that an appropriately fitted piecewise–uniform mesh suffices for such problems. We illustrate this by constructing an ε–uniform method for Problem Class 6.1 based on a piecewise–uniform fitted mesh. Because the domain is rectangular, an appropriate fitted mesh $\overline{\Omega}_\varepsilon^N$ is the tensor product of two one–dimensional fitted meshes. Our method is described formally in the following definition.

Method 6.2.   Upwind finite difference operator on piecewise–uniform fitted mesh for Problem Class 6.1.

$$[\varepsilon(\delta_x^2 + \delta_y^2) + a_1(x_i, y_j)D_x^+]U_\varepsilon = f(x_i, y_j) \quad (x_i, y_j) \in \Omega_\varepsilon^N \qquad (6.17a)$$

$$U_\varepsilon = g, \quad (x_i, y_j) \in \Gamma^N \qquad (6.17b)$$

where

$$\overline{\Omega}_\varepsilon^N = \overline{\Omega}_{\sigma_1}^N \times \overline{\Omega}_{\sigma_2}^N \qquad (6.17c)$$

$$\overline{\Omega}_{\sigma_1}^N = \{x_i : 0 \le i \le N\} \quad \text{and} \quad \overline{\Omega}_{\sigma_2}^N = \{y_j : 0 \le j \le N\} \qquad (6.17d)$$

$$x_i = \begin{cases} 2i\sigma_1/N, & \text{for } 0 \le i \le N/2 \\ \sigma_1 + 2(i - (N/2))(1 - \sigma_1)/N, & \text{for } N/2 \le i \le N \end{cases} \qquad (6.17e)$$

$$y_j = \begin{cases} 4i\sigma_2/N, & \text{for } 0 \le j \le N/4 \\ \sigma_2 + 2(i - (N/4))(1 - 2\sigma_2)/N, & \text{for } N/4 \le j \le 3N/4 \\ 1 - \sigma_2 + 4(i - (3N/4))\sigma_2/N, & \text{for } 3N/4 \le j \le N \end{cases} \qquad (6.17f)$$

$$\sigma_1 = \min\{1/2, \, c\varepsilon \ln N\}, \quad c \ge \alpha_1^{-1} \qquad (6.17g)$$

$$\sigma_2 = \min\{1/4, \, \varepsilon^{\frac{1}{2}} \ln N\}. \qquad (6.17h)$$

We remark that the value of c in (6.17g) is not critical, since it turns out that the errors in the numerical solutions generated by Method 6.2 are not particularly sensitive to its choice. This can be shown by numerical experiments and it is similar to the situation for the analogous one–dimensional case discussed in §3.6. For all of the numerical examples in this chapter we take $c = \alpha_1^{-1}$.

We now apply Method 6.2 to the following specific convection–diffusion problem from Problem Class 6.1 on the unit square Ω

$$\varepsilon \Delta u_\varepsilon + \frac{\partial u_\varepsilon}{\partial x} = 0, \quad (x, y) \in \Omega \qquad (6.18a)$$

$$u_\varepsilon(x, 0) = 64x^3(1 - x)^3, \quad (x, y) \in \Gamma_B, \qquad (6.18b)$$

$$u_\varepsilon(1, y) = 64y^3(1 - y)^3, \quad (x, y) \in \Gamma_R, \qquad (6.18c)$$

$$u_\varepsilon(x, y) = 0, \quad (x, y) \in \Gamma \setminus (\Gamma_B \cup \Gamma_R). \qquad (6.18d)$$

Regular, parabolic and corner boundary layers are present in its solution. The regular boundary layer is on the edge Γ_L, the parabolic boundary layers are on the edges Γ_B and Γ_T and the corner boundary layers are at C_{BL} and C_{TL}. The resulting numerical solutions for this problem are displayed in Tables 6.3 and 6.4. It is clear from these tables that Method 6.2 is ε–uniform for this problem. This demonstrates once again that an appropriate piecewise–uniform fitted mesh combined with a standard upwind finite difference operator suffice for the construction of an ε–uniform method.

It is important to observe that there is not a lot of leeway in the choice of the fitted mesh. To see this we consider, for example, two other potentially

Table 6.3 *Computed maximum pointwise errors E_ε^N and E^N generated by the fitted mesh Method 6.2 applied to problem (6.18) for various values of ε and N.*

	Number of intervals N					
ε	8	16	32	64	128	256
1	0.185D-01	0.737D-02	0.301D-02	0.126D-02	0.514D-03	0.169D-03
2^{-2}	0.325D-01	0.168D-01	0.836D-02	0.399D-02	0.174D-02	0.599D-03
2^{-4}	0.991D-01	0.504D-01	0.251D-01	0.128D-01	0.624D-02	0.235D-02
2^{-6}	0.116D+00	0.682D-01	0.379D-01	0.200D-01	0.960D-02	0.354D-02
2^{-8}	0.150D+00	0.883D-01	0.484D-01	0.251D-01	0.118D-01	0.428D-02
2^{-10}	0.163D+00	0.943D-01	0.515D-01	0.267D-01	0.126D-01	0.454D-02
2^{-12}	0.167D+00	0.959D-01	0.524D-01	0.272D-01	0.128D-01	0.462D-02
2^{-14}	0.168D+00	0.963D-01	0.526D-01	0.273D-01	0.128D-01	0.464D-02
2^{-16}	0.168D+00	0.964D-01	0.526D-01	0.273D-01	0.129D-01	0.465D-02
2^{-18}	0.168D+00	0.965D-01	0.527D-01	0.273D-01	0.129D-01	0.465D-02
.	.	.	.	.	.	.
.	.	.	.	.	.	.
2^{-34}	0.168D+00	0.965D-01	0.527D-01	0.273D-01	0.129D-01	0.465D-02
E^N	0.168D+00	0.965D-01	0.527D-01	0.273D-01	0.129D-01	0.465D-02

Table 6.4 *Computed orders of convergence p_ε^N and p^N generated by the fitted mesh Method 6.2 applied to problem (6.18) for various values of ε and N.*

	Number of intervals N				
ε	8	16	32	64	128
1	1.42	1.34	1.21	1.12	1.03
2^{-2}	0.95	0.95	0.97	0.98	0.93
2^{-4}	0.93	1.12	0.96	0.79	0.72
2^{-6}	0.64	0.93	0.76	0.79	0.78
2^{-8}	0.69	0.79	0.81	0.82	0.82
2^{-10}	0.74	0.80	0.81	0.81	0.83
2^{-12}	0.75	0.81	0.82	0.81	0.83
2^{-14}	0.76	0.81	0.82	0.81	0.83
2^{-16}	0.76	0.81	0.82	0.81	0.83
2^{-18}	0.76	0.81	0.82	0.81	0.82
.	.	.	.	.	.
.	.	.	.	.	.
2^{-34}	0.76	0.81	0.82	0.81	0.82
p^N	0.76	0.81	0.82	0.81	0.82

promising fitted meshes, both of which fail to produce an ε–uniform method for this problem. The first candidate is a piecewise–uniform partially fitted mesh, which is fitted to the parabolic boundary layers but not to the regular boundary layer. This is the mesh

$$\Omega_u^N \times \Omega_{\sigma_2}^N \tag{6.19}$$

where Ω_u^N is a uniform mesh along the x–axis and $\Omega_{\sigma_2}^N$ is the one–dimensional fitted mesh in (6.17). Using the numerical method based on the same upwind finite difference operator as in (6.17a) and the fitted mesh (6.19) to solve problem (6.18), we obtain the numerical results shown in Table 6.5. From the maximum values in each column of this table it is clear that this monotone method is not ε–uniform. As before there is a diagonal in the table along which the maximum pointwise error has a magnitude of at least 0.12 .

The second candidate is a piecewise–uniform partially fitted mesh, which is fitted to the regular boundary layer but not to the parabolic boundary layers. This is the mesh

$$\Omega_{\sigma_1}^N \times \Omega_u^N \tag{6.20}$$

where Ω_u^N is a uniform mesh along the y–axis and $\Omega_{\sigma_1}^N$ is the one–dimensional fitted mesh in (6.17). We use the numerical method based on the upwind finite difference operator in (6.17a) and the fitted mesh (6.20), to solve problem (6.18). The numerical results are given in Table 6.6. It is less obvious here, than in the previous case, that this monotone method is not ε–uniform. As before, we see the error ridge behaviour along a diagonal of the table typical of a non ε–uniform method, but in this case the values of E_ε^N along this diagonal appear to decrease

Table 6.5 *Computed maximum pointwise error* $E_{\varepsilon,\text{comp}}^N$ *generated by the upwind finite difference operator and the partially fitted mesh (6.19) with no refinement in the regular layer applied to problem (6.18) for various values of* ε *and* N.

ε	Number of intervals N					
	8	16	32	64	128	256
1	0.185D-01	0.737D-02	0.301D-02	0.126D-02	0.514D-03	0.169D-03
2^{-2}	0.325D-01	0.168D-01	0.836D-02	0.399D-02	0.174D-02	0.599D-03
2^{-4}	**0.123D+00**	0.846D-01	0.486D-01	0.257D-01	0.122D-01	0.480D-02
2^{-6}	0.964D-01	**0.124D+00**	**0.160D+00**	0.124D+00	0.706D-01	0.371D-01
2^{-8}	0.748D-01	0.539D-01	0.861D-01	**0.151D+00**	**0.189D+00**	0.145D+00
2^{-10}	0.765D-01	0.544D-01	0.275D-01	0.474D-01	0.899D-01	**0.157D+00**
2^{-12}	0.765D-01	0.545D-01	0.275D-01	0.124D-01	0.245D-01	0.475D-01
2^{-14}	0.763D-01	0.545D-01	0.275D-01	0.109D-01	0.626D-02	0.124D-01
2^{-16}	0.763D-01	0.545D-01	0.275D-01	0.109D-01	0.329D-02	0.312D-02
2^{-18}	0.762D-01	0.545D-01	0.275D-01	0.109D-01	0.329D-02	0.855D-03
.	.	.	.	.	.	.
.	.	.	.	.	.	.
.	.	.	.	.	.	.
2^{-34}	0.762D-01	0.545D-01	0.275D-01	0.109D-01	0.329D-02	0.855D-03

Table 6.6 *Computed maximum pointwise error $E^N_{\varepsilon,\text{comp}}$ generated by the upwind finite difference operator and the partially fitted mesh (6.20) with no refinement in the parabolic layers applied to problem (6.18) for various values of ε and N.*

ε	Number of intervals N					
	8	16	32	64	128	256
1	0.185D-01	0.737D-02	0.301D-02	0.126D-02	0.514D-03	0.169D-03
2^{-2}	0.325D-01	0.168D-01	0.836D-02	0.399D-02	0.174D-02	0.599D-03
2^{-4}	0.991D-01	0.504D-01	0.251D-01	0.128D-01	0.624D-02	0.235D-02
2^{-6}	0.116D+00	0.682D-01	0.379D-01	0.200D-01	0.960D-02	0.354D-02
2^{-8}	0.152D+00	0.886D-01	0.484D-01	0.251D-01	0.118D-01	0.428D-02
2^{-10}	0.164D+00	0.945D-01	**0.650D-01**	0.323D-01	0.126D-01	0.455D-02
2^{-12}	0.167D+00	0.960D-01	0.568D-01	**0.590D-01**	0.267D-01	0.725D-02
2^{-14}	0.168D+00	0.963D-01	0.526D-01	0.556D-01	**0.558D-01**	0.238D-01
2^{-16}	0.168D+00	0.964D-01	0.526D-01	0.273D-01	0.551D-01	0.540D-01
2^{-18}	0.168D+00	0.965D-01	0.527D-01	0.273D-01	0.216D-01	**0.547D-01**
2^{-20}	0.168D+00	0.965D-01	0.527D-01	0.273D-01	0.129D-01	0.216D-01
2^{-22}	0.168D+00	0.965D-01	0.527D-01	0.273D-01	0.129D-01	0.561D-02
.	.	.	.	.	.	.
2^{-34}	0.168D+00	0.965D-01	0.527D-01	0.273D-01	0.129D-01	0.465D-02

with increasing N. However, the rate of this decrease tends almost to zero as N increases.

It is when this method, based on the partially fitted mesh (6.20), is used to generate approximations to a scaled first derivative of the solution, that its inadequacy becomes indisputable. This observation is of practical importance, because often the physical quantities of interest involve scaled first order derivatives of the solution, such as the flux . A small error in the approximation of the solution may be tolerable, but in practice we want comparably small errors in the numerical approximations of, for example, the scaled derivatives $(\varepsilon u_x, \sqrt{\varepsilon} u_y)$. We show now with a method based on the partially fitted mesh (6.20) the error in the numerical approximations to the scaled derivatives can be much greater than the error in the numerical approximations to the solution itself.

For problem (6.18), from the known theoretical bounds on the first order derivatives, we know that the scaled derivatives $(\varepsilon u_x, \sqrt{\varepsilon} u_y)$ remain bounded as $\varepsilon \to 0$. Therefore, the above scaled derivatives are appropriate in the singularly perturbed case and it is natural to approximate them by the following scaled discrete derivatives $(\varepsilon D_x^+ U_\varepsilon, \sqrt{\varepsilon} D_y^+ U_\varepsilon)$. The computed maximum pointwise errors and computed orders of convergence of the numerical approximation to the scaled derivatives, generated by the fully fitted Method 6.2, are given in Tables 6.7, 6.8, 6.9 and 6.10. These results indicate that the scaled discrete derivatives are ε–uniform approximations to the scaled derivatives. On the other hand, the results in Tables 6.11 and 6.12 are computed using the partially fitted method based on the upwind finite difference operator and the partially fitted mesh (6.20). These results demonstrate that the numerical approximations of

Table 6.7 *Computed maximum pointwise errors* $E_\varepsilon^N(\varepsilon D_x^+ U)$ *and* $E^N(\varepsilon D_x^+ U)$ *generated by Method 6.2 applied to problem (6.18) for various values of* ε *and* N.

ε	Number of intervals N					
	8	16	32	64	128	256
1	0.444D+00	0.204D+00	0.989D-01	0.473D-01	0.204D-01	0.681D-02
2^{-2}	0.375D+00	0.219D+00	0.104D+00	0.481D-01	0.205D-01	0.684D-02
2^{-4}	0.142D+00	0.101D+00	0.536D-01	0.284D-01	0.134D-01	0.480D-02
2^{-6}	0.120D+00	0.805D-01	0.488D-01	0.266D-01	0.128D-01	0.465D-02
2^{-8}	0.101D+00	0.767D-01	0.463D-01	0.254D-01	0.123D-01	0.452D-02
2^{-10}	0.942D-01	0.755D-01	0.459D-01	0.251D-01	0.122D-01	0.448D-02
2^{-12}	0.934D-01	0.748D-01	0.457D-01	0.250D-01	0.121D-01	0.446D-02
2^{-14}	0.932D-01	0.744D-01	0.455D-01	0.250D-01	0.121D-01	0.446D-02
2^{-16}	0.931D-01	0.742D-01	0.453D-01	0.250D-01	0.121D-01	0.446D-02
2^{-18}	0.931D-01	0.741D-01	0.453D-01	0.250D-01	0.121D-01	0.445D-02
.	.	.	.	.	.	.
2^{-34}	0.931D-01	0.740D-01	0.453D-01	0.250D-01	0.121D-01	0.445D-02
$E^N(\varepsilon D_x^+ U)$	0.444D+00	0.219D+00	0.104D+00	0.481D-01	0.205D-01	0.684D-02

Table 6.8 *Computed orders of convergence* $p_\varepsilon^N(\varepsilon D_x^+ U)$ *and* $p^N(\varepsilon D_x^+ U)$ *generated by Method 6.2 applied to problem (6.18) for various values of* ε *and* N.

ε	Number of intervals N				
	8	16	32	64	128
1	0.73	0.79	0.85	0.92	0.96
2^{-2}	0.52	0.74	0.87	0.93	0.97
2^{-4}	0.35	0.79	0.83	0.66	0.74
2^{-6}	0.71	0.46	0.37	0.63	0.70
2^{-8}	0.60	0.09	0.36	0.60	0.69
2^{-10}	0.85	0.05	0.35	0.61	0.69
2^{-12}	0.94	0.04	0.34	0.61	0.69
2^{-14}	0.98	0.02	0.34	0.60	0.69
2^{-16}	0.99	0.02	0.34	0.60	0.69
2^{-18}	1.00	0.01	0.34	0.60	0.69
.	.	.	.	.	.
2^{-34}	1.00	0.01	0.34	0.60	0.69
$p^N(\varepsilon D_x^+ U)$	0.73	0.79	0.85	0.92	0.96

the scaled derivatives are satisfactory in the boundary layer regions, where the mesh is fitted, but that they do not converge ε–uniformly in the boundary layer regions where the mesh is not fitted. In Fig. 6.6 and Fig. 6.7 we present graphs of the approximation to $\sqrt{\varepsilon}u_y$ generated using, respectively, Method 6.2 and the partially fitted method based on the upwind finite difference operator and the partially fitted mesh (6.20). It is important to note the different scales in these figures and the fact that the approximation on the fitted mesh correctly shows

Table 6.9 *Computed maximum pointwise errors $E_\varepsilon^N(\sqrt{\varepsilon}D_y^+U)$, $E^N(\sqrt{\varepsilon}D_y^+U)$ generated by Method 6.2 applied to problem (6.18) for various values of ε and N.*

	Number of intervals N					
ε	8	16	32	64	128	256
1	0.430D+00	0.210D+00	0.102D+00	0.469D-01	0.205D-01	0.690D-02
2^{-2}	0.430D+00	0.230D+00	0.129D+00	0.627D-01	0.275D-01	0.925D-02
2^{-4}	0.650D+00	0.305D+00	0.154D+00	0.742D-01	0.325D-01	0.110D-01
2^{-6}	0.934D+00	0.482D+00	0.230D+00	0.113D+00	0.498D-01	0.169D-01
2^{-8}	0.736D+00	0.587D+00	0.370D+00	0.219D+00	0.999D-01	0.346D-01
2^{-10}	0.680D+00	0.562D+00	0.347D+00	0.208D+00	0.108D+00	0.411D-01
2^{-12}	0.672D+00	0.560D+00	0.347D+00	0.208D+00	0.108D+00	0.410D-01
2^{-14}	0.671D+00	0.560D+00	0.347D+00	0.208D+00	0.108D+00	0.410D-01
.	.	.	.	.	.	.
.	.	.	.	.	.	.
.	.	.	.	.	.	.
2^{-34}	0.671D+00	0.560D+00	0.347D+00	0.208D+00	0.108D+00	0.410D-01
$E^N(\sqrt{\varepsilon}D_y^+U)$	0.934D+00	0.587D+00	0.370D+00	0.219D+00	0.108D+00	0.411D-01

Table 6.10 *Computed orders of convergence $p_\varepsilon^N(\sqrt{\varepsilon}D_y^+U)$ and $p^N(\sqrt{\varepsilon}D_y^+U)$ generated by Method 6.2 applied to problem (6.18) for various values of ε and N.*

	Number of intervals N				
ε	8	16	32	64	128
1	0.88	0.97	0.95	0.94	0.98
2^{-2}	0.63	0.80	0.90	0.95	0.98
2^{-4}	0.65	0.83	0.87	0.93	0.97
2^{-6}	0.25	0.65	0.77	0.88	0.94
2^{-8}	-0.49	-0.25	0.01	0.77	0.88
2^{-10}	-0.55	-0.11	0.22	0.44	0.60
.	.	.	.	.	.
.	.	.	.	.	.
.	.	.	.	.	.
2^{-34}	-0.56	-0.11	0.22	0.44	0.60
$p^N(\sqrt{\varepsilon}D_y^+U)$	0.88	0.75	0.01	0.77	0.65

significant positive derivatives in the parabolic layer, while the approximation on the partially fitted mesh (6.20) completely misses this behaviour. A graph of the approximate solution of problem (6.18) with $\varepsilon = 10^{-6}$ generated by Method 6.2 with $N = 32$ is displayed in Fig. 6.8.

In Table 6.6 we see that, for all values of ε, the computed maximum pointwise errors in the approximations to the solution of problem (6.18) decrease to about 5% for N sufficiently large. But this is not an indication that the method gives ε–uniform approximations to the scaled derivatives, because in Table 6.12 we see that there is a large persistent error in the approximation of one of the scaled derivatives for all values of ε and N.

Table 6.11 *Computed maximum pointwise errors $E^N_{\varepsilon,\text{comp}}(\varepsilon D^+_x U)$ generated by the upwind finite difference operator and the partially fitted mesh (6.20) applied to problem (6.18) for various values of ε and N.*

	Number of intervals N					
ε	8	16	32	64	128	256
1	0.444D+00	0.204D+00	0.989D-01	0.473D-01	0.204D-01	0.681D-02
2^{-2}	0.375D+00	0.219D+00	0.104D+00	0.481D-01	0.205D-01	0.684D-02
2^{-4}	0.142D+00	0.101D+00	0.536D-01	0.284D-01	0.134D-01	0.480D-02
2^{-6}	0.120D+00	0.805D-01	0.488D-01	0.266D-01	0.128D-01	0.465D-02
2^{-8}	0.109D+00	0.759D-01	0.463D-01	0.254D-01	0.123D-01	0.452D-02
2^{-10}	0.106D+00	0.745D-01	0.456D-01	0.251D-01	0.122D-01	0.448D-02
2^{-12}	0.105D+00	0.742D-01	0.455D-01	0.336D-01	0.121D-01	0.446D-02
2^{-14}	0.105D+00	0.741D-01	0.455D-01	0.250D-01	0.307D-01	0.953D-02
2^{-16}	0.105D+00	0.740D-01	0.455D-01	0.250D-01	0.121D-01	0.292D-01
2^{-18}	0.105D+00	0.740D-01	0.455D-01	0.250D-01	0.168D-01	0.659D-02
2^{-20}	0.105D+00	0.740D-01	0.455D-01	0.250D-01	0.121D-01	0.168D-01
2^{-22}	0.105D+00	0.740D-01	0.455D-01	0.250D-01	0.121D-01	0.446D-02
.	.	.	.	.	.	.
.	.	.	.	.	.	.
2^{-34}	0.105D+00	0.740D-01	0.454D-01	0.250D-01	0.121D-01	0.445D-02

Table 6.12 *Computed maximum pointwise errors $E^N_{\varepsilon,\text{comp}}(\sqrt{\varepsilon}D^+_y U)$ generated by the upwind finite difference operator and the partially fitted mesh (6.20) applied to problem (6.18) for various values of ε and N.*

	Number of intervals N					
ε	8	16	32	64	128	256
1	0.430D+00	0.210D+00	0.102D+00	0.469D-01	0.205D-01	0.690D-02
2^{-2}	0.430D+00	0.230D+00	0.129D+00	0.627D-01	0.275D-01	0.925D-02
2^{-4}	0.650D+00	0.305D+00	0.154D+00	0.742D-01	0.325D-01	0.110D-01
2^{-6}	0.934D+00	0.482D+00	0.230D+00	0.113D+00	0.498D-01	0.169D-01
2^{-8}	0.123D+01	0.775D+00	0.433D+00	0.219D+00	0.999D-01	0.346D-01
2^{-10}	**0.124D + 01**	0.102D+01	0.756D+00	0.423D+00	0.210D+00	0.846D-01
2^{-12}	0.123D+01	**0.104D+01**	0.988D+00	0.754D+00	0.428D+00	0.215D+00
2^{-14}	0.123D+01	0.104D+01	**0.101D+01**	0.983D+00	0.754D+00	0.431D+00
2^{-16}	0.123D+01	0.103D+01	0.100D+01	**0.100D+01**	0.986D+00	0.755D+00
2^{-18}	0.123D+01	0.103D+01	0.100D+01	0.100D+01	**0.100D+01**	0.985D+00
2^{-20}	0.123D+01	0.103D+01	0.100D+01	0.100D+01	0.100D+01	**0.100D+01**
2^{-22}	0.123D+01	0.103D+01	0.100D+01	0.100D+01	0.100D+01	0.100D+01
.	.	.	.	.	.	.
.	.	.	.	.	.	.
2^{-34}	0.123D+01	0.103D+01	0.100D+01	0.100D+01	0.100D+01	0.100D+01

6.5   Simple parameter–uniform analytic approximations

It seems natural to assume that a problem, the exact solution of which has a complicated boundary layer structure, requires a complicated numerical method to produce numerical approximations of guaranteed accuracy. That this is a

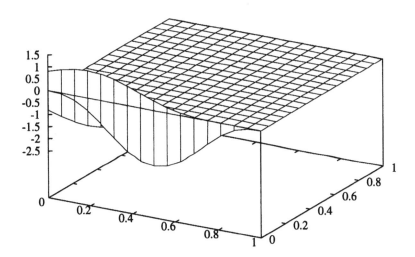

FIG. 6.6. Scaled discrete derivative $\sqrt{\varepsilon} D_y^+ U$ generated by Method 6.2 applied to problem (6.18) for $\varepsilon = 10^{-6}$ and with $N = 32$.

misconception we now illustrate by examining the analytic structure of the exact solution of problem (6.18). On the one hand, we demonstrate in this section that this structure is complicated, while on the other hand, the numerical results in the previous section show that the simple piecewise–uniform fitted mesh Method 6.2 generates numerical approximations converging ε–uniformly not only to u_ε but also to its scaled derivatives εu_x and $\sqrt{\varepsilon} u_y$. Furthermore, these numerical solutions can be extended to ε–uniform approximations at each point of the domain $\overline{\Omega}$ by piecewise bilinear interpolation.

That the solution u_ε to problem (6.18) has a complicated analytic form can be inferred from an asymptotic expansion of the solution of a related, but simpler, problem, which we now describe. This is the constant coefficient convection–diffusion problem

$$-\varepsilon \Delta u_\varepsilon + a \frac{\partial u_\varepsilon}{\partial x} = f(x, y), \quad (x, y) \in \Omega \qquad (6.21\text{a})$$

$$u_\varepsilon(x, y) = g(x, y), \quad (x, y) \in \Omega \qquad (6.21\text{b})$$

where $a > 0$ is a constant. Shih and Kellogg (1987) give an asymptotic expansion u_{asy} for the solution of this problem. They show that the solution u_ε can be

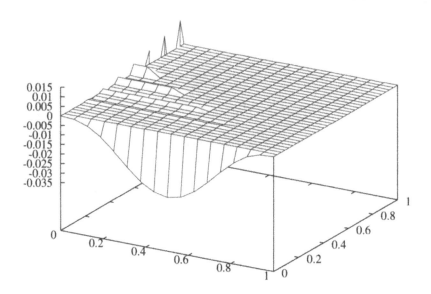

FIG. 6.7. *Scaled discrete derivative* $\sqrt{\varepsilon}D_y^+U$ *generated by the upwind finite difference operator and the partially fitted mesh* (6.20) *applied to problem* (6.18) *with* $\varepsilon = 10^{-6}$ *and with* $N = 32$.

written in the form

$$u_\varepsilon = u_{\text{asy}} + R$$

where

$$\begin{aligned}
u_{\text{asy}}(x,y) = {} & v_0(x,y) + w(X_1,y;\varepsilon) + v(X,Y;\varepsilon) + z(x,\tilde{Y};\varepsilon) + \\
& + W(X_1,\tilde{Y};\varepsilon) + V(X_1,Y;\varepsilon) + v^T(X,Y_1^T;\varepsilon) + \\
& + z^T(x,\tilde{Y}^T;\varepsilon) + W^T(X_1,\tilde{Y}^T;\varepsilon) + V^T(X_1,Y_1^T;\varepsilon)
\end{aligned}$$

and the remainder term satisfies the condition

$$|R(x,y;\varepsilon)| \leq C\varepsilon \quad \text{for all} \quad (x,y) \in \overline{\Omega} \quad \text{and} \quad \varepsilon \in (0,1].$$

The individual terms in the asymptotic expansion are defined as follows:
(i) $v_0(x,y)$ is the reduced solution, that is the solution of the initial value problem

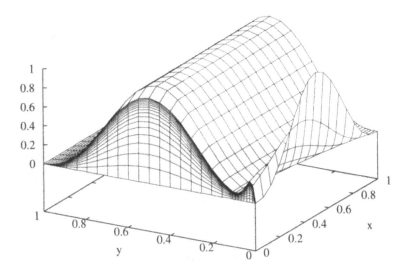

FIG. 6.8. Exact solution generated by Method 6.2 applied to problem (6.18) for $\varepsilon = 10^{-6}$ and with $N = 32$.

$$a\frac{\partial v_0}{\partial x} = f(x,y), \quad x > 0$$
$$v_0(0,y) = g(0,y)$$

(ii) $w(X_1, y; \varepsilon)$, $X_1 = (1-x)/\varepsilon$, is the regular boundary layer function near the outflow side Γ_R, which is the solution of the two–point boundary value problem

$$\frac{\partial^2 w}{\partial X_1^2} + a\frac{\partial w}{\partial X_1} = 0 \quad X_1 \in (0, \infty)$$
$$w(0, y) = g(1, y) - v_0(1, y), \quad w(X_1, y) \to 0 \text{ as } X_1 \to \infty$$

(iii) $v(X, Y; \varepsilon)$, $X = x/\varepsilon, Y = y/\varepsilon$, is an elliptic boundary layer function defined in a neighbourhood of the characteristic edge Γ_B near the inflow corner C_{BL}. It satisfies the elliptic differential equation

$$-\frac{\partial^2 v}{\partial X^2} - \frac{\partial^2 v}{\partial Y^2} + a\frac{\partial v}{\partial X} = 0 \quad (X, Y) \in (0, \infty) \times (0, \infty)$$

with the boundary conditions

$$v(0, Y_1; \varepsilon) = 0, \quad v(X, Y_1; \varepsilon) \to 0 \text{ as } X^2 + Y^2 \to \infty$$

$$\text{and} \quad v(X, 0; \varepsilon) = \varepsilon X \big(\frac{\partial g}{\partial y}(0,0) - \frac{\partial v_0}{\partial x}(0,0) \big)$$

(iv) $z(x, \tilde{Y}; \varepsilon), \tilde{Y} = y/\sqrt{\varepsilon}$, is a parabolic boundary layer function near the characteristic edge Γ_B. It satisfies the parabolic differential equation

$$-\frac{\partial^2 z}{\partial \tilde{Y}^2} + a\frac{\partial z}{\partial x} = 0, \quad x > 0, \quad \tilde{Y} \in (0, \infty)$$

with the boundary conditions

$$z(0, \tilde{Y}) = 0, \quad z(x, 0) = g(x, 0) - v_0(x, 0) - v(x, 0)$$

$$\text{and} \quad z(x, \tilde{Y}) \to 0 \text{ as } \tilde{Y} \to \infty$$

(v) $W(X_1, \tilde{Y}; \varepsilon), X_1 = (1 - x)/\varepsilon, \tilde{Y} = y/\sqrt{\varepsilon}$, is an ordinary corner boundary layer function near the outflow corner C_{BR}. It is the solution of the two–point boundary value problem

$$\frac{\partial^2 W}{\partial X_1^2} + a\frac{\partial W}{\partial X_1} = 0, \quad X_1 \in (0, \infty)$$

$$W(0, \tilde{Y}) = -z(1, \tilde{Y}), \quad W(X_1, \tilde{Y}) \to 0 \text{ as } X_1 \to \infty$$

(vi) $V(X_1, Y; \varepsilon), \quad X_1 = (1 - x)/\varepsilon, \quad Y = y/\varepsilon$, is an elliptic corner boundary layer function near the outflow corner C_{BR}, which satisfies the elliptic differential equation

$$-\frac{\partial^2 V}{\partial X_1^2} - \frac{\partial^2 V}{\partial Y^2} - a\frac{\partial V}{\partial X_1} = 0, \quad (X_1, Y) \in (0, \infty) \times (0, \infty)$$

with the boundary conditions

$$V(0, Y; \varepsilon) = -v(1/\varepsilon, Y; \varepsilon), \quad v(X_1, Y; \varepsilon) \to 0 \text{ as } X_1^2 + Y^2 \to \infty$$

$$\text{and} \quad v(X_1, 0; \varepsilon) = -w(X_1, 0) - W(X_1, 0).$$

The remaining boundary layer functions

$$v^T(X, Y_1^T; \varepsilon), \quad z^T(x, \tilde{Y}_1^T; \varepsilon), \quad W^T(X_1, \tilde{Y}_1^T; \varepsilon), \quad V^T(X_1, Y_1^T; \varepsilon)$$

are the analogous elliptic, parabolic, ordinary corner and elliptic corner boundary layer functions near the side Γ_T, where $Y_1^T = (1 - y)/\varepsilon$, $\tilde{Y}_1^T = (1 - y)/\sqrt{\varepsilon}$.

The complicated boundary layer structure in this asymptotic expansion is for the constant coefficient problem (6.21). For the variable coefficient problem (6.18) from Problem Class 6.1 we can expect that the analytic representation of its solution is at least as complicated, and probably more so. The same is likely to be the case for analytic representations of the derivatives of the solutions of such problems.

It is therefore remarkable that the monotone Method 6.2, consisting of a standard upwind finite difference operator and a simple fitted piecewise–uniform mesh, resolves all of the regular, parabolic and corner boundary layers present in the solutions of the singular perturbation problems from Problem Class 6.1. Moreover, the simple piecewise bilinear interpolants of the corresponding finite difference solutions (and their scaled discrete derivatives), given by Method 6.2, give ε–uniform pointwise accurate approximations to the solutions (and their scaled derivatives) of these problems over the whole domain. We remark also that these piecewise bilinear interpolants provide simple analytic approximations to the functions u_ε, and to their scaled derivatives. This is particularly noteworthy because the u_ε and their scaled derivatives have complicated boundary layer structures.

CHAPTER 7

Convection–diffusion problems with no slip boundary conditions

7.1 No–slip boundary conditions

In this chapter we again consider convection–diffusion problems for a substance in a fluid, which is flowing parallel to a wall at some points of the boundary of the domain. But, in contrast to the situation in the previous chapter, in this case the wall is not frictionless, and so we assume that on the wall the velocity of the fluid relative to the wall is zero. Therefore, if $\mathbf{v}$ denotes the relative velocity profile, we have $\mathbf{v} = \mathbf{0}$ at each point of the wall (see Fig. 7.1). This is the no slip boundary condition and its physical relevance is confirmed by extensive experimental observations. If the velocity profile $\mathbf{v}$ satisfies the no slip condition at a point on the boundary, then the tangential component of the velocity vanishes at that point on the boundary, and the corresponding parabolic boundary layer is said to be degenerate at that point. Suppose now that $\mathbf{x}^*$ is a point on the boundary, at which there is a degenerate parabolic boundary layer. Then $\mathbf{v}(\mathbf{x}^*) = \mathbf{0}$ and consequently $\mathbf{v}$ has the form $\mathbf{v}(x) = \mathbf{a}(x)|\mathbf{x} - \mathbf{x}^*|^\beta$, where each component of $\mathbf{a}(\mathbf{x}^*)$ is strictly positive and β is real and positive.

We are interested in flow in the unit channel Ω, such that the edges Γ_L and Γ_R are respectively the inflow and outflow boundaries, while the edges Γ_B and Γ_T are the walls on which the no slip condition is satisfied. Comparing experimentally

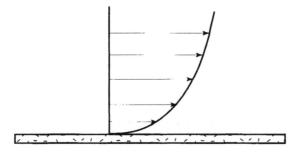

FIG. 7.1. Velocity profile for flow parallel to a wall with a no–slip boundary condition at the wall.

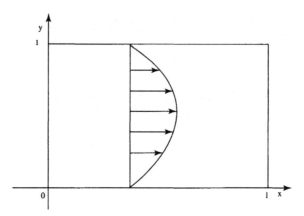

FIG. 7.2. Laminar velocity profile $\mathbf{v}(x,y) = (y^\beta(1-y)^\beta, 0)$ for $\beta = 1$

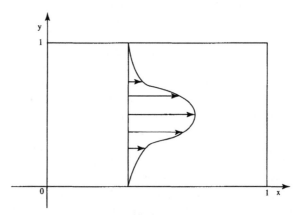

FIG. 7.3. Separating velocity profile $\mathbf{v}(x,y) = (y^\beta(1-y)^\beta, 0)$ for $\beta > 1$

determined velocity profiles of flows in such channels with the velocity profiles of the form $\mathbf{v}(\mathbf{x}) = (y^\beta(1-y)^\beta, 0)$, illustrated in Figs. 7.2–7.4, we conclude that values $\beta < 1$ give a shape similar to the velocity profile of a turbulent boundary layer, $\beta = 1$ similar to that of a laminar boundary layer and $\beta > 1$ similar to that of a separating boundary layer.

In this chapter we examine model problems with these three kinds of degenerate boundary layers in their solutions. We show that a monotone numerical method based on a standard upwind finite difference operator and a piecewise–uniform fitted mesh is ε–uniform for a typical problem from a wide problem class. We also show, through numerical experiments, that it is essential to use the correct expression for the width of the boundary layer in the construction of the mesh. Problems with these three kinds of degenerate parabolic boundary

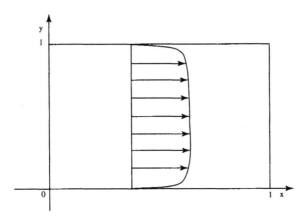

FIG. 7.4. Turbulent velocity profile $\mathbf{v}(x,y) = (y^\beta(1-y)^\beta, 0)$ for $\beta < 1$

layers are included in the following problem class, which also contains more general degenerate parabolic boundary layers on the walls Γ_B and Γ_T corresponding to the velocity profiles $\mathbf{v}(x,y) = (y^{\beta_1}(1-y)^{\beta_2}, 0)$.

Problem Class 7.1. Linear convection–diffusion in two–dimensions with degenerate parabolic boundary layers.

$$\varepsilon \Delta u_\varepsilon + y^{\beta_1}(1-y)^{\beta_2} \frac{\partial u_\varepsilon}{\partial x} = f, \quad (x,y) \in \Omega \tag{7.1a}$$

$$\beta_1 > 0, \quad \beta_2 > 0 \tag{7.1b}$$

$$\frac{\partial u_\varepsilon}{\partial x} = \psi, \quad (x,y) \in \Gamma_L \tag{7.1c}$$

$$u_\varepsilon = \phi, \quad (x,y) \in \Gamma_R \cup \Gamma_B \cup \Gamma_T \tag{7.1d}$$

$$f \in C^{0,\alpha}(\bar{\Omega}), \quad \phi, \psi \in C^\alpha(\Gamma), \quad \alpha > 0 \tag{7.1e}$$

$$\sup_{(x,y) \in \bar{\Omega}} \left| \frac{f(x,y)}{y^{\beta_1}(1-y)^{\beta_2}} \right| \le C \tag{7.1f}$$

The parabolic boundary layers in the solutions of problems in Problem Class 7.1 are all degenerate, because the coefficient of the first derivative term in the differential equation vanishes at the bottom and top edges Γ_B and Γ_T of Ω. Furthermore, from the discussion above, it is clear that for various choices of the parameters β_1 and β_2 the solutions of such problems mimic some of the features of real flows. Note that the restriction (7.1f) on the inhomogeneous term guarantees that the solution u_ε is ε–uniformly bounded on $\bar{\Omega}$.

7.2 Width of degenerate parabolic boundary layers

In order to construct a piecewise–uniform fitted mesh, for problems from Problem Class 7.1, we need to know the width of each of the boundary layers in the solution u_ε on the edges Γ_B and Γ_T. We illustrate how to find the width of the boundary layer on the edge Γ_B. We stress that it is not necessary to develop a complete asymptotic expansion of u_ε; in fact it suffices to determine only its first two terms. We refer the reader to Tobiska (1976) for a complete asymptotic expansion for the solution of a problem with degenerate parabolic boundary layers. We write simply

$$u_\varepsilon = v_0 + w_{0,B} + \quad \text{other terms}$$

where the leading term v_0 is the solution of the reduced problem

$$y^{\beta_1}(1-y)^{\beta_2}\frac{\partial v_0}{\partial x} = f(x,y), \quad (x,y) \in \Omega$$
$$v_0 = \phi, \quad (x,y) \in \Gamma_R$$

and the parabolic boundary layer function $w_{0,B}$, associated with the edge Γ_B, is defined to be the solution of the degenerate parabolic problem

$$\frac{\partial^2 \tilde{w}_{0,B}}{\partial \eta^2} - \eta^{\beta_1}\frac{\partial \tilde{w}_{0,B}}{\partial x_1} = 0, \quad (\eta, x_1) \in (0,\infty) \times (a,b]$$
$$\tilde{w}_{0,B}(0,x_1) = 0 \quad, \quad \tilde{w}_{0,B}(\infty,x_1) = 0$$
$$\tilde{w}_{0,B}(\eta,a) = v_0(a,\eta), \quad 0 < a < b < 1$$

where $x_1 = 1 - x$, $\eta = y/\varepsilon^\gamma$ is a stretched variable and the value of γ has to be determined. To find the correct choice of γ, we transform equation (7.1a) using the change of variable $(x,y) \to (x,\eta)$ where $\eta = y/\varepsilon^\gamma$. This gives the equation

$$\varepsilon^{1-2\gamma}\frac{\partial^2 u(x,y(\eta))}{\partial \eta^2} + \varepsilon^{\beta_1\gamma}\eta^{\beta_1}(1-\varepsilon^\gamma\eta)^{\beta_1}\frac{\partial u(x,y(\eta))}{\partial x} = f(x,y(\eta)) - \varepsilon\frac{\partial^2 u(x,y(\eta))}{\partial x^2}$$

from which it is easy to see that the two terms on the left hand side are of the same order in ε, provided that we choose γ so that $1-2\gamma = \beta_1\gamma$ or, equivalently, $\gamma = \frac{1}{2+\beta_1}$. This shows that the width of the corresponding degenerate parabolic boundary layer on the edge Γ_B is of order

$$\varepsilon^{\frac{1}{2+\beta_1}}. \tag{7.2}$$

Applying an analogous technique on Γ_T enables us to determine the width of the degenerate boundary layer there. This information enables us to construct a suitable piecewise–uniform fitted mesh in the next section.

7.3 Monotone fitted mesh method

To construct a monotone ε–uniform fitted mesh method for Problem Class 7.1 we use the standard upwind finite difference operator on a simple piecewise–uniform fitted mesh. Because the domain is a rectangle and the two boundary layers are on edges parallel to just one of the coordinate axes, we can take the piecewise–uniform fitted mesh to be the tensor product of a uniform mesh in one dimension and a piecewise–uniform fitted mesh in the other. The precise definition of the numerical method is the following.

Method 7.1. Upwind finite difference operator on piecewise–uniform fitted mesh for Problem Class 7.1.

$$[\varepsilon(\delta_x^2 + \delta_y^2) + y_j^{\beta_1}(1 - y_j)^{\beta_2} D_x^+]U_\varepsilon(x_i, y_j) = f(x_i, y_j), \quad (x_i, y_j) \in \Omega_\varepsilon^N \quad (7.3a)$$

$$U_\varepsilon = \phi, \quad (x_i, y_j) \in \Gamma \setminus \Gamma_{\mathrm{L}}, \quad D_x^+ U_\varepsilon = \psi, \quad (x_i, y_j) \in \Gamma_{\mathrm{L}} \quad (7.3b)$$

$$\overline{\Omega}_\varepsilon^N = \overline{\Omega}_u^N \times \overline{\Omega}_\sigma^N \quad (7.3c)$$

$$\overline{\Omega}_u^N = \{x_i = i/N : 0 \leq i \leq N\}, \quad \text{and} \quad \overline{\Omega}_\sigma^N = \{y_j : 0 \leq j \leq N\} \quad (7.3d)$$

with

$$y_j = \begin{cases} 4j\sigma_1/N, & \text{for } 0 \leq j \leq N/4 \\ \sigma_1 + 2(j - (N/4))(1 - \sigma_1 - \sigma_2)/N, & \text{for } N/4 \leq j \leq 3N/4 \\ 1 - \sigma_2 + 4(j - (3N/4))\sigma_2/N, & \text{for } 3N/4 \leq j \leq N \end{cases} \quad (7.3e)$$

and the transition parameters

$$\sigma_1 = \min\{1/4, \varepsilon^{\frac{1}{2+\beta_1}} \ln N\}, \quad \sigma_2 = \min\{1/4, \varepsilon^{\frac{1}{2+\beta_2}} \ln N\}. \quad (7.3f)$$

In an analogous fashion to Shishkin (1991a, 1992a, 2000), it can be shown that Method 7.1 applied to Problem Class 7.1 satisfies the following theoretical ε–uniform error estimate for all $N \geq N_0$ and for all $\beta_1, \beta_2 > 0$

$$\sup_{0 < \varepsilon \leq 1} \|\overline{U}_\varepsilon - u_\varepsilon\|_{\overline{\Omega}} \leq CN^{-p}, \quad p > 0, \quad (7.4)$$

where the constants N_0, p and C are independent of ε and N.

In the next section we apply Method 7.1 to the solution of specific problems from Problem Class 7.1, for various values of β_1 and β_2, and we derive computed ε–uniform orders of convergence which are significantly larger than the theoretical orders. We also study the adverse effect of using an incorrect expression for the width of the degenerate parabolic boundary layer in the construction of the piecewise–uniform fitted mesh in Method 7.1.

7.4 Numerical results

We apply Method 7.1 to problems from Problem Class 7.1 of the particular form

$$\varepsilon \Delta u_\varepsilon + y^{\beta_1}(1-y)^{\beta_2}\frac{\partial u_\varepsilon}{\partial x} = xy^{\beta_1}(1-y)^{\beta_2}, \quad (x,y) \in \Omega \qquad (7.5a)$$

$$u_\varepsilon(1,y) = 16y^2(1-y)^2 \quad \text{on } \Gamma_R, \quad \frac{\partial u_\varepsilon}{\partial x}(0,y) = 8 \quad \text{on } \Gamma_L, \qquad (7.5b)$$

$$u_\varepsilon(x,0) = 0 \quad \text{on } \Gamma_B, \quad u_\varepsilon(x,1) = 0 \quad \text{on } \Gamma_T. \qquad (7.5c)$$

for various choices of the parameters β_1 and β_2. These choices of β_1 and β_2 include the three kinds of degenerate parabolic boundary layers discussed in §7.1. The resulting computed orders of ε–uniform convergence p^N for different values of N, over the range $\varepsilon = 1$ to 2^{-30}, corresponding to various choices of β_1 and β_2 are given in Table 7.1.

The numerical values of p^N in Table 7.1 demonstrate experimentally that Method 7.1 is ε–uniform of first order for problem (7.5) for each choice of β_1 and β_2 in this table. This shows that a monotone method, using an upwind finite difference operator on a piecewise–uniform fitted mesh, gives ε–uniform accurate approximations to the solutions of problems containing a wide range of degenerate parabolic boundary layers.

In the remainder of this section we show that in the definition of the mesh it is essential to use the correct expression for the width of the degenerate parabolic boundary layer. We do this by studying a numerical method with a fitted piecewise–uniform mesh, which is based on an incorrect expression for this width. We show experimentally that the corresponding monotone numerical method is not ε–uniform. We obtain this piecewise–uniform mesh by replacing the correct expressions for the transition parameters σ_1 and σ_2 (7.3g) in Method 7.1 by the simpler, but incorrect, definitions

$$\sigma_1 = \sigma_2 = \min[1/4, \varepsilon^{\frac{1}{2}} \ln N] \qquad (7.6)$$

These correspond to the non–degenerate parabolic boundary layer width used in Method 6.2 for problems from Problem Class 6.1. The resulting variant of

Table 7.1 *Computed order of ε–uniform convergence p^N generated by Method 7.1 applied to problem (7.5) for various values of β_1 and β_2.*

β_1	β_2	$N = 8$	16	32	64	128
1.0	1.0	1.07	1.04	1.02	1.01	1.01
0.5	0.5	1.10	1.05	1.03	1.01	1.01
0.1	0.1	1.07	1.05	1.03	1.01	1.01
2.0	2.0	1.07	1.05	1.02	1.01	1.01
4.0	4.0	1.09	1.05	1.03	1.01	1.01
10.0	10.0	1.67	1.11	1.22	1.01	1.01

Table 7.2 *Computed maximum pointwise error $E_{\varepsilon,\text{comp}}^N$ generated by a variant of Method 7.1, with an incorrect choice of σ_1 and σ_2, applied to problem (7.5) with $\beta_1 = 1$ and $\beta_2 = 1$, for various values of N and ε.*

ε	\multicolumn{6}{c}{Number of intervals N}					
	8	16	32	64	128	256
1	0.235D-01	0.960D-02	0.422D-02	0.188D-02	0.785D-03	0.258D-03
2^{-2}	0.289D-01	0.119D-01	0.524D-02	0.234D-02	0.984D-03	0.330D-03
2^{-4}	0.506D-01	0.218D-01	0.987D-02	0.444D-02	0.185D-02	0.597D-03
2^{-6}	0.857D-01	0.398D-01	0.187D-01	0.863D-02	0.367D-02	0.122D-02
2^{-8}	0.988D-01	0.443D-01	0.205D-01	0.942D-02	0.401D-02	0.133D-02
2^{-10}	0.747D-01	0.352D-01	0.166D-01	0.769D-02	0.328D-02	0.110D-02
2^{-12}	0.668D-01	0.348D-01	0.161D-01	0.725D-02	0.304D-02	0.101D-02
2^{-14}	0.629D-01	0.346D-01	0.190D-01	0.793D-02	0.324D-02	0.106D-02
2^{-16}	0.824D-01	0.590D-01	0.308D-01	0.137D-01	0.395D-02	0.122D-02
2^{-18}	0.106D+00	0.847D-01	0.524D-01	0.243D-01	0.104D-01	0.230D-02
2^{-22}	0.141D+00	0.127D+00	0.942D-01	0.577D-01	0.324D-01	0.163D-01
2^{-26}	0.163D+00	0.155D+00	0.126D+00	0.886D-01	0.536D-01	0.367D-01
2^{-30}	0.177D+00	0.173D+00	0.147D+00	0.111D+00	0.751D-01	0.447D-01
2^{-34}	0.186D+00	0.185D+00	0.161D+00	0.127D+00	0.906D-01	0.588D-01
2^{-38}	0.191D+00	0.192D+00	0.170D+00	0.137D+00	0.101D+00	0.687D-01
2^{-42}	0.195D+00	0.197D+00	0.175D+00	0.143D+00	0.108D+00	0.754D-01
2^{-46}	0.197D+00	0.199D+00	0.179D+00	0.147D+00	0.112D+00	0.798D-01
2^{-50}	0.198D+00	0.201D+00	0.181D+00	0.150D+00	0.115D+00	0.826D-01
2^{-54}	0.199D+00	0.202D+00	0.183D+00	0.151D+00	0.117D+00	0.843D-01
2^{-58}	0.200D+00	0.203D+00	0.183D+00	0.152D+00	0.118D+00	0.855D-01
2^{-62}	0.200D+00	0.204D+00	0.184D+00	0.153D+00	0.118D+00	0.862D-01
2^{-68}	0.200D+00	0.204D+00	0.184D+00	0.153D+00	0.119D+00	0.866D-01

Method 6.2 is then applied to solve problem (7.5) from Problem Class 7.1, again in the case $\beta_1 = \beta_2 = 1$. The computed maximum pointwise errors E_ε^N are displayed in Table 7.2. The results indicate that the method is not ε–uniform for this problem. In particular we remark that in Table 7.2 the computed maximum pointwise errors E_ε^N increase as $\varepsilon \to 0$.

7.5 Slip versus no–slip

In this final section we examine the relation between slip and no–slip boundary conditions and the widths of the corresponding boundary layers. To do so we compare the exact solutions of the following two similar looking linear convection–diffusion problems.

$$\varepsilon \Delta u_\varepsilon + (y + 0.1)^2 \frac{\partial u_\varepsilon}{\partial x} = y^2 \quad (x, y) \in \Omega \tag{7.7a}$$

$$u_\varepsilon(x\,, 0) = u_\varepsilon(x, 1) = 0, \quad u_\varepsilon(1, y) = 16y^2(1 - y)^2, \quad \frac{\partial u_\varepsilon}{\partial x}(0, y) = 8 \tag{7.7b}$$

and

$$\varepsilon \Delta u_\varepsilon + y^2 \frac{\partial u_\varepsilon}{\partial x} = y^2, \quad (x,y) \in \Omega \tag{7.8a}$$

$$u_\varepsilon(x,0) = u_\varepsilon(x,1) = 0, \quad u_\varepsilon(1,y) = 16y^2(1-y)^2, \quad \frac{\partial u_\varepsilon}{\partial x}(0,y) = 8 \tag{7.8b}$$

The only difference between these problems is the small difference in the coefficient of the first derivative term in the differential equation. With this choice of coefficients it is clear that on the boundary Γ_B the solution of problem (7.7) satisfies a slip boundary condition, while the solution of problem (7.8) satisfies a no–slip boundary condition. The solutions of both problems have a parabolic boundary layer on Γ_B, which in the case of problem (7.7) is non–degenerate, and for problem (7.8) is degenerate. We use the expressions $\varepsilon^{1/2}$ and $\varepsilon^{1/4}$ from (6.17h) and (7.2) respectively for the width of the parabolic boundary layer in the non–degenerate and degenerate cases. We conclude that the widths of the boundary layers in the solutions of these two problems are radically different, because if, for example, $\varepsilon = 10^{-12}$ then the width of the non–degenerate parabolic boundary layer in problem (7.7) is $O(10^{-6})$, while the width of the degenerate parabolic boundary layer in problem (7.8) is $O(10^{-3})$. That is, the boundary layer is wider by a factor of $1,000$ with the no–slip boundary condition compared with the slip boundary condition. This suggests that the thinner non–degenerate parabolic boundary layers in the solutions of problems with slip boundary conditions may be significantly harder to resolve numerically, than the wider degenerate parabolic boundary layers in problems with no–slip boundary conditions. Moreover, the appropriate meshes to use in the construction of ε–uniform numerical methods are radically different for these two kinds of problems. They have the same form but the transition parameters differ by several orders of magnitude. We observe the significant differences between the graphs of the exact solutions displayed in Fig. 7.5 and Fig. 7.6, especially near the outflow corner C_{BL}. In Fig. 7.5, where there is a slip condition on the boundary Γ_B, the inflow profile specified on Γ_R is not significantly changed as we approach the corner C_{BL} along the outflow boundary Γ_L. On the other hand, in Fig. 7.6 there is a significant change in the solution as we move from the inflow edge Γ_R to the outflow edge Γ_L, especially near the corner C_{BL}. The principal point of these examples is to show that a small disturbance of the coefficient of the differential equation can cause a large change in the physical boundary layer, and, furthermore, that this difference between the solutions of these two problems can be detected numerically, if we use ε–uniform numerical methods to solve the problems.

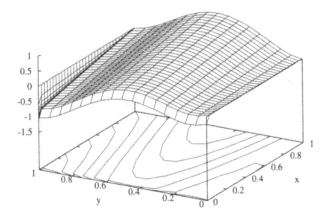

FIG. 7.5. Surface and contour plots of the solution of problem (7.7) for $\varepsilon = 10^{-6}$ showing the non–degenerate parabolic boundary layer on Γ_B.

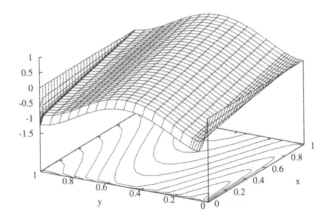

FIG. 7.6. Surface and contour plots of the solution of problem (7.8) for $\varepsilon = 10^{-6}$ showing the degenerate parabolic boundary layer on Γ_B.

CHAPTER 8

Experimental estimation of errors

8.1 Theoretical error estimates

Numerical methods for finding approximate solutions of partial differential equations are usually applied to problems for which the exact solution is unknown. However, the key step of estimating the errors in the resulting numerical approximations is often glossed over. The standard criterion for assessing the quality of a numerical method for solving problems from a class of non–singularly perturbed problems is a theoretical error estimate of the following form: there exist positive constants N_0, $C = C(N_0)$ and $p = p(N_0)$, all independent of N, such that for all $N \geq N_0$

$$\|U^N - u\|_{\Omega^N} \leq CN^{-p} \tag{8.1}$$

where u is the exact solution of any problem in the problem class, U^N is a numerical approximation on the mesh Ω^N and N is the number of mesh points in each coordinate direction of the discrete problem associated with the numerical method. The maximum pointwise error $\|U^N - u\|_{\Omega^N}$ of the numerical method on the mesh Ω^N is bounded above by the error bound CN^{-p}, which is determined, for a given N, by the error constant C and the order of convergence p. The quantities p and C are called the error parameters.

A common maxim for the ranking of numerical methods is: the higher the order of convergence p the better the numerical method. Therefore, for a particular numerical method, numerical analysts work hard to establish a sharp estimate of p, while the error constant C is normally left unspecified. It is clear that without an estimate of the error constant C, an error bound of this form gives information only about the rate of convergence of the numerical solutions. In other words, it tells us that if we use more mesh points then the error decreases like N^{-p}. However, if the value of the error constant C is unknown, we cannot compute the magnitude of the right hand side of (8.1) and, consequently, we have no estimate of the magnitude of the errors in our numerical approximations.

To see that the matter is not trivial, suppose that, for example, the following theoretical pointwise error bounds are known to be valid for some numerical method applied to a particular problem:

$$C_1 N^{-p_1},\ C_2 N^{-p_2},\ C_3 N^{-p_3}, \tag{8.2a}$$

where $\quad C_1 < C_2 < C_3 \quad$ and $\quad p_1 < p_2 < p_3. \tag{8.2b}$

It is not immediately obvious which of these bounds should be used. But an examination of the graphs of three such functions in Fig. 8.1 shows that the best bound to use depends on the range of values of N involved, because for different ranges of N different error bounds are better, in the sense that they are smaller in size over the relevant range. One consequence of this is that, for a given value of N, a second order ($p = 2$) error bound is not necessarily sharper than a first order ($p = 1$) error bound, because the error bound depends not only on the order p but also on the error constant C.

Therefore to choose a good error bound for a given range of values of N, it is helpful to examine the graphs of the available error bounds. In practice, of course, this cannot be done precisely for several reasons. For example, a sharp estimate of the order of convergence p may not be known, because of technical difficulties in proving the theoretical result. Furthermore, such theoretical results are established for a class of problems, while for a specific problem the order of convergence may be higher. An even more important reason is that theoretical estimates of the error constant C are rarely available, and when available they are usually wildly pessimistic.

It is worth noting that if $\|u\|_{\Omega^N} \leq C_1$ and $\|U^N\|_{\Omega^N} \leq C_1$, then the constraint $N \geq N_0$ in (8.1) is not a practical restriction, since in this case, for all N satisfying $1 \leq N \leq N_0$, we have

$$\|U^N - u\|_{\Omega^N} \leq 2C_1 \leq 2C_1 \frac{N_0^p}{N^p} = C' N^{-p}$$

where $C' = 2C_1 N_0^p$. Combining this inequality with (8.1) we obtain, for all $N \geq 1$, an error bound of the same form as (8.1), but with the new error constant $C_{new} = \max\{C, C'\}$.

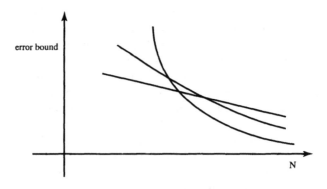

FIG. 8.1. Error bound CN^{-p} as a function of N for various values of p.

Note also that, given (8.1), we can always obtain a new error bound of the same form with a smaller error constant but a lower order of convergence, since for each p' satisfying $0 \leq p' < p$ we have

$$\|U^N - u\|_{\Omega^N} \leq CN^{-p} = CN^{-p'}N^{-(p-p')} < C'N^{-p'}$$

where $C' = \sup_N CN^{-(p-p')} < C$. In both of the above cases the new error constant depends on p, which shows that it is more appropriate to write error estimates in the form

$$\|U^N - u\|_{\Omega^N} \leq C_p N^{-p} \tag{8.3}$$

than in the original form (8.1), where now the dependence of the error constant on the order of convergence p is explicitly indicated in the notation.

When a numerical method is applied to a specific problem with a known exact solution u, we define the order of local convergence for the value N by

$$p_{\text{exact}}^N = \log_2 \frac{\|U^N - u\|_{\Omega^N}}{\|U^{2N} - u\|_{\Omega^{2N}}} \tag{8.4}$$

and the asymptotic order of convergence by

$$p_{\text{exact}} = \lim_{N \to \infty} p_{\text{exact}}^N. \tag{8.5}$$

The theoretical order of convergence is the value p in (8.1) and (8.3), which is assumed to be valid for a whole problem class. It may be much smaller than the asymptotic order of convergence p_{exact} in (8.5) for a particular problem in this problem class. Moreover, for a specific problem, the asymptotic order of convergence can be quite different from the order of local convergence for a given value of N. The order of local convergence is also, in general, different from both the asymptotic order of convergence and the theoretical order of convergence. This is illustrated in Fig. 8.2, where the theoretical error bound is larger than the error bound based on the order of local convergence and the same is true of the error bound based on the asymptotic rate of convergence for small values of N.

In practice, for a specific computation, the values of N lie in a finite range $R_N = \{N : \underline{N} \leq N \leq \overline{N}\}$, and so we want to choose p and its associated C_p in such a way that, for all $N \in R_N$, the resulting error bound $C_p N^{-p}$ is as small as possible. In other words, we want to obtain error bounds that are both practical and useful for the range of N involved. Of course an error bound of the form (8.3) gives only an upper–bound for the error. In practice, the actual error for a particular problem could, for example, have the form shown in Fig. 8.2.

The main point of the present chapter is to describe a number of computational techniques for determining realistic estimates of p and C_p and to use

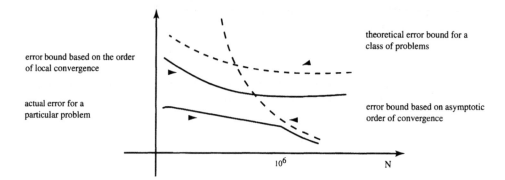

F IG. 8.2. Local, asymptotic and theoretical error bounds and actual errors as
    functions of N.

them to estimate the maximum pointwise error in numerical approximations to
the solution and its scaled derivatives of the singular perturbation problems considered in this book. For non–singularly perturbed problems the error bound
$C_p N^{-p}$ is a function of the three variables p, N and C_p, but for a singularly
perturbed problem an extra variable is added, because it is essential to study
the behaviour of the numerical method for a range of values $R_\varepsilon = [\underline{\varepsilon}, \overline{\varepsilon}]$ of the
singular perturbation parameter ε. It is clear that the actual maximum pointwise
error $\|U_\varepsilon^N - u_\varepsilon\|_{\Omega^N}$ in the numerical solution of a singularly perturbed problem
depends on the two variables N and ε. If the numerical method is ε–uniform for
the particular problem being solved, then the ε–uniform nature of this dependence on these variables is of the form shown in Fig. 8.3. On the other hand, for
any standard non–ε–uniform numerical method, the error constant C_p increases
as ε decreases. Worse still, given a range R_N, for such numerical methods there
are always values of ε for which the errors grow as N increases, that is the errors become larger as the mesh is refined. This is the fundamental reason for
constructing ε–uniform numerical methods, because such behaviour is normally
regarded as unacceptable.

    Throughout this book, in order to estimate the error in the numerical solution
of a singularly perturbed problem, we assume that either the numerical method
is ε–uniform, of some positive but unknown order or approximate values of the
solution at all points of the domain can be determined to any prescribed accuracy.
That is, we require knowledge of the existence of a theoretical parameter–uniform
pointwise global error bound of the following form: there exist positive constants
N_0, $\bar{C}_{\bar{p}} = \bar{C}_{\bar{p}}(N_0)$ and $\bar{p} = \bar{p}(N_0)$, all independent of N and ε, such that for all
$N \geq N_0$

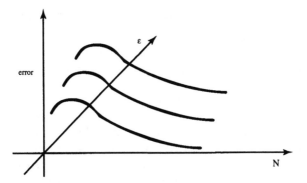

FIG. 8.3. Maximum pointwise error $E_\varepsilon^N = \|U_\varepsilon^N - u_\varepsilon\|_{\overline{\Omega}}$ as a function of N and ε for an ε–uniform method.

$$\sup_{0<\varepsilon\leq 1} \|\overline{U}_\varepsilon^N - u_\varepsilon\|_{\overline{\Omega}} \leq \bar{C}_{\bar{p}} N^{-\bar{p}} \tag{8.6}$$

where $\overline{U}_\varepsilon^N$ is the piecewise linear interpolant of the numerical solution U_ε^N on Ω^N to $\overline{\Omega}$. Note that here we take the maximum norm over the whole domain $\overline{\Omega}$ and not just over the mesh Ω^N, as is usually the case for standard numerical methods. It is clear that if the global estimate (8.6) holds, then *a fortiori* at the mesh points we have the estimate

$$\sup_{0<\varepsilon\leq 1} \|U_\varepsilon^N - u_\varepsilon\|_{\overline{\Omega}_\varepsilon^N} \leq \bar{C}_{\bar{p}} N^{-\bar{p}} \tag{8.7}$$

where $\bar{C}_{\bar{p}}$ and $\bar{p}$ are called the global error parameters and $\overline{\Omega}_\varepsilon^N$ is the mesh, which normally depends on both N and ε in the case of an ε–uniform method. It is worth pointing out that for many problems, for which the global error estimate (8.6) holds, it is possible to establish directly a sharper estimate than (8.7) of the form

$$\sup_{0<\varepsilon\leq 1} \|U_\varepsilon^N - u_\varepsilon\|_{\overline{\Omega}_\varepsilon^N} \leq C_p N^{-p}, \tag{8.8}$$

where $p \geq \bar{p}$ and $C_p \leq \bar{C}_{\bar{p}}$.

In what follows a family of computational algorithms of increasing sophistication is described. These algorithms give various *a posteriori* approximations to the values of the error parameters of C_p and p in (8.8) and of $\bar{C}_{\bar{p}}$ and $\bar{p}$ in (8.6). Realistic error bounds are then obtained from these by evaluating the right–hand side of (8.6) and (8.8). These bounds can be used in practice to estimate the error in a numerical solution for a given value of N, or to estimate the value of N required to achieve a specified accuracy. Examples of these uses are given in §8.5.

8.2   Quick algorithms for estimating the error parameters

In this section three simple algorithms are described, which provide quick computed approximations to the ε–uniform error parameters p and C_p when an ε–uniform numerical method is applied to a specific singularly perturbed problem.

It is assumed throughout that the numerical method is known to be ε–uniform in the sense of (8.6). In such circumstances usually only a rough estimate of the ε–uniform order of convergence p and no estimate of the ε–uniform error constant C_p are available. Our task here is to describe simple algorithms to obtain realistic computed approximations to the ε–uniform error parameters p and C_p which are valid for all values of ε and N in given ranges R_ε and R_N respectively.

More precisely, the situation considered is as follows. Suppose that a given numerical method is used to solve a given singularly perturbed problem on a sequence of piecewise uniform fitted meshes for several values of $N \in R_N$ and $\varepsilon \in R_\varepsilon$. Suppose also that the exact solution u_ε of the problem is unknown, which is the usual situation. This means that the error $U_\varepsilon^N - u_\varepsilon$ in the numerical solution U_ε^N cannot be determined directly, and therefore an indirect estimate of its magnitude is all that can be computed.

The first algorithm, which we now describe, is useful in cases where two numerical solutions can be computed on two different meshes for a single value of ε. It is assumed that the ε–uniform order of convergence is known to be approximately p. The algorithm then provides an approximate value for the error constant C_p in the following way. Let u_ε denote the exact solution of the problem and R_ε, R_N the ranges of ε and N respectively in which numerical solutions U_ε^N can be computed. Choose any convenient values $\varepsilon_1 \in R_\varepsilon$ and N_1 such that $N_1, 4N_1 \in R_N$. The first step is to use the numerical method to compute the two numerical solutions of the problem for the chosen value ε_1 on the two fitted piecewise uniform meshes $\Omega_{\varepsilon_1}^{N_1}$ and $\Omega_{\varepsilon_1}^{4N_1}$. Denoting these approximate solutions by $U_{\varepsilon_1}^{N_1}$ and $U_{\varepsilon_1}^{4N_1}$ respectively, we then compute the maximum pointwise two–mesh difference on the mesh $\Omega_{\varepsilon_1}^{N_1}$ between $U_{\varepsilon_1}^{N_1}$ and the piecewise linear interpolant $\overline{U}_{\varepsilon_1}^{4N_1}$, namely

$$D = \|U_{\varepsilon_1}^{N_1} - \overline{U}_{\varepsilon_1}^{4N_1}\|_{\Omega_{\varepsilon_1}^{N_1}}. \tag{8.9}$$

Using the triangle inequality and (8.6) we obtain

$$D \geq \|U_{\varepsilon_1}^{N_1} - u_{\varepsilon_1}\|_{\Omega_{\varepsilon_1}^{N_1}} - \|\overline{U}_{\varepsilon_1}^{4N_1} - u_{\varepsilon_1}\|_{\overline{\Omega}}$$

$$\geq \|U_{\varepsilon_1}^{N_1} - u_{\varepsilon_1}\|_{\Omega_{\varepsilon_1}^{N_1}} - C_p(4N_1)^{-p} \approx C_p(1 - 4^{-p})N_1^{-p}.$$

This motivates our taking

$$C_p^* = \frac{DN_1^p}{1 - 4^{-p}} \tag{8.10}$$

to be the computed approximation to the unknown ε–uniform error constant.

The second algorithm is applicable in cases when three numerical solutions can be computed on three different meshes for a single value of ε. It provides us with approximations of both the ε–uniform order of convergence p and the ε–uniform error constant C_p. A computed estimate of p is obtained first by computing the three numerical solutions $U_{\varepsilon_1}^{N_1}, U_{\varepsilon_1}^{2N_1}$ and $U_{\varepsilon_1}^{4N_1}$ on the meshes $\Omega_{\varepsilon_1}^{N_1}, \Omega_{\varepsilon_1}^{2N_1}$ and $\Omega_{\varepsilon_1}^{4N_1}$ respectively, where ε_1 and N_1 are chosen so that $\varepsilon_1 \in R_\varepsilon$ and $N_1, 2N_1, 4N_1 \in R_N$. Let $\overline{U}_{\varepsilon_1}^{2N_1}, \overline{U}_{\varepsilon_1}^{4N_1}$ denote the piecewise linear interpolants of $U_{\varepsilon_1}^{2N_1}, U_{\varepsilon_1}^{4N_1}$ on $\overline{\Omega}$. We then compute the maximum pointwise two–mesh differences

$$D_1 = \|U_{\varepsilon_1}^{N_1} - \overline{U}_{\varepsilon_1}^{2N_1}\|_{\Omega_{\varepsilon_1}^{N_1}}, \quad D_2 = \|U_{\varepsilon_1}^{2N_1} - \overline{U}_{\varepsilon_1}^{4N_1}\|_{\Omega_{\varepsilon_1}^{2N_1}}$$

and

$$D = \|U_{\varepsilon_1}^{N_1} - \overline{U}_{\varepsilon_1}^{4N_1}\|_{\Omega_{\varepsilon_1}^{N_1}}$$

on the appropriate meshes. Using the triangle inequality (8.6), and a similar argument to that leading to (8.10), we obtain

$$\frac{D_1}{D_2} \approx \frac{C_p N_1^{-p}(1 + 2^{-p})}{C_p(2N_1)^{-p}(1 - 2^{-p})} \approx 2^p$$

which motivates our taking

$$p^* = \log_2 \frac{D_1}{D_2}$$

to be the computed approximation to the unknown value of the ε–uniform order of convergence. Using this computed value of p and the above value of D we then apply the previous algorithm to obtain, from (8.10), the computed approximation

$$C_{p^*}^* = \frac{DN_1^{p^*}}{1 - 4^{-p^*}} \tag{8.11}$$

Without computing further numerical solutions of the problem, it is now possible to test the sensitivity, relative to changes in N, of the computed error constant $C_{p^*}^*$, produced by this second algorithm, in the following way. Using an analogous argument to that used to obtain (8.10) from (8.6), we see that each of the quantities

$$C_1^* = \frac{D_1 N_1^{p^*}}{1 - 2^{-p^*}} \quad \text{and} \quad C_2^* = \frac{D_2(2N_1)^{p^*}}{1 - 2^{-p^*}}$$

may be taken as computed approximates to the ε–uniform error constant. The values C_1^* and C_2^* may not be close to the value $C_{p^*}^*$ in (8.11), but, if $C_1^* \approx C_2^*$

then it can be concluded, as we see later in the discussion of Table 8.3, that p^* and $C^*_{p^*}$ are insensitive to variations in N between N_1 and $4N_1$.

Our third algorithm also provides us with approximates to p and C_p and allows us to test their robustness. In this case we must compute six numerical solutions corresponding to two different values of ε on three different meshes. We begin by computing the six numerical solutions $U^{N_1}_{\underline{\varepsilon}}, U^{2N_1}_{\underline{\varepsilon}}, U^{4N_1}_{\underline{\varepsilon}}, U^{N_1}_{\hat{\varepsilon}}, U^{2N_1}_{\hat{\varepsilon}}, U^{4N_1}_{\hat{\varepsilon}}$, where $N_1, 2N_1, 4N_1 \in R_N$ as before, $\underline{\varepsilon}, \hat{\varepsilon} \in R_\varepsilon$, and we construct the required piecewise linear interpolants. Then, for all $\varepsilon \in R_\varepsilon$, we define the maximum pointwise two–mesh differences

$$D_1(\varepsilon) = ||U^{N_1}_\varepsilon - \overline{U}^{2N_1}_\varepsilon||_{\Omega^{N_1}_\varepsilon} \quad D_2(\varepsilon) = ||U^{2N_1}_\varepsilon - \overline{U}^{4N_1}_\varepsilon||_{\Omega^{2N_1}_\varepsilon} \qquad (8.12)$$

and

$$D(\varepsilon) = ||U^{N_1}_\varepsilon - \overline{U}^{4N_1}_\varepsilon||_{\Omega^{N_1}_\varepsilon} \qquad (8.13)$$

on the appropriate meshes. We compute the ε–uniform maximum pointwise two–mesh differences

$$D_1 = \max\{D_1(\underline{\varepsilon}), D_1(\hat{\varepsilon})\}, \quad D_2 = \max\{D_2(\underline{\varepsilon}), D_2(\hat{\varepsilon})\}$$

and

$$D = \max\{D(\underline{\varepsilon}), D(\hat{\varepsilon})\}.$$

Using an analogous argument to that in the second algorithm, we obtain the computed approximations to the ε–uniform error parameters

$$p^* = \log_2 \frac{D_1}{D_2} \quad \text{and} \quad C^*_{p^*} = \frac{DN^{p^*}_1}{1 - 4^{-p^*}}. \qquad (8.14)$$

Calculating the quantities

$$C^*_1 = \frac{D_1 N^{p^*}_1}{1 - 2^{-p^*}}, \quad C^*_2 = \frac{D_2 (2N_1)^{p^*}}{1 - 2^{-p^*}}$$

we can expect, for the same reason as before, that if $C_1 \approx C_2$ then the values of p^* and $C^*_{p^*}$ are insensitive to changes in N for all $N \in R_N$. Without computing additional numerical solutions, it is also possible to test the robustness of p^* and $C^*_{p^*}$ with respect to $\varepsilon \in R_\varepsilon$. To do this we compute $D_1(\underline{\varepsilon}), D_2(\underline{\varepsilon})$ and $D(\underline{\varepsilon})$ from (8.12) and (8.13). Then, as in the second algorithm, we put

$$\underline{p} = \log_2 \frac{D_1(\underline{\varepsilon})}{D_2(\underline{\varepsilon})}$$

and

$$C_{\underline{p}} = \frac{D(\underline{\varepsilon})N^{\underline{p}}_1}{1 - 4^{-\underline{p}}}$$

Similarly, computing $D_1(\hat{\varepsilon}), D_2(\hat{\varepsilon})$ and $D(\hat{\varepsilon})$ we obtain the analogous quantities $\hat{p}$ and $C_{\hat{p}}$. The values $\underline{p}$ and $\hat{p}$ may not be close to the value p^* in (8.14). However, if $\underline{p} \approx \hat{p}$ then p^* is robust for all $\varepsilon \in R_\varepsilon$ and if $C_{\underline{p}} \approx C_{\hat{p}}$ then $C_{p^*}^*$ is robust for all $\varepsilon \in R_\varepsilon$.

We now use the three algorithms of this section to obtain computed approximations to the error parameters p and C_p for Method 6.2 applied to the two–dimensional convection–diffusion problem (6.18), which has both regular and parabolic boundary layers. We note that this numerical method is known to be ε–uniform for this problem with an ε–uniform order of convergence $p \approx 1$. We consider the solution for all values $\varepsilon \in R_\varepsilon = [2^{-8}, 2^{-1}]$ and we choose $\varepsilon_1 = 2^{-8}$. We assume that, in practice, we can compute numerical solutions for all values of $N \in R_N = [2^2, 2^9]$ and we choose $N_1 = 2^7$.

The maximum pointwise two–mesh difference (8.9) is $D = 0.0118$. Assuming now, for example, that Method 6.2 is ε–uniform of order $p = 0.5$, the first algorithm gives the corresponding error constant $C_{0.5}^* = 0.267$. With the second algorithm the three maximum pointwise two–mesh differences are $D_1 = 0.00756, D_2 = 0.00428$ and $D = 0.0118$, and so the corresponding computed error parameters are $p^* = 0.821$ and $C_{0.821}^* = 0.933$. The insensitivity with respect to N for $N \in R_N$ of this value of the error constant is a consequence of the fact that $|C_1^* - C_2^*| = |0.9356231 - 0.9357687| = 0.00015$ is small. For the third algorithm the six maximum pointwise two–mesh differences are $D_1(\underline{\varepsilon}) = 0.00823$, $D_2(\underline{\varepsilon}) = 0.00465$ and $D(\underline{\varepsilon}) = 0.0129$ and $D_1(\hat{\varepsilon}) = 0.00756$, $D_2(\hat{\varepsilon}) = 0.00428$ and $D(\hat{\varepsilon}) = 0.0118$. From these we calculate $\underline{p} = 0.824$, $C_{\underline{p}} = 1.032$ and $\hat{p} = 0.821$, $C_{\hat{p}} = 0.933$. The corresponding computed error parameters are $p^* = 0.824$ and $C_{0.824}^* = 1.032$. The insensitivity with respect to N, for all $N \in R_N$, of this value of the error constant is a consequence of the fact that $|C_1^* - C_2^*| = |1.0306837 - 1.0309254| = 0.00024$ is small. The robustness with respect to ε for all $\varepsilon \in R_\varepsilon$ of these values of both error parameters follows from the observation that the two quantities $|\underline{p} - \hat{p}| = 0.003$ and $|C_{\underline{p}} - C_{\hat{p}}| = 0.099$ are small. The above results are summarized in Table 8.1.

It is an obvious inference, from the discussion in this section, that if numerical solutions are computed on more than three meshes and for several values of ε, then it should be possible to obtain more accurate estimates of the ε–uniform error parameters. This is the topic of the next section.

Table 8.1 *Computed error parameters, sensitivity, and robustness criteria generated by the quick algorithms of §8.2 for the fitted mesh Method 6.2 applied to problem (6.18), with $N_1 = 2^7$ and $\varepsilon_1 = 2^{-8}$.*

Algorithm	p^*	$C_{p^*}^*$	$\lvert C_1^* - C_2^* \rvert$	$\lvert \underline{p} - \hat{p} \rvert$	$\lvert C_{\underline{p}} - C_{\hat{p}} \rvert$
First	0.5	0.267			
Second	0.821	0.933	0.00015		
Third	0.824	1.032	0.00024	0.003	0.099

8.3 General algorithm for estimating error parameters

A systematic technique for the determination of *a posteriori* approximations to the ε–uniform error constant C_p, and the ε–uniform order of convergence p, of a numerical method for solving a given singularly perturbed problem, is now discussed for cases when the numerical solutions can be computed for several values of N and ε. As in the previous section the arguments are heuristic rather than rigorous, which is typical for *a posteriori* techniques of this type.

We assume that, on the appropriate meshes Ω_ε^N, the piecewise linear inter-polants $\overline{U}_\varepsilon^N$ of the numerical solutions U_ε^N have been determined. Then, for all integers N satisfying $N, 2N \in R_N$ and for a finite set of values $\varepsilon \in R_\varepsilon$, we compute the maximum pointwise two–mesh differences

$$D_\varepsilon^N = \|U_\varepsilon^N - \overline{U}_\varepsilon^{2N}\|_{\Omega_\varepsilon^N}. \tag{8.15}$$

From these values the ε–uniform maximum pointwise two–mesh differences

$$D^N = \max_{\varepsilon \in R_\varepsilon} D_\varepsilon^N \tag{8.16}$$

are formed for each available value of N satisfying $N, 2N \in R_N$. Approximations to the ε–uniform order of local convergence are defined, for all $N, 4N \in R_N$, by

$$p^N = \log_2 \frac{D^N}{D^{2N}} \tag{8.17}$$

and we take the computed ε–uniform order of convergence to be

$$p^* = \min_N p^N. \tag{8.18}$$

Corresponding to this value of p^* we calculate the quantities

$$C_{p^*}^N = \frac{D^N N^{p^*}}{1 - 2^{-p^*}} \tag{8.19}$$

and we take the computed ε–uniform error constant to be

$$C_{p^*}^* = \max_N C_{p^*}^N. \tag{8.20}$$

The results of the above procedure are summarized in Table 8.2 for the two–dimensional convection–diffusion problem (6.18) and the numerical Method 6.2. The resulting computed estimates of p and C_p are $p^* = 0.76$ from (8.18) and $C_{0.76}^* = 0.897$ from (8.20). These values should be compared with those in Table 8.1, obtained by the three quick algorithms of the previous section.

The robustness for $\varepsilon \in R_\varepsilon$ and $N \in R_N$ of the computed values of p and C_p, given by (8.18) and (8.20), is now examined for problem (6.18) and Method 6.2

Table 8.2 *Values of $D_\varepsilon^N, D^N, p^N, p^*$ and $C_{p^*}^N$ generated by the general algorithm of §8.3 for the fitted mesh Method 6.2 applied to problem (6.18) for various values of ε and N.*

	Number of intervals N					
ε	8	16	32	64	128	256
1	0.117D-01	0.439D-02	0.174D-02	0.751D-03	0.345D-03	0.169D-03
2^{-2}	0.163D-01	0.847D-02	0.439D-02	0.225D-02	0.114D-02	0.599D-03
2^{-4}	0.538D-01	0.282D-01	0.130D-01	0.670D-02	0.388D-02	0.235D-02
2^{-6}	0.525D-01	0.337D-01	0.177D-01	0.105D-01	0.607D-02	0.354D-02
2^{-8}	0.650D-01	0.403D-01	0.233D-01	0.133D-01	0.756D-02	0.428D-02
2^{-10}	0.723D-01	0.434D-01	0.249D-01	0.142D-01	0.806D-02	0.454D-02
2^{-12}	0.746D-01	0.443D-01	0.253D-01	0.144D-01	0.819D-02	0.462D-02
2^{-14}	0.752D-01	0.446D-01	0.254D-01	0.144D-01	0.822D-02	0.464D-02
2^{-16}	0.754D-01	0.446D-01	0.254D-01	0.144D-01	0.823D-02	0.465D-02
.	.	.	.	.	.	.
.	.	.	.	.	.	.
2^{-34}	0.754D-01	0.446D-01	0.254D-01	0.144D-01	0.823D-02	0.465D-02
D^N	0.754D-01	0.446D-01	0.254D-01	0.144D-01	0.823D-02	0.465D-02
p^N	0.76	0.81	0.82	0.81	0.82	$p^* = 0.76$
$C_{0.76}^N$	0.894	0.897	0.865	0.832	0.803	0.768

by a more elaborate technique than that used in the previous section. Starting from the values of D^N for $N = 8, 16, 32, 64, 128, 256$, given in Table 8.2, the quantities C_p^N are computed from the formula

$$C_p^N = \frac{D^N N^p}{1 - 2^{-p}}$$

for the same values of N and for a convenient range of values of p surrounding and including the computed value p^* given by (8.18). The results are set out in Table 8.3. The rows of the table contain the values of C_p^N for fixed p and the columns correspond to the values for fixed N. The bold entry in each row of the table is the maximum value in the row. We take the computed ε–uniform order of convergence p^* to be the maximum value of p for which this bold entry is the first entry of the row, and we take the value of that entry to be the computed ε–uniform error constant $C_{p^*}^*$. Thus $p^* = 0.75$ and $C_{0.75}^* = 0.885$ for values of N in the range $R_N = [8, 512]$ and $p^* = 0.81$, $C_{0.81}^* = 0.981$ for values of N in the smaller range $R_N = [16, 512]$. If the computed approximations p^* and $C_{p^*}^*$, obtained by this tabular technique, are either equal to or close to the computed values obtained with the general algorithm, this indicates their insensitivity to changes in N and their robustness with respect to ε for all $\varepsilon \in R_\varepsilon$ and all $N \in R_N$. This is indeed the case for the numerical solutions of problem (6.18) obtained by Method 6.2, because the general algorithm yields $p^* = 0.76$ and $C_{0.76}^* = 0.897$ and the tabular technique yields $p^* = 0.75$ and $C_{0.75}^* = 0.885$ for

Table 8.3 *Values of C_p^N defined by (8.19) generated by the fitted mesh Method 6.2 applied to problem (6.18) for various values of p and N.*

p	Number of intervals N					
	8	16	32	64	128	256
0.1	**1.39**	0.880	0.537	0.327	0.200	0.121
0.2	**0.883**	0.600	0.393	0.256	0.168	0.109
0.3	**0.750**	0.546	0.383	0.268	0.188	0.131
0.5	**0.728**	0.610	0.491	0.395	0.318	0.254
0.7	**0.841**	0.809	0.748	0.691	0.639	0.587
0.75	**0.885**	0.881	0.844	0.806	0.773	0.734
0.76	0.894	**0.897**	0.865	0.832	0.803	0.768
0.8	0.935	**0.964**	0.956	0.946	0.938	0.922
0.81	0.946	**0.981**	0.979	0.973	0.975	0.966
0.9	1.06	1.17	1.24	1.31	1.40	**1.47**
1.0	1.21	1.43	1.63	1.85	2.11	**2.38**

$N \in R_N = [8, 512]$. We then have the option of taking the values $p^* = 0.75$ and $C_{0.75}^* = 0.897$ as the 'safest' approximations to the error parameters.

It is instructive to compare these computed values of p^* and $C_{p^*}^*$ with the computed values obtained by the three quick algorithms of the previous section given in Table 8.1. We see that both the second and third quick algorithms yield reasonable approximations to p.

We can also introduce the order of local convergence p_ε^N using the maximum pointwise two–mesh differences D_ε^N, defined in (8.15), and

$$p_\varepsilon^N = \log_2 \frac{D_\varepsilon^N}{D_\varepsilon^{2N}}. \qquad (8.21)$$

It might then be expected that the minimum over ε of these quantities would coincide with p^N defined in (8.17). However, in general, $p^N \neq \min_{\varepsilon \in R_\varepsilon} p_\varepsilon^N$, which is confirmed by the numerical results given in Table 5.4.

Throughout this chapter, we develop our algorithms to approximate the error parameters corresponding to theoretical ε–uniform error estimates of the form

$$\|U_\varepsilon^N - u_\varepsilon\| \leq CN^{-p}.$$

However, current theoretical research on ε–uniform numerical methods with piecewise–uniform fitted meshes, suggests that ε–uniform error estimates of the form

$$\|U_\varepsilon^N - u_\varepsilon\| \leq CN^{-p}(\ln N)^q$$

may be more appropriate. In such cases, the computed order of convergence obtained from the general two–mesh algorithm with

$$D^N = \max_\varepsilon \|U_\varepsilon^N - \overline{U}_\varepsilon^{2N}\|_{\Omega_\varepsilon^N} \approx CN^{-p}(\ln N)^q - C(2N)^{-p}(\ln 2N)^q,$$

Table 8.4 *Orders of local convergence p^N corresponding to different theoretical error bounds for various values of N.*

	Number of intervals N										
	8	16	32	64	128	256	512	1024	2048	4096	8192
$N^{-1}\ln N$	0.42	0.58	0.68	0.74	0.78	0.81	0.83	0.85	0.86	0.87	0.88
$N^{-1}(\ln N)^2$	-0.81	0.00	0.28	0.44	0.53	0.60	0.65	0.69	0.72	0.75	0.77
$(N^{-1}\ln N)^2$	1.04	1.29	1.43	1.53	1.59	1.64	1.68	1.71	1.74	1.76	1.78

is given by

$$p^N = \log_2 \frac{D^N}{D^{2N}} \approx p - \log_2\left(\frac{2^p(\ln 2N)^q - (\ln 4N)^q}{2^p(\ln N)^q - (\ln 2N)^q}\right) < p. \qquad (8.22)$$

This shows that, for moderate values of N, the two–mesh orders of local convergence p^N underestimate the correct value of p. Specific examples of this phenomena may be seen in Tables 5.2, 5.4 and 5.10. More generally, the order of local convergence p^N corresponding to different theoretical error behaviour for sample values of p and q are tabulated in Table 8.4 for various values of N.

8.4   Validation

In this section we apply the general algorithm, described in the previous section, to obtain a computed approximations to the ε–uniform error parameters for a one dimensional problem. We compute the corresponding ε–uniform maximum pointwise error bounds for the numerical solutions. We choose a problem for which the exact solution is available in a simple closed form, which allows us to compute the exact ε–uniform errors. We then compare our computed ε–uniform error bounds with these exact ε–uniform errors. This enables us to judge, in this particular case, whether or not the experimental techniques used in this chapter to compute the ε–uniform error bounds yield realistic estimates of the true errors.

In particular, we use Method 3.1 to solve problem (2.3) and we compute the exact ε–uniform maximum pointwise error E^N_{exact} for various values of N using the known exact solution. Then, using the computed values of the ε–uniform error parameters $p^*, C^*_{p^*}$ found with the general algorithm, we compute an upper bound $C^*_{p^*}.N^{-p^*}$ for the ε–uniform error for various values of N. The results are displayed in Table 8.5, where the exact ε–uniform maximum pointwise errors and the computed ε–uniform error bounds are compared. It is seen that, for all values of N, the computed ε–uniform error bounds are realistic upper bounds for the exact ε–uniform maximum pointwise error, and that they are not wildly pessimistic, in contrast to the usual situation when the upper bounds are obtained by theoretical arguments.

Table 8.5 *Exact ε–uniform maximum pointwise errors E^N_{exact} and computed ε–uniform maximum pointwise error bounds $C^*_{p^*} N^{-p^*}$ generated by the general algorithm of §8.3 for Method 3.1 applied to problem (2.3) for various values of N.*

	Number of intervals N						
	8	16	32	64	128	256	512
D^N	0.0776	0.0501	0.0293	0.0163	0.0089	0.0049	0.0027
p^N	**0.63**	0.78	0.85	0.86	0.87	0.87	$p^* = 0.63$
$C^N_{0.63}$	**0.813**	0.812	0.735	0.633	0.535	0.456	0.389
$C^*_{0.63} N^{-0.63}$	0.219	0.142	0.092	0.059	0.038	0.025	0.016
E^N_{exact}	0.191	0.115	0.065	0.036	0.020	0.011	0.006

8.5 Practical uses of ε–uniform error parameters

In this section some important uses of the computed ε–uniform error parameters are illustrated. It is assumed that, for the problem under consideration, the numerical solutions are computed on a sequence of meshes Ω^N_ε, with $N \in R_N$ and $\varepsilon \in R_\varepsilon$, and that the numerical method is known to be ε–uniform of some positive, but unknown, order p. The first obvious use of the computed ε–uniform error parameters is to compute estimates of the ε–uniform maximum pointwise error bound for a range of values of N. For example, for Method 6.2 applied to problem (6.18), using the computed ε–uniform error parameters $p^* = 0.821, C^*_{0.821} = 0.983$ obtained from the second quick algorithm in §8.2, the computed ε–uniform error bound is $0.983 N^{-0.821}$. Values of this for various values of N are in Table 8.6. These computed ε–uniform error bounds are likely to be realistic approximations to the true ε–uniform errors for all $\varepsilon \in R_\varepsilon = [2^{-18}, 2^{-8}]$, and $N \in R_N = [128, 512]$, because it is known theoretically that Method 6.2 is ε–uniform. It should be noted that if a similar attempt is made to obtain computed error bounds for a non–ε–uniform method, then the resulting approximate errors are unlikely to be realistic.

We now use the general algorithm to obtain more accurate approximations p^* and $C^*_{p^*}$ of the ε–uniform error parameters , and from these we find a more reliable ε–uniform error bound $C^*_{p^*} N^{-p^*}$ for $N \in R_N$. The resulting computed ε–uniform error bound for Method 6.2 applied to problem (6.18) is $0.897 N^{-0.76}$ for $N \in R_N = [8, 512]$ or $0.981 N^{-0.81}$ for $N \in R_N = [16, 512]$, values of which are given in Table 8.7 for various values of N.

Table 8.6 *Computed ε–uniform maximum pointwise error bound $C^*_{p^*} N^{-p^*}$ generated by the second quick algorithm of §8.2 for the fitted mesh Method 6.2 applied to problem (6.18) for various values of N.*

	Number of intervals N						
	8	16	32	64	128	256	512
$C^*_{0.821} N^{-0.821}$	0.169	0.096	0.054	0.031	0.017	0.010	0.006

Table 8.7 *Computed ε-uniform maximum pointwise error bound $C_{p^*}^* . N^{-p^*}$ generated by the general algorithm of §8.3 for the fitted mesh Method 6.2 applied to problem (6.18) for various values of N.*

	Number of intervals N						
	8	16	32	64	128	256	512
$0.897N^{-0.76}$	0.185	0.110	0.064	0.038	0.022	0.013	0.008
$0.981N^{-0.81}$	–	0.104	0.059	0.034	0.019	0.011	0.006

We can also use the computed ε-uniform error parameters to answer questions of the following kind: suppose that we require a guaranteed ε-uniform maximum pointwise error no greater than δ, say, what is the number N of mesh points that must be used? It follows immediately from (8.6) that this accuracy is attained if

$$N \geq \max\{N_0, (\frac{C_{\bar{p}}}{\delta})^{1/\bar{p}}\}.$$

To apply this in a specific case we now consider the solution of problem (6.18) by Method 6.2. With the values $p^* = 0.76$ and $C_{0.76}^* = 0.897$ obtained above, and taking $\delta = 0.01$, it suffices to choose N such that

$$N \geq \left[\left(\frac{0.897}{0.01} \right)^{1/0.76} \right] + 1 = 373$$

where the notation $[x]$ denotes the greatest integer less than or equal to the real number x. This shows that, in this case, if we take $N = 373$, we are likely to have an ε-uniform maximum pointwise error no greater than 0.01, for all values of ε in the range R_ε.

Finally, we make the following important observation about the use in general of ε-uniform methods. Suppose that we have a numerical method that is ε-uniform of some positive order for a class of problems. We can compute an approximation at each point of the domain $\overline{\Omega}$ to the exact solution, of a specific problem from this problem class, to whatever accuracy we wish by using a sufficiently large value of N. We can then use this approximate solution as a benchmark solution for testing the performance of any other numerical method proposed for solving this problem, whether or not any theoretical error estimates are available for the method. We illustrate the use of this approach in the next chapter.

8.6   Global error parameters

To estimate the global error constant $\bar{C}_{\bar{p}}$ in (8.6) we modify the general algorithm in §8.3 as follows. Because we use piecewise linear or bilinear interpolation to form $\overline{U}_\varepsilon^N$, the maximum and minimum values of $|\overline{U}_\varepsilon^N|$ occur at mesh points in $\overline{\Omega}_\varepsilon^N$. It follows that

$$\|\overline{U}_\varepsilon^N - \overline{U}_\varepsilon^{2N}\|_\Omega = \|\overline{U}_\varepsilon^N - \overline{U}_\varepsilon^{2N}\|_{\Omega_\varepsilon^N \cup \Omega_\varepsilon^{2N}}$$
$$= \max\{\|U_\varepsilon^N - \overline{U}_\varepsilon^{2N}\|_{\Omega_\varepsilon^N}, \|\overline{U}_\varepsilon^N - U_\varepsilon^{2N}\|_{\Omega_\varepsilon^{2N}}\}.$$

We first compute the global maximum pointwise two–mesh differences

$$\bar{D}_\varepsilon^N = \|\overline{U}_\varepsilon^N - \overline{U}_\varepsilon^{2N}\|_{\Omega_\varepsilon^N \cup \Omega_\varepsilon^{2N}} \tag{8.23}$$

for all integers N satisfying $N, 2N \in R_N$ and for a finite set of values $\varepsilon \in R_\varepsilon$. Then, from these values, we find the global ε–uniform maximum pointwise two–mesh differences

$$\bar{D}^N = \max_{\varepsilon \in R_\varepsilon} \bar{D}_\varepsilon^N \tag{8.24}$$

for each available value of N satisfying $N, 2N \in R_N$. We define the computed ε–uniform order of local convergence, for all $N, 4N \in R_N$, by

$$\bar{p}^N = \log_2 \frac{\bar{D}^N}{\bar{D}^{2N}} \tag{8.25}$$

and we take the computed global ε–uniform order of convergence to be

$$\bar{p}^* = \min_N \bar{p}^N. \tag{8.26}$$

Corresponding to this value of $\bar{p}^*$ we define

$$\bar{C}_{\bar{p}^*}^N = \frac{\bar{D}^N N^{\bar{p}^*}}{1 - 2^{-\bar{p}^*}} \tag{8.27}$$

and we take the computed global ε–uniform error constant to be

$$\bar{C}_{\bar{p}^*}^* = \max_N \bar{C}_{\bar{p}^*}^N. \tag{8.28}$$

We now apply this procedure to find the computed global maximum pointwise error bounds when Method 3.1 is used to solve the one–dimensional problem

Table 8.8 *Exact global maximum pointwise error $\bar{E}_{\text{exact}}^N$ and computed global maximum pointwise error bound $\bar{C}_{\bar{p}^*}^* . N^{-\bar{p}^*}$ generated by the general algorithm of §8.3 for Method 3.1 applied to problem (2.3) for various values of N.*

		Number of intervals N						
		8	16	32	64	128	256	512
	$\bar{D}^N$	0.1578	0.0861	0.0439	0.0208	0.0097	0.0050	0.0027
	$\bar{p}^N$	**0.87**	0.97	1.07	1.10	0.96	0.88	$\bar{p}^* = 0.87$
	$\bar{C}_{0.87}^N$	**2.128**	2.122	1.977	1.716	1.459	1.367	1.356
$\bar{C}_{\bar{p}^*}^* . N^{-\bar{p}^*}$		0.349	0.191	0.104	0.057	0.031	0.017	0.009
Global error		0.316	0.169	0.083	0.039	0.020	0.011	0.006

(2.3). Since problem (2.3) has a known solution we can also compute the exact errors. The results for various values of N are compared in Table 8.8. Again we see from the numerical results in the table that for all values of N, the computed error bounds are realistic upper bounds for the true error.

CHAPTER 9

Non–monotone methods in two dimensions

9.1 Non–monotone methods

In chapters 2 and 4 we studied the use of non–monotone finite difference methods for finding numerical solutions of singular perturbation problems in one dimension. We saw in chapter 2 that, on uniform meshes with $\varepsilon N < 1$, the numerical solutions have non–physical numerical oscillations with amplitudes, which are either unbounded or comparable to the magnitude of the exact solution. In chapter 4 we showed that these oscillations are stabilized by using an appropriate piecewise–uniform fitted mesh, and are of negligible amplitude if the mesh is tuned by making a careful choice of the transition parameter σ. With an appropriately tuned mesh the non–monotone Method 4.3 gives numerical solutions with smaller ε–uniform maximum pointwise errors than the corresponding errors given by the monotone Method 3.1. A choice must be made therefore between the less accurate but physically correct and robust behaviour of monotone methods, and the more accurate but non–physical and non–robust behaviour of non–monotone methods. These issues are even more critical when numerical approximations of appropriately scaled derivatives of the exact solutions are required.

In this chapter we examine the analogous situation in two dimensions. We want to know if it is similar to the one–dimensional case, or if there are further complications. We can anticipate that the numerical solutions have non–physical numerical oscillations and that we must be careful in our choice of the transition parameters. Even if we are prepared to accept small numerical oscillations in the solution, as the price for the anticipated higher accuracy, we still need to know if there are other issues that should concern us in two or more dimensions.

9.2 Tuned non–monotone method

In this section we consider non–monotone numerical methods for solving singularly perturbed convection–diffusion problems from Problem Class 6.1 which are of the form

$$\varepsilon \Delta u_\varepsilon + a_1(x,y)\frac{\partial u_\varepsilon}{\partial x} = f(x,y), \quad (x,y) \in \Omega$$

$$u_\varepsilon = g, \quad (x,y) \in \Gamma$$

$$a_1(x,y) \geq \alpha_1 > 0, \quad (x,y) \in \overline{\Omega}.$$

The exact solution u_ε of such a problem typically has a regular boundary layer of width $O(\varepsilon)$ on the edge Γ_L and a non–degenerate parabolic boundary layer of width $O(\sqrt{\varepsilon})$ on each of the edges Γ_B and Γ_T, because the velocity profile $\mathbf{a} = (a_1, 0)$ is parallel to these edges and $a_1 \geq \alpha_1 > 0$.

We introduce the following non–monotone numerical method with a centred finite difference operator on a piecewise–uniform fitted mesh to compute numerical approximations U_ε to the solution u_ε of a typical problem from Problem Class 6.1.

Method 9.1. Centred finite difference operator on piecewise–uniform fitted mesh for Problem Class 6.1.

$$[\varepsilon(\delta_x^2 + \delta_y^2) + a_1(x_i, y_j)D_x^0]U_\varepsilon = f(x_i, y_j) \quad (x_i, y_j) \in \Omega_\varepsilon^N \qquad (9.1a)$$

$$U_\varepsilon = g, \quad (x_i, y_j) \in \Gamma^N \qquad (9.1b)$$

where

$$\overline{\Omega}_\varepsilon^N = \overline{\Omega}_{\sigma_1}^N \times \overline{\Omega}_{\sigma_2}^N \qquad (9.1c)$$

$$\overline{\Omega}_{\sigma_1}^N = \{x_i : 0 \leq i \leq N\} \quad \text{and} \quad \overline{\Omega}_{\sigma_2}^N = \{y_j : 0 \leq j \leq N\} \qquad (9.1d)$$

$$x_i = \begin{cases} 2i\sigma_1/N, & \text{for } 0 \leq i \leq N/2 \\ \sigma_1 + 2(i - (N/2))(1 - \sigma_1)/N, & \text{for } N/2 \leq i \leq N \end{cases} \qquad (9.1e)$$

$$y_j = \begin{cases} 4i\sigma_2/N, & \text{for } 0 \leq j \leq N/4 \\ \sigma_2 + 2(i - (N/4))(1 - 2\sigma_2)/N, & \text{for } N/4 \leq j \leq 3N/4 \\ 1 - \sigma_2 + 4(i - (3N/4))\sigma_2/N, & \text{for } 3N/4 \leq j \leq N \end{cases} \qquad (9.1f)$$

and the transition parameter σ_1 for $\overline{\Omega}_{\sigma_1}^N$ is

$$\sigma_1 \equiv \min\{1/2, c\varepsilon \ln N\}, \quad c \geq \frac{1}{\alpha_1} \qquad (9.1g)$$

where c is a tuning parameter and the transition parameter σ_2 for $\bar{\Omega}_{\sigma_2}^N$ is

$$\sigma_2 = \min\{1/4, \ \varepsilon^{\frac{1}{2}} \ln N\}. \tag{9.1h}$$

We note that with the choice $c = 1/\alpha_1$ for the tuning parameter the resulting mesh $\bar{\Omega}_\varepsilon^N$ coincides with the piecewise–uniform fitted mesh in Method 6.2. In the present section we use the non–monotone finite difference operator (9.1a) on the piecewise–uniform fitted mesh $\bar{\Omega}_\varepsilon^N$ for various choices of the tuning parameter c.

In the one dimensional case, studied in chapter 4, an appropriate choice of the tuning parameter c is critical for the damping of the non–physical numerical oscillations arising from the use of a non–monotone finite difference operator. In two dimensions we anticipate similar oscillatory behaviour of the numerical solutions, and also that the choice of the tuning parameter c is critical for their control. To see if these expectations are correct, we apply the non–monotone Method 8.1 to the solution of the specific problem (6.18) from Problem Class 6.1. Graphs of the numerical solutions of problem (6.18) for $\varepsilon = 0.0001$ obtained from Method 9.1 with $c = 2/\alpha_1$, the 'optimal' choice in chapter 4, and for two values of N, are displayed in Fig. 9.1. We observe that, with this choice of the tuning parameter c, no numerical oscillations are visible.

We now estimate the ε–uniform maximum pointwise errors in the numerical solutions of problem (6.18). Since we do not have an explicit expression for the exact solution we use u_{comp} as an approximation, where u_{comp} is the solution computed using the monotone Method 6.2 with $N = 512$. It is important to observe that the ε–uniform accuracy of u_{comp} is guaranteed by the theoretical ε–uniform error estimate for the monotone Method 6.2 applied to problems from

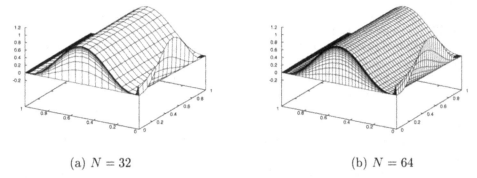

(a) $N = 32$ (b) $N = 64$

FIG. 9.1. *Numerical solutions generated by Method 9.1 with $c = 2/\alpha_1$ applied to problem (6.18) for $\varepsilon = 0.0001$, with $N = 32$ and $N = 64$.*

Problem Class 6.1, which is given in Shishkin (1992b). We note that no comparable theoretical results are available for the non–monotone Method 9.1 or for
the other non–monotone methods, based on the centred finite difference operators, which are discussed in this chapter. The computed estimates of the error
parameters p and C_p, obtained in §8.3 for the numerical solutions generated by
Method 6.2, are given in Table 8.7. From the entry corresponding to $p = 0.81$
and $N = 512$ we obtain the following computed error bound for the ε–uniform
maximum pointwise error

$$\max_{\varepsilon} \|u_{\text{comp}} - u_{\varepsilon}\|_{\bar{\Omega}_{\varepsilon}^{512}} \leq 0.006.$$

This upper bound shows that u_{comp} is a good approximation to u_{ε}.

We approximate the exact maximum pointwise errors $E^N_{\varepsilon,\text{exact}}$ and E^N_{exact} for
Method 9.1 applied to problem (6.18) by $E^N_{\varepsilon,\text{comp}}$ and E^N_{comp}, which are defined
by

$$E^N_{\varepsilon,\text{comp}} = \|U^N_{\varepsilon} - \bar{u}_{\text{comp}}\|_{\Omega^N_{\varepsilon}} \qquad (9.2a)$$

and

$$E^N_{\text{comp}} = \max_{\varepsilon} E^N_{\varepsilon,\text{comp}}, \qquad (9.2b)$$

where $\bar{u}_{\text{comp}}$ denotes the piecewise bilinear interpolant of u_{comp}. The values of
$E^N_{\varepsilon,\text{comp}}$ and E^N_{comp} for problem (6.18), using Method 9.1 with $c = 2/\alpha_1$, for
various values of ε and N are presented in Table 9.1. In this table we see that,
for each fixed N, the errors stabilize at a constant value as ε decreases, and that
these stabilized values decrease with increasing N. However, unlike the situation
for Method 6.2, which we know is an ε–uniform method for problem (6.18),
we cannot infer any strong conclusions about the ε–uniform convergence of the
numerical solutions generated by the non–monotone Method 9.1, with the choice
of tuning parameter $c = 2/\alpha_1$, to the exact solution of problem (6.18).

We now use the two–mesh technique described in §8.3 to find the computed
orders of local convergence p^N_{ε}, defined in (8.21) and the computed ε–uniform
orders of convergence p^N, defined in (8.17), for various values of ε and N. The
results are presented in Table 9.2, where we see that the computed ε–uniform
orders p^N suggest that this non–monotone method is ε–uniform of second order
for all sufficiently large N. More precisely, comparing the values of p^N in Table
9.2 with the values in the last row of Table 8.4 leads us to the conjecture that the
correct ε-uniform maximum pointwise error bound is of the form $C(N^{-1} \ln N)^2$,
where C is some constant independent of N and ε.

We now test the sensitivity of the non–monotone Method 9.1, to a change
in the value of the tuning parameter c, by using Method 9.1 with $c = 1/\alpha_1$ to
compute numerical solutions of problem (6.18) with $\varepsilon = 0.0001$. Graphs of the

Table 9.1 *Computed maximum pointwise errors* $E^N_{\varepsilon,\mathrm{comp}}$ *and* E^N_{comp} *generated by Method 9.1 with* $c = 2/\alpha_1$ *applied to problem (6.18) for various values of* ε *and* N.

ε	8	16	32	64	128	256
	\multicolumn{6}{c}{Number of intervals N}					
1	0.519D-02	0.142D-02	0.372D-03	0.946D-04	0.228D-04	0.457D-05
2^{-2}	0.118D-01	0.354D-02	0.994D-03	0.261D-03	0.642D-04	0.130D-04
2^{-4}	0.521D-01	0.265D-01	0.947D-02	0.313D-02	0.750D-03	0.149D-03
2^{-6}	0.695D-01	0.344D-01	0.132D-01	0.495D-02	0.177D-02	0.782D-03
2^{-8}	0.106D+00	0.462D-01	0.215D-01	0.936D-02	0.282D-02	0.937D-03
2^{-10}	0.130D+00	0.571D-01	0.215D-01	0.918D-02	0.331D-02	0.106D-02
2^{-12}	0.143D+00	0.632D-01	0.220D-01	0.922D-02	0.333D-02	0.114D-02
2^{-14}	0.150D+00	0.664D-01	0.234D-01	0.923D-02	0.334D-02	0.119D-02
2^{-16}	0.153D+00	0.680D-01	0.242D-01	0.923D-02	0.334D-02	0.122D-02
2^{-18}	0.155D+00	0.688D-01	0.246D-01	0.923D-02	0.334D-02	0.124D-02
2^{-20}	0.156D+00	0.692D-01	0.247D-01	0.923D-02	0.334D-02	0.125D-02
2^{-22}	0.156D+00	0.694D-01	0.248D-01	0.923D-02	0.334D-02	0.125D-02
2^{-24}	0.156D+00	0.695D-01	0.249D-01	0.923D-02	0.334D-02	0.125D-02
2^{-26}	0.156D+00	0.696D-01	0.249D-01	0.923D-02	0.334D-02	0.125D-02
2^{-28}	0.156D+00	0.696D-01	0.249D-01	0.923D-02	0.334D-02	0.125D-02
2^{-30}	0.156D+00	0.696D-01	0.249D-01	0.923D-02	0.334D-02	0.125D-02
E^N_{comp}	0.156D+00	0.696D-01	0.249D-01	0.923D-02	0.334D-02	0.125D-02

Table 9.2 *Computed orders of convergence* p^N_ε *and* p^N *generated by Method 9.1 with* $c = 2/\alpha_1$ *applied to problem (6.18) for various values of* ε *and* N.

ε	8	16	32	64	128
	\multicolumn{5}{c}{Number of intervals N}				
1	1.87	1.92	1.95	1.98	2.00
2^{-2}	1.72	1.78	1.90	1.95	1.98
2^{-4}	0.73	1.35	2.39	2.00	2.01
2^{-6}	0.78	1.17	1.49	1.46	1.60
2^{-8}	1.01	1.25	1.53	1.46	1.57
2^{-10}	1.03	1.30	1.50	1.49	1.59
2^{-12}	1.03	1.29	1.51	1.49	1.59
2^{-14}	1.02	1.28	1.51	1.50	1.59
.	.	.	.	.	.
.	.	.	.	.	.
2^{-30}	1.02	1.26	1.50	1.50	1.59
p^N	1.00	1.25	1.50	1.50	1.59

numerical solutions for $N = 32$ and $N = 64$ are displayed in Fig. 9.2 where, in contrast to the case with $c = 2/\alpha_1$, we observe that now there are non–physical numerical oscillations on the coarse mesh outside the boundary layers. However, further numerical experiments show that if c is fixed at $1/\alpha_1$ and N is increased, the amplitude of these oscillations decreases towards zero as N increases.

Because of this promising behaviour, it is of interest to determine whether or not the non–monotone Method 9.1 with $c = 1/\alpha_1$ is ε–uniform and, if so,

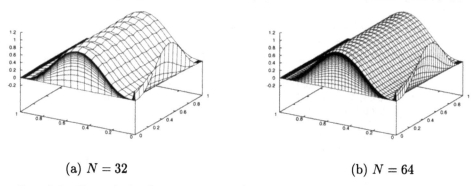

(a) $N = 32$ (b) $N = 64$

FIG. 9.2. *Numerical solutions generated by Method 9.1 with $c = 1/\alpha_1$ applied to problem (6.18) for $\varepsilon = 0.0001$, with $N = 32$ and $N = 64$.*

whether or not the presence of these oscillations affects the order of ε–uniform convergence. The answers are obtained from the numerical results displayed in Tables 9.3 and 9.4. In Table 9.3 we present the computed maximum pointwise errors $E_{\varepsilon,\text{comp}}^N$ and E_{comp}^N in the numerical solutions of problem (6.18) using Method 9.1 with $c = 1/\alpha_1$. The results indicate numerically that, as in Table 9.1 for the value $c = 2/\alpha_1$, the errors stabilize for each fixed N as ε decreases,

Table 9.3 *Computed maximum pointwise errors $E_{\varepsilon,\text{comp}}^N$ and E_{comp}^N generated by Method 9.1 with $c = 1/\alpha_1$ applied to problem (6.18) for various values of ε and N.*

ε	Number of intervals N					
	8	16	32	64	128	256
1	0.519D-02	0.142D-02	0.372D-03	0.946D-04	0.228D-04	0.457D-05
2^{-2}	0.118D-01	0.354D-02	0.994D-03	0.261D-03	0.642D-04	0.130D-04
2^{-4}	0.338D-01	0.135D-01	0.431D-02	0.118D-02	0.297D-03	0.139D-03
2^{-6}	0.102D+00	0.291D-01	0.111D-01	0.365D-02	0.112D-02	0.253D-03
2^{-8}	0.187D+00	0.767D-01	0.224D-01	0.851D-02	0.209D-02	0.961D-03
2^{-10}	0.235D+00	0.105D+00	0.407D-01	0.134D-01	0.563D-02	0.225D-02
2^{-12}	0.263D+00	0.123D+00	0.555D-01	0.219D-01	0.803D-02	0.441D-02
2^{-14}	0.276D+00	0.130D+00	0.622D-01	0.293D-01	0.126D-01	0.559D-02
2^{-16}	0.283D+00	0.134D+00	0.650D-01	0.323D-01	0.159D-01	0.743D-02
2^{-18}	0.286D+00	0.135D+00	0.654D-01	0.324D-01	0.163D-01	0.816D-02
2^{-20}	0.287D+00	0.135D+00	0.652D-01	0.320D-01	0.160D-01	0.799D-02
2^{-22}	0.288D+00	0.135D+00	0.651D-01	0.319D-01	0.158D-01	0.789D-02
2^{-24}	0.288D+00	0.136D+00	0.651D-01	0.318D-01	0.158D-01	0.786D-02
2^{-26}	0.288D+00	0.136D+00	0.651D-01	0.318D-01	0.158D-01	0.785D-02
2^{-28}	0.288D+00	0.136D+00	0.651D-01	0.318D-01	0.158D-01	0.785D-02
2^{-30}	0.288D+00	0.136D+00	0.651D-01	0.318D-01	0.158D-01	0.785D-02
E_{comp}^N	0.288D+00	0.136D+00	0.654D-01	0.324D-01	0.163D-01	0.816D-02

Table 9.4 *Computed orders of convergence p_ε^N and p^N generated by Method 9.1 with $c = 1/\alpha_1$ applied to problem (6.18) for various values of ε and N.*

ε	Number of intervals N				
	8	16	32	64	128
1	1.87	1.92	1.95	1.98	2.00
2^{-2}	1.72	1.78	1.90	1.95	1.98
2^{-4}	1.47	1.43	1.45	1.50	1.65
2^{-6}	2.04	1.79	1.62	1.60	1.76
2^{-8}	1.28	1.76	1.67	1.84	1.67
2^{-10}	1.08	1.25	1.55	1.39	1.58
2^{-12}	1.08	1.06	1.19	1.44	1.23
2^{-14}	1.09	1.07	1.02	1.15	1.38
2^{-16}	1.09	1.08	1.03	0.99	1.13
2^{-18}	1.09	1.09	1.04	1.00	0.97
2^{-20}	1.10	1.09	1.05	1.01	0.99
2^{-22}	1.10	1.09	1.05	1.02	1.00
2^{-24}	1.10	1.09	1.05	1.02	1.01
2^{-26}	1.10	1.09	1.05	1.02	1.01
2^{-28}	1.10	1.09	1.05	1.02	1.01
2^{-30}	1.10	1.09	1.05	1.02	1.01
p^N	1.10	1.09	1.04	0.99	1.00

and that these stabilized values decrease with increasing N. However, we observe that the values of E_{comp}^N in this table are larger than the corresponding values in Table 9.1, which shows that, for this particular problem, $c = 1/\alpha_1$ may not be as good a choice as $c = 2/\alpha_1$.

The corresponding computed orders of convergence p_ε^N and p^N are presented in Table 9.4. These indicate numerically that this non–monotone method has an ε–uniform order of convergence of approximately 1, which is significantly lower than the corresponding ε–uniform order of convergence indicated by the entries in Table 9.2 for the value $c = 2/\alpha_1$.

The computed ε–uniform order of convergence p^N is presented in Table 9.5 for Method 9.1 applied to problem (6.18) for $\varepsilon = 2^{-i}, i = 1, \dots, 30$, for various values of N and for several additional values of the tuning parameter c. The numerical results there suggest that a good value of c for this particular problem is $2/\alpha_1$. They also show that the order of ε–uniform convergence is sensitive to the choice of the tuning parameter.

It should be noted that the linear solver used for all the computations in the present section is BiCGStab with MILU preconditioning. The reader may consult van der Vorst (1992) for details. The criterion for the convergence of its iterates is the reduction of the maximum norm of the residual to less than 10^{-8}. In the case when Method 9.1 is used to solve problem (6.18) it is apparent, from the entries in Table 9.6, that the number of iterations required for convergence of the corresponding iterates is independent of ε. This establishes experimentally the fact that if the non–monotone Method 9.1 is applied to the particular problem

Table 9.5 *Computed ε–uniform order of convergence p^N generated by Method 9.1 applied to problem (6.18) for various values of c and N.*

	Number of intervals N				
c	8	16	32	64	128
0.5	0.71	0.64	0.57	0.50	0.63
1.0	1.10	1.09	1.04	0.99	1.00
2.0	1.00	1.25	1.50	1.50	1.59
4.0	0.75	0.97	1.29	1.56	1.54
8.0	0.91	0.65	0.99	1.32	1.60

Table 9.6 *Number of BiCGStab solver iterations required for convergence for Method 9.1 with $c = 2/\alpha_1$ applied to problem (6.18) for various values of ε and N.*

	Number of intervals N						
ε	8	16	32	64	128	256	512
1	3	5	10	19	31	61	119
2^{-2}	3	5	10	19	36	68	113
2^{-4}	3	5	9	16	33	66	132
2^{-6}	2	4	5	9	20	49	46
2^{-8}	2	3	5	6	12	22	18
2^{-10}	2	3	5	9	12	19	47
2^{-12}	2	4	6	10	16	26	37
2^{-14}	1	4	6	10	21	34	54
2^{-16}	1	4	6	11	21	48	51
2^{-18}	1	4	6	11	25	50	65
2^{-20}	1	4	6	11	25	50	59
2^{-22}	1	4	6	11	25	48	54

(6.18) then variations in the choice of the tuning parameter c have no obvious deleterious effect on the performance of the linear solver, despite the fact that the condition number of the linear system is highly ε–dependent. But, in contrast to this satisfactory situation, we show, in the next section, that when the same non–monotone numerical method is applied to a different problem, the behaviour of the linear solver is highly sensitive to the value of the tuning parameter.

9.3   Difficulties in tuning non–monotone methods

To illustrate some of the difficulties encountered in tuning non–monotone methods we consider their use for solving singularly perturbed convection–diffusion problems from Problem Class 5.1, which are of the form

$$\varepsilon \Delta u_\varepsilon + a_1(x,y)\frac{\partial u_\varepsilon}{\partial x} + a_2(x,y)\frac{\partial u_\varepsilon}{\partial y} = f(x,y), \quad (x,y) \in \Omega$$

$$u_\varepsilon = g, \quad (x,y) \in \Gamma$$

$$a_1(x,y) \geq \alpha_1 > 0, \ a_2(x,y) \geq \alpha_2 > 0 , \quad (x,y) \in \overline{\Omega}.$$

The exact solution u_ε of such a problem typically has regular boundary layers on the edges Γ_L and Γ_B but, unlike problems in Problem Class 6.1, has no parabolic boundary layer.

To compute numerical approximations U_ε to the solution u_ε of this problem, we introduce the following non–monotone method with a centred finite difference operator on a piecewise–uniform fitted mesh with two tuning parameters.

Method 9.2.   Centred finite difference operator on piecewise–uniform fitted mesh for Problem Class 5.1.

$$[\varepsilon(\delta_x^2 + \delta_y^2) + a_1(x_i, y_j)D_x^0 + a_2(x_i, y_j)D_y^0]U_\varepsilon = f(x_i, y_j) \quad (x_i, y_j) \in \Omega_\varepsilon^N \quad (9.3a)$$

$$U_\varepsilon = g, \quad (x_i, y_j) \in \Gamma^N \tag{9.3b}$$

where

$$\overline{\Omega}_\varepsilon^N = \overline{\Omega}_{\sigma_1}^N \times \overline{\Omega}_{\sigma_2}^N \tag{9.3c}$$

$$\overline{\Omega}_{\sigma_1}^N = \{x_i : 0 \le i \le N\} \quad \text{and} \quad \overline{\Omega}_{\sigma_2}^N = \{y_j : 0 \le j \le N\} \tag{9.3d}$$

$$x_i = \begin{cases} 2i\sigma_1/N, & \text{for } 0 \le i \le N/2 \\ \sigma_1 + 2(i - (N/2))(1 - \sigma_1)/N, & \text{for } N/2 \le i \le N \end{cases} \tag{9.3e}$$

$$y_j = \begin{cases} 2j\sigma_2/N, & \text{for } 0 \le j \le N/2 \\ \sigma_2 + 2(j - (N/2))(1 - \sigma_2)/N, & \text{for } N/2 \le j \le N \end{cases} \tag{9.3f}$$

and

$$\sigma_i \equiv \min\{1/2, c_i\varepsilon \ln N\}, \quad c_i \ge \frac{1}{\alpha_i}, \quad i = 1, 2 \tag{9.3g}$$

where c_1 and c_2 are tuning parameters.

We apply Method 9.2 with the values $c_1 = 2/\alpha_1, c_2 = 2/\alpha_2$ for the tuning parameters, to the particular problem (5.17) with $\varepsilon = 0.0001$. Graphs of the resulting numerical solutions for $N = 32$ and $N = 64$ are displayed in Fig. 9.3. We see that these numerical solutions have non–physical oscillations outside the boundary layers. The numerical solutions of the same problem using the same method, but with the tuning parameters $c_1 = 1/\alpha_1$, $c_2 = 1/\alpha_2$, are shown in the

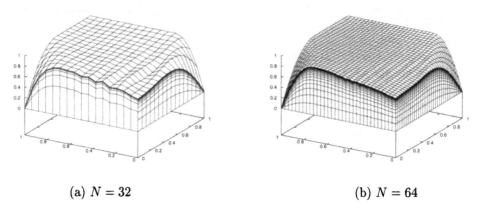

(a) $N = 32$ (b) $N = 64$

FIG. 9.3. Numerical solutions generated by Method 9.2 with $c_1 = 2/\alpha_1$, $c_2 = 2/\alpha_2$ applied to problem (5.17) for $\varepsilon = 0.0001$, with $N = 32$ and $N = 64$.

graphs in Fig. 9.4. We see that we now have non–physical oscillations not only on the coarse mesh but also on part of the fine mesh in one of the boundary layers. This shows that, at least for this specific problem, the occurrence of non–physical oscillations depends critically on the tuning parameters.

Since we do not have an explicit expression for the exact solution of problem (5.17), we cannot compute the errors exactly. We approximate the error, as before, by replacing the exact solution by u_{comp} in the expression for the error, where u_{comp} is obtained by applying the monotone ε–uniform Method 5.1

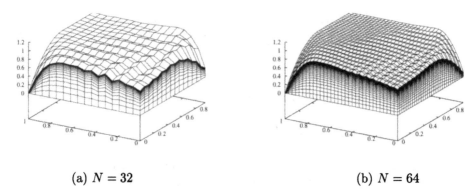

(a) $N = 32$ (b) $N = 64$

FIG. 9.4. Numerical solutions generated by Method 9.2 with $c_1 = 1/\alpha_1$, $c_2 = 1/\alpha_2$ applied to problem (5.17) for $\varepsilon = 0.0001$, with $N = 32$ and $N = 64$.

Table 9.7 *Values of $D_\varepsilon^N, D^N, p^N, p^*$ and $C_{p^*}^N$ generated by the general algorithm of §8.3 for the fitted mesh Method 5.1 applied to problem (5.17) for various values of N.*

	Number of intervals N						
	8	16	32	64	128	256	512
D^N	0.142	0.107	0.0786	0.0521	0.0316	0.0181	
p^N	**0.41**	0.45	0.59	0.72	0.80	$p^* = 0.41$	
$C_{0.41}^N$	1.346	**1.348**	1.316	1.159	0.934	0.711	
$C_{0.72}^N$	1.615	2.005	2.426	**2.649**	2.646	2.496	
$C_{0.41}^* N^{-0.41}$	0.576	0.433	0.326	0.245	0.185	0.139	0.105
$C_{0.72}^* N^{-0.72}$	–	–	–	0.133	0.081	0.049	0.030

to problem (5.17). We use the general algorithm from §8.3, to compute 'safe' values of the error parameters p^* and $C_{p^*}^*$, from which we obtain the computed maximum pointwise error bound $C_{p^*}^* N^{-p^*}$ for Method 5.1 applied to problem (5.17). The results are given in Table 9.7 where the ε–uniform error parameters, for $N \in R_N = [64, 512]$, are $p^* = 0.72$ and $C_{0.72}^* = 2.649$. From the entry corresponding to $N = 512$ we obtain the following computed ε–uniform maximum pointwise error bound

$$\max_\varepsilon \|u_{\text{comp}} - u_\varepsilon\|_{\bar{\Omega}_\varepsilon^{512}} \leq 0.030$$

which shows that u_{comp} is a good approximation to u_ε. In Table 9.8 we give the computed maximum pointwise errors $E_{\varepsilon,\text{comp}}^N$ and E_{comp}^N for Method 9.2 with $c_1 = 2/\alpha_1, c_2 = 2/\alpha_2$ applied to problem (5.17) for various values of N and ε. The numerical results in this table suggest that the method is ε–uniform for this specific problem. The computed orders of convergence p_ε^N for various

Table 9.8 *Computed maximum pointwise errors $E_{\varepsilon,\text{comp}}^N$ and E_{comp}^N generated by Method 9.2 with $c_1 = 2/\alpha_1$, $c_2 = 2/\alpha_2$ applied to problem (5.17) for various values of ε and N.*

ε	Number of intervals N					
	8	16	32	64	128	256
1	0.138D-01	0.384D-02	0.995D-03	0.250D-03	0.597D-04	0.119D-04
2^{-2}	0.266D-01	0.621D-02	0.154D-02	0.379D-03	0.903D-04	0.180D-04
2^{-4}	0.502D-01	0.156D-01	0.582D-02	0.197D-02	0.621D-03	0.258D-03
2^{-6}	0.161D+00	0.644D-01	0.176D-01	0.417D-02	0.896D-03	0.267D-03
2^{-8}	0.278D+00	0.152D+00	0.511D-01	0.115D-01	0.232D-02	0.422D-03
2^{-10}	0.319D+00	0.186D+00	0.732D-01	0.193D-01	0.506D-02	0.106D-02
2^{-12}	0.330D+00	0.195D+00	0.817D-01	0.237D-01	0.711D-02	0.178D-02
2^{-14}	0.332D+00	0.197D+00	0.839D-01	0.257D-01	0.816D-02	0.222D-02
2^{-16}	0.333D+00	0.198D+00	0.845D-01	0.263D-01	0.863D-02	0.282D-02
2^{-18}	0.333D+00	0.198D+00	0.847D-01	0.264D-01	0.876D-02	0.305D-02
E_{comp}^N	0.333D+00	0.198D+00	0.847D-01	0.265D-01	0.878D-02	0.308D-02

Table 9.9 *Computed orders of convergence p_ε^N and p^N generated by Method 9.2 with $c_1 = 2/\alpha_1, c_2 = 2/\alpha_2$ applied to problem (5.17).*

	\multicolumn{5}{c}{Number of intervals N}			
ε	8	16	32	64	128
1	1.80	1.93	1.97	2.00	2.00
2^{-2}	2.12	2.01	2.01	2.00	2.00
2^{-4}	1.23	1.43	1.49	1.52	1.66
2^{-6}	1.54	2.15	2.03	2.02	1.65
2^{-8}	1.23	1.74	2.22	2.32	2.27
2^{-10}	1.09	1.57	2.05	1.95	2.08
2^{-12}	1.06	1.53	1.85	1.81	1.80
2^{-14}	1.05	1.53	1.80	1.69	1.69
2^{-16}	1.05	1.52	1.79	1.64	1.41
2^{-18}	1.05	1.52	1.79	1.63	1.31
p^N	1.05	1.52	1.79	1.63	1.30

values of N and ε, and the computed ε–uniform orders of convergence p^N for various values of N, are presented in Table 9.9. At first sight, the entries in this table suggest that the non–monotone Method 9.2 is ε–uniform of order greater than 1. But this experimental evidence is not convincing, because the values of p^N in this table are not monotonically increasing with increasing N, which suggests that the values $c_1 = 2/\alpha_1$, $c_2 = 2/\alpha_2$ of the tuning parameters in Method 9.2 are not the 'optimal' choice for this specific problem. This shows that the ' optimal' choice of the tuning parameters for a general problem class may be impossible, and that each problem has to be considered separately. The necessity to determine appropriate values of one or more tuning parameters, for each specific problem, is an unsatisfactory feature of non–monotone methods in comparison with monotone methods.

Table 9.10 *Number of BiCGStab solver iterations required for convergence for Method 9.2 with $c_1 = 2/\alpha_1, c_2 = 2/\alpha_2$ applied to problem (5.17) for various values of ε and N.*

	\multicolumn{7}{c}{Number of intervals N}					
ε	8	16	32	64	128	256	512
1	3	5	7	10	15	21	30
2^{-2}	2	4	5	8	12	16	22
2^{-4}	3	4	5	7	9	13	21
2^{-6}	5	8	8	11	17	25	38
2^{-8}	5	12	14	15	17	21	42
2^{-10}	6	14	25	33	34	38	39
2^{-12}	6	14	34	55	73	77	86
2^{-14}	6	17	37	74	146	184	209
2^{-16}	6	17	38	87	168	331	428
2^{-18}	7	17	40	93	209	471	930
2^{-20}	7	17	43	101	217	527	*
2^{-22}	7	19	45	108	246	588	*

When the non–monotone Method 9.2 is used to solve problem (5.17) with $\varepsilon = 0.0001$ the determination of the number of solver iterations required for convergence of the iterates, corresponding to the solver BiCGStab with MILU preconditioning, is no longer straightforward. This can be seen in Table 9.10, in which the number of iterations required to reduce the maximum norm of the residual to less than 10^{-7} is presented. The entry $*$ in this table indicates that the corresponding iteration count exceeds 1000, which we interpret as a failure of the iterates to converge. It is evident from this table that the solver behaves quite differently on this problem than on problem (6.18), which we considered in the last section. In particular, we see that, for each fixed N, the number of solver iterations required for convergence increases as ε approaches zero. For $N = 256$ a graph of the number of iterations required for convergence versus $\log_2 1/\varepsilon$ is given in Fig. 9.5. It is clear that as ε approaches zero the required number of solver iterations increases rapidly, which shows that the amount of work needed to solve problem (5.17) with Method 9.2 is not ε–uniform, at least in the case when this solver is used. A plot of $\log_{10}$ (residual) versus the number of solver iterations required for convergence, for $N = 256$ and $\varepsilon = 2^{-24}$, is given in Fig. 9.6. The highly oscillatory behaviour of the residuals is clearly not desirable.

The convergence difficulties described above when the non–monotone Method 9.2 is used to solve problem (5.17) can be much worse when similar non–monotone methods are used to solve other problems. To illustrate this we modify the co-efficient of the first derivative in problem (5.17) from $2 + x^2 y$ to $1 + x^2 y$. The resulting problem is

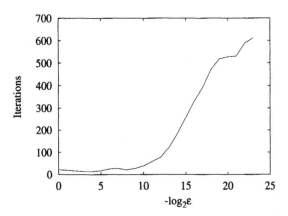

FIG. 9.5. Number of BiCGStab solver iterations required for convergence of Method 9.2 with $c_1 = 2/\alpha_1$, $c_2 = 2/\alpha_2$ and $N = 256$, versus $\log_2 1/\varepsilon$ applied to problem (5.17).

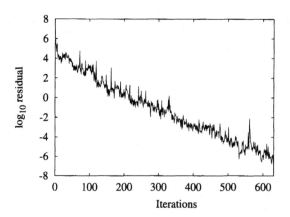

FIG. 9.6. $\log_{10}$ (residual) versus number of BiCGStab solver iterations for Method 9.2 with $c_1 = 2/\alpha_1$, $c_2 = 2/\alpha_2$ and $N = 256$ applied to problem (5.17) for $\varepsilon = 2^{-24}$.

$$\varepsilon\Delta u + (1 + x^2 y)\frac{\partial u}{\partial x} + (1 + xy)\frac{\partial u}{\partial y} = x^2 + y^3 + \cos(x + 2y) \quad (x, y) \in \Omega \text{(9.4a)}$$

$$u(x, 0) = 0, \quad u(x, 1) = \begin{cases} 4x(1 - x), x < 1/2 \\ 1, x \geq 1/2 \end{cases} \quad \text{(9.4b)}$$

$$u(0, y) = 0, \quad u(1, y) = \begin{cases} 8y(1 - 2y), \ y < 1/4 \\ 1, \ y \geq 1/4 \end{cases} \quad \text{(9.4c)}$$

which has the same form as problem (5.17). The locations of the boundary layers in its solution are the same as in problem (5.17). We now solve problem (9.4) using the non–monotone method Method 9.2 in conjunction with the solver BiCGStab(2). We present the number of BiCGStab(2) iterations required for convergence in Table 9.11. Comparing the entries in Tables 9.11 and 9.10 we see that for problem (9.4) the number of iterations required for convergence using the more powerful BiCGStab(2) solver is always greater than that for problem (5.17) using the BiCGStab solver. We see also that the computational effort required for convergence of the iterates is highly ε–dependent, and that a failure to converge seems to be unpredictable. These results demonstrate conclusively that the non–monotone Method 9.2 is not robust, because an apparently trivial change in the data of a problem leads to unpredictable behaviour.

If we insist on using a non–monotone method at all costs, then we can either let the iterations carry on for as long as is necessary and hope for convergence or, slightly more realistically, we can try using a more powerful linear solver. A family of such solvers, called BiCGStab(ℓ), is described in Sleijpen and Fokkema (1993). Note that BiCGStab(1) is equivalent to BiCGStab. With these solvers

Table 9.11 *Number of BiCGStab(2) solver iterations required for convergence for Method 9.2 with $c_1 = 2/\alpha_1, c_2 = 2/\alpha_2$ applied to problem (9.4) for various values of ε and N.*

	\multicolumn{6}{c}{Number of intervals N}
ε	8	16	32	64	128	256
1	4	6	8	12	19	27
2^{-2}	3	4	6	10	14	19
2^{-4}	3	3	4	7	10	14
2^{-6}	5	7	8	9	13	18
2^{-8}	6	16	15	15	14	17
2^{-10}	6	20	44	36	35	30
2^{-12}	6	22	79	126	92	84
2^{-14}	6	22	90	285	387	254
2^{-16}	7	22	98	385	962	965
2^{-18}	7	23	103	419	*	*
2^{-20}	7	23	100	466	*	*
2^{-22}	7	24	104	516	*	*

we obtain convergence for any given ε by using BiCGStab(ℓ), provided that we choose ℓ sufficiently large. However, the number of matrix products required in each iteration is proportional to ℓ. Taking into account the fact that these iterations themselves take longer as ℓ increases, we have a situation where, even on a powerful computer, it can take days to solve a problem for a small value of ε. This demonstrates that the amount of computational work required to solve problems of the form (9.4) using Method 9.2 is not ε–uniform. Therefore, it is highly questionable whether it is really worth this effort to compute a numerical solution which is, at best, only slightly more accurate than that obtained with a monotone method.

9.4 Weaknesses of non–monotone ε–uniform methods

If we use a non–monotone method we need to choose appropriate values of the tuning parameters c_1 and c_2 in order to suppress non–physical oscillations in the numerical solutions. Appropriate choices of these tuning parameters are not obvious and *ad hoc* choices must be made for the particular problem in question. Moreover, even with an appropriate choice of these parameters, the convergence behaviour of the iterates generated by the linear solver may be highly ε–dependent. Thus, we are led to the conclusion that any reduction in the error, gained by using a non–monotone method, as opposed to a monotone method, has to be paid for by an unpredictable increase in the number of solver iterations required for convergence. Furthermore, few theoretical ε–uniform error estimates in the maximum norm are known for non–monotone methods for problems in more than one dimension. On the other hand, ε–uniform error estimates in the maximum norm are available for monotone methods on fitted piecewise–uniform meshes. Furthermore, we can achieve any desired accuracy by using a sufficiently large number of mesh points, the required number of which can be estimated by the experimental techniques described in §8.5.

It should be observed that, for some special cases, Hemker et al. (1997) show that higher order ε–uniform convergence can be obtained using defect correction techniques, provided that the underlying numerical method is monotone and ε–uniform. We believe that higher order ε–uniform convergence can also be obtained for more general problem classes by developing appropriate defect correction techniques in conjunction with monotone ε–uniform methods.

The numerical results and theoretical discussions in chapter 4 for one–dimensional problems illustrate in a simple, easily comprehended setting, some of the dangers inherent in the use of non–monotone ε–uniform numerical methods for singular perturbation problems. In the present chapter the numerical results using non–monotone methods for two–dimensional problems confirm our worst fears. They demonstrate decisively that we cannot guarantee the physical relevance, the pointwise accurate or the robustness of the numerical solutions, if they are generated by a non–monotone method.

We have reached the above conclusion by comparing the performance of monotone and non–monotone ε–uniform numerical methods on relatively simple one– and two– dimensional linear stationary singularly perturbed problems. Once we move away from this familiar territory, into the realm of nonstationary nonlinear problems in three spatial dimensions, for example, it is essential to use robust numerical methods, because we now know that, with non–monotone methods, we can have no confidence in the physical relevance of their numerical solutions. Nor can we be sure that we can compute them with a reasonable amount of computational effort. It is precisely for such complicated problems that numerical methods are employed as the main tool for uncovering the unknown structure of the exact solution, and therefore it is essential that we use robust numerical methods in such cases.

CHAPTER 10

Linear and nonlinear reaction–diffusion problems

10.1 Linear reaction diffusion problems

Because nonlinear phenomena with boundary layers abound in nature, we need to consider problems involving singularly perturbed nonlinear differential equations, with boundary layers in their solutions, and to construct ε–uniform numerical methods for approximating these solutions. To keep things reasonably simple we restrict the discussion in this chapter to reaction–diffusion problems, but in the final two chapters we consider more complicated problems. In this section we introduce a class of singularly perturbed linear reaction–diffusion problems in one dimension on the unit interval $\Omega = (0, 1)$, and we examine several numerical methods on a uniform mesh for approximating their solutions. Then, in subsequent sections, we introduce singularly perturbed semilinear reaction–diffusion problems.

Problem Class 10.1. Linear reaction–diffusion in one dimension.

$$\varepsilon u_\varepsilon'' - c(x)u_\varepsilon = f(x), \quad x \in \Omega, \tag{10.1a}$$
$$u_\varepsilon(0) = A, \quad u_\varepsilon(1) = B, \tag{10.1b}$$
$$c, f \in C^2(\overline{\Omega}) \tag{10.1c}$$
$$c(x) \geq \alpha > 0, \quad x \in \overline{\Omega}. \tag{10.1d}$$

It is well known that the following ε–explicit bound for all $x \in \overline{\Omega}$ is valid for the derivatives of the solution of a typical problem in Problem Class 10.1

$$|u_\varepsilon^{(k)}(x)| \leq C(1 + \varepsilon^{-k/2}e^{-\sqrt{\alpha/\varepsilon}(x)} + \varepsilon^{-k/2}e^{-\sqrt{\alpha/\varepsilon}(1-x)}),$$

where C is a constant independent of ε. This shows that, in general, regular boundary layers of width $O(\sqrt{\varepsilon})$ appear at each boundary point $x = 0$ and $x = 1$.

Taking account of the discussions in chapters 1 and 2, it should come as no surprise to the reader that the following standard centred finite difference method on a uniform mesh is not suitable for solving these problems.

Method 10.1. Centred finite difference method on uniform mesh for Problem Class 10.1.

$$\varepsilon \delta^2 U_\varepsilon(x_i) - c(x_i)U_\varepsilon(x_i) = f(x_i), \ x_i \in \Omega_u^N, \qquad (10.2a)$$
$$U_\varepsilon(0) = u_\varepsilon(0), \ U_\varepsilon(1) = u_\varepsilon(1), \qquad (10.2b)$$

$$\overline{\Omega}_u^N = \{x_i | x_i = i/N, i = 0, \dots, N\}. \qquad (10.2c)$$

To demonstrate the inadequacy of this method for solving problems from Problem Class 10.1 we now apply it to the following specific linear reaction–diffusion problem

$$\varepsilon u_\varepsilon''(x) - (1 + (x-1)^2)u_\varepsilon(x) = 0, \qquad (10.3a)$$
$$u_\varepsilon(0) = 1, \quad u_\varepsilon(1) = 0. \qquad (10.3b)$$

We observe that the boundary conditions in this problem are such that a boundary layer arises only at the boundary point $x = 0$. We make this choice for the sake of simplicity.

In Table 10.1 we present the resulting computed maximum pointwise errors E_ε^N, for various values of ε and N. It is evident from the numerical results in this table that Method 10.1 is not ε–uniform for this problem, because there is a persistent error of almost constant magnitude along one of the diagonals. We remark that the dip in the error ridge for $N = 256$ is due to the fact that, in the expression for the error, the unknown exact solution is replaced by the numerical solution computed on the finest available uniform mesh Ω_u^{1024}, which is itself inaccurate near the layer regions when ε is small.

On the other hand, it is easy to find numerical methods that are ε–uniform at the mesh points for this problem class, as we now show. We begin by introducing

Table 10.1 *Computed maximum pointwise error E_ε^N generated by Method 10.1 applied to problem (10.3) for various values of ε and N.*

ε	Number of intervals N					
	8	16	32	64	128	256
2^{-2}	0.002400	0.000621	0.000156	0.000039	0.000010	0.000002
2^{-4}	0.008410	0.002307	0.000582	0.000146	0.000036	0.000009
2^{-6}	0.027382	0.007671	0.002099	0.000528	0.000131	0.000031
2^{-8}	**0.045372**	0.026088	0.007294	0.001988	0.000495	0.000118
2^{-10}	0.028812	**0.043604**	0.025431	0.007087	0.001913	0.000459
2^{-12}	0.008679	0.027342	**0.042703**	0.025032	0.006909	0.001796
2^{-14}	0.002203	0.008166	0.026612	**0.042115**	0.024539	0.006522
2^{-16}	0.000552	0.002070	0.007919	0.026184	**0.041254**	0.023132
2^{-18}	0.000138	0.000519	0.002007	0.007796	0.025699	**0.038616**
2^{-20}	0.000035	0.000130	0.000503	0.001976	0.007725	0.024361

two fitted–operator methods on a uniform mesh Ω_u^N. The first has a frozen fitting factor and the second has a variable fitting factor.

Method 10.2. **Fitted finite difference operator with frozen fitting factor on uniform mesh for Problem Class 10.1.**

$$\varepsilon \gamma_i \delta^2 U_\varepsilon(x_i) - c(x_i)U_\varepsilon(x_i) = f(x_i), \quad x_i \in \Omega_u^N, \quad (10.4a)$$
$$U_\varepsilon(0) = u_\varepsilon(0), \quad U_\varepsilon(1) = u_\varepsilon(1), \quad (10.4b)$$

with the frozen fitting factor

$$\gamma_i = \begin{cases} \left(\frac{\rho(0)}{\sinh \rho(0)}\right)^2, & 0 \leq i \leq N/2, \\ \left(\frac{\rho(1)}{\sinh \rho(1)}\right)^2, & N/2 < i \leq N \end{cases} \quad (10.4c)$$

where $\rho(y) = \frac{1}{2N}\sqrt{\frac{c(y)}{\varepsilon}}$.

Method 10.3. **Fitted finite difference operator with variable fitting factor on uniform mesh for Problem Class 10.1.**

$$\varepsilon \gamma_i \delta^2 U_\varepsilon(x_i) - c(x_i)U_\varepsilon(x_i) = f(x_i), \quad x_i \in \Omega_u^N, \quad (10.5a)$$
$$U_\varepsilon(0) = u_\varepsilon(0), \quad U_\varepsilon(1) = u_\varepsilon(1), \quad (10.5b)$$

with the variable fitting factor

$$\gamma_i = \left(\frac{\rho(x_i)}{\sinh \rho(x_i)}\right)^2, \quad \text{where } \rho(y) = \frac{1}{2N}\sqrt{\frac{c(y)}{\varepsilon}}.$$

In Tables 10.2 and 10.3 we present the computed ε–uniform maximum pointwise error E^N, and computed ε–uniform order of convergence p^N, for Methods 10.2 and 10.3 respectively applied to problem (10.3) for various values of N. The numerical results given in these tables confirm experimentally that both of these

Table 10.2 *Computed ε–uniform maximum pointwise error E^N and order of ε–uniform convergence p^N generated by Method 10.2 applied to problem (10.3) for various values of N.*

N	8	16	32	64	128	256
E^N	0.001625	0.000799	0.000397	0.000198	0.000097	0.000046
p^N	1.02	1.01	1.00	1.00	1.00	1.00

Table 10.3 *Computed ε–uniform maximum pointwise error E^N and order of ε–uniform convergence p^N generated by Method 10.3 applied to problem (10.3) for various values of N.*

N	8	16	32	64	128	256
E^N	0.004711	0.002316	0.001146	0.000568	0.000279	0.000131
p^N	1.03	1.00	1.00	1.00	1.00	1.00

methods are ε–uniform at the mesh points for the typical problem (10.3) from Problem Class 10.1.

It is not hard to prove rigorously, see for example Doolan et al. (1980), that for the linear problems of Problem Class 10.1 we can construct fitted operator methods on uniform meshes that are ε–uniform at least at the mesh points $\overline{\Omega}_u^N$. Furthermore, using exponential basis functions, we can construct a piecewise exponential interpolant from Ω_u^N to $\overline{\Omega}$, which is an ε–uniform global approximation to the solution (see for example O'Riordan and Stynes (1986)). In the following sections we show that analogous results are not valid for problems involving singularly perturbed semilinear ordinary differential equations. We also demonstrate that ε–uniform methods for such semilinear problems can be constructed by using a piecewise–uniform rather than a uniform mesh.

10.2   Semilinear reaction–diffusion problems

For the first time in this book we now consider a singularly perturbed problem involving a nonlinear differential equation. We introduce the following class of singularly perturbed semilinear reaction–diffusion problems in one dimension on the unit interval $\Omega = (0, 1)$.

Problem Class 10.2.   Semilinear reaction–diffusion in one dimension.

$$\varepsilon u_\varepsilon'' - c(u_\varepsilon(x))u_\varepsilon = 0, \quad x \in \Omega, \tag{10.6a}$$

$$u_\varepsilon(0) = A, \quad u_\varepsilon(1) = B, \tag{10.6b}$$

$$c(y) \geq \alpha > 0, \quad \text{for all} \quad y \in \Re \tag{10.6c}$$

It can be shown that the nature of the boundary layers in the solutions u_ε of the semilinear problems in Problem Class 10.2 is similar to that of the regular boundary layers that occur in the solutions of linear problems in Problem Class 10.1. If $0 < A, B < 1$ these boundary layers occur at the boundary points $x = 0$ and $x = 1$. It is also worth observing that for the linear differential equation

$$\varepsilon v_\varepsilon'' + \lambda v_\varepsilon = 0, \quad \lambda > 0$$

the oscillatory function

$$v_\varepsilon(x) = C \sin \sqrt{\lambda/\varepsilon}(x - \phi)$$

is a solution for any constants C and ϕ, and that the number of oscillations in v_ε, occurring in the interval Ω, tends to infinity as $\varepsilon \to 0$. This shows that condition (10.6c) is a natural restriction if we want to avoid oscillatory behaviour of this kind in the solutions of problems from Problem Class 10.2.

In what follows we compute numerical solutions to problems from Problem Class 10.2 using numerical methods based on fitted nonlinear finite difference operators of the form

$$\varepsilon\gamma_i\delta^2 U_\varepsilon(x_i) - c(U_\varepsilon(x_i))U_\varepsilon(x_i) = 0, \ x_i \in \Omega^N \qquad (10.7)$$

where Ω^N is a uniform or piecewise–uniform mesh and $\gamma_i = \gamma_i(x_i, U_\varepsilon(x_i))$ is a fitting factor. If $\gamma_i = 1$ this finite difference operator reduces to the standard centred finite difference operator. Since (10.7) is a nonlinear finite difference operator it is necessary to find a suitable way to solve the resulting nonlinear finite difference method. This is the topic of the next section.

10.3 Nonlinear solvers

When $\varepsilon = 1$ the nonlinear finite difference operator (10.7) has no singularities and the standard procedure is to use Newton's method as our nonlinear solver. However, when ε is small, (10.7) is a singularly perturbed operator and the radius of convergence of Newton's method may depend on a positive power of ε, which renders it useless as $\varepsilon \to 0$. For this reason we use a continuation method of the following form as our nonlinear solver. For $j = 1, \ldots K$,

$$(\varepsilon\gamma_i(x_i, U_\varepsilon(x, t_{j-1}))\delta_x^2 - c(U_\varepsilon(x, t_{j-1})) - D_t^-)U_\varepsilon(x, t_j) = 0, \qquad (10.8a)$$
$$U_\varepsilon(0, t_j) = u_\varepsilon(0), \ U_\varepsilon(1, t_j) = u_\varepsilon(1) \text{ for all } j, \qquad (10.8b)$$
$$U_\varepsilon(x, 0) = u_{\text{init}}(x). \qquad (10.8c)$$

We can interpret (10.8) as a discretization of the following time–dependent version of a problem from Problem Class 10.2

$$\varepsilon u_{xx} - c(u(x, t))u - u_t = 0, \quad (x, t) \in (0, 1) \times (0, T] \qquad (10.9a)$$
$$u(0, t) = u_\varepsilon(0) \quad u(1, t) = u_\varepsilon(1) \quad t \geq 0 \qquad (10.9b)$$
$$u(x, 0) = u_{\text{init}}(x) \quad 0 < x < 1. \qquad (10.9c)$$

In the numerical experiments that follow, in order to test the robustness of the continuation algorithm (10.8), we choose the starting values $u_{\text{init}}(x)$ in a variety of ways. The number of iterations K and the choice of the uniform time step $h_t = t_j - t_{j-1}$ are determined as follows. At each time level t_j we compute the maximum of the discrete derivatives with respect to t of U_ε as follows

$$d(j) \equiv \max_{1 \leq i \leq N} |U_\varepsilon(x_i, t_j) - U_\varepsilon(x_i, t_{j-1})|/h_t, \text{ for } j = 1, 2, \ldots, K.$$

We then choose the time step h_t to be sufficiently small, so that these discrete derivatives do not grow with increasing j. That is, we require that

$$d(j) \leq d(j-1), \text{ for all } j, \ 1 < j \leq K. \tag{10.10}$$

The number of iterations K is then made sufficiently large so that

$$d(K) \leq \ tol \tag{10.11}$$

where *tol* is some prescribed tolerance. We consider the effect of different choices for *tol* later in the chapter.

In the initial experiments we set $tol = 0.00001$. More precisely, we compute the steady–state numerical solution as follows: we start with $h_t = 1.0$, for example, and then compute $d(j)$ for successive values of j starting from $j = 1$. If, at some value of j, (10.10) is not satisfied by $d(j)$, we halve the time step, and recompute the iterates repeatedly until (10.10) is satisfied by $d(j)$. We continue the iterations until either (10.11) is satisfied or $K = 90$. If (10.11) is not satisfied, we repeat the entire process starting with the smaller value $h_t = 0.5$. We continue in this way until (10.11) is satisfied. The transient solution of (10.9) converges to the steady–state solution, and for h_t sufficiently small the discrete counterpart to (10.9) inherits this property. From this we know that it is possible, eventually, to satisfy (10.11). We take the resulting values of $\{U_\varepsilon(x_i, T) : 1 \leq i \leq N, \ T = Kh_t\}$ as the required computed solution of the nonlinear finite difference method arising from (10.7)

The numerical results in this chapter are computed for $N = 8, 16, 32, 64,$ 128, 256, 512, 1024 and $\varepsilon = 2^{-j}, j = 1, 2, \ldots j_{red}$, where j_{red} is chosen so that $\varepsilon = 2^{-j_{red}}$ is a value at which the order of convergence stabilizes. This normally occurs when ε is so sufficiently small that, to machine accuracy, we are solving essentially the reduced problem. The exact pointwise errors $|U_\varepsilon(x_i, T) - u_\varepsilon(x_i)|$ are approximated on each mesh, for successive values of ε, by the computed pointwise errors $e_\varepsilon^N(x_i) = |U_\varepsilon(x_i, T) - \overline{U}(x_i, T)|$, where the piecewise linear interpolant $\overline{U}(x, T)$ is defined on the subinterval $[x_{i-1}, x_i]$ for each i, $1 \leq i \leq 1024$, by

$$\overline{U}(x, T) = U^{1024}(x_{i-1}, T) + (U^{1024}(x_i, K) - U^{1024}(x_{i-1}, T))\frac{x - x_{i-1}}{x_i - x_{i-1}}.$$

Here the set $\{U^{1024}(x_i, T)\}_{i=0}^{1024}$ contains the nodal values of the approximate solution, of the nonlinear finite difference method with $N = 1024$, generated by the continuation method (10.8). As is the case for linear problems, the maximum pointwise error $E_{\varepsilon,\text{exact}}^N$ is approximated for each ε and each N by the computed maximum pointwise error

$$E_\varepsilon^N = \max_{0 \leq i \leq N} e_\varepsilon^N(x_i)$$

and the ε–uniform maximum pointwise error E^N_{exact} is approximated by the computed ε–uniform maximum pointwise error

$$E^N = \max_\varepsilon E^N_\varepsilon.$$

10.4 Numerical methods on uniform meshes

We now examine the effectiveness of three nonlinear finite difference methods, composed of a finite difference operator of the form (10.7) on a uniform mesh, for solving two typical semilinear reaction–diffusion problems from Problem Class 10.2. These finite difference methods are analogous to the Methods 10.1 – 10.3 described in §10.1 for linear problems. We solve all of the nonlinear finite difference methods in this section using the continuation algorithm in § 10.3 with *tol* = 0.00001 and an initial guess which is the linear interpolant of the boundary values, that is, $u_{\text{init}}(x) = u_\varepsilon(0) + x(u_\varepsilon(1) - u_\varepsilon(0))$, $0 \le x \le 1$.

The first semilinear problem from Problem Class 10.2 is

$$\varepsilon u''_\varepsilon(x) - (1 + u_\varepsilon)u_\varepsilon(x) = 0, \tag{10.12a}$$
$$u_\varepsilon(0) = 1, \quad u_\varepsilon(1) = 0 \tag{10.12b}$$

which has a boundary layer only at $x = 0$. The second of these semilinear problems is

$$\varepsilon u''_\varepsilon(x) - (1 + u_\varepsilon)u_\varepsilon(x) = 0, \tag{10.13a}$$
$$u_\varepsilon(0) = 0.5, \quad u_\varepsilon(1) = 0.7 \tag{10.13b}$$

in which the boundary conditions are chosen to ensure that its solution has a boundary layer at each of the boundary points $x = 0$ and $x = 1$.

The first numerical method is analogous to Method 10.1 for the linear reaction–diffusion problems from Problem Class 10.1. It has a centred finite difference operator on a uniform mesh.

Method 10.4. Centred finite difference operator on uniform mesh for Problem Class 10.2.

$$\varepsilon\delta^2 U_\varepsilon(x_i) - c(U_\varepsilon(x_i))U_\varepsilon(x_i) = 0, \quad x_i \in \Omega^N_u, \tag{10.14a}$$
$$U_\varepsilon(0) = u_\varepsilon(0), \; U_\varepsilon(1) = u_\varepsilon(1). \tag{10.14b}$$

In Table 10.4 we present the computed maximum pointwise error E^N_ε for various values of ε and N, obtained when Method 10.4 applied to the semilinear problem (10.12). Since the computed ε–uniform maximum pointwise errors in this table do not decrease significantly as N increases, it is clear that Method 10.4 is not ε–uniform for this semilinear problem.

Table 10.4 *Computed maximum pointwise error E_ε^N generated by Method 10.4 applied to problem (10.12) for various values of ε and N.*

ε	Number of intervals N					
	8	16	32	64	128	256
2^{-4}	0.008796	0.002368	0.000612	0.000153	0.000038	0.000009
2^{-6}	0.027901	0.008808	0.002372	0.000611	0.000152	0.000036
2^{-8}	**0.048220**	0.027895	0.008801	0.002365	0.000604	0.000144
2^{-10}	0.039177	**0.048207**	0.027869	0.008773	0.002336	0.000575
2^{-12}	0.014682	0.039166	**0.048153**	0.027765	0.008659	0.002222
2^{-14}	0.003861	0.014681	0.039123	**0.047937**	0.027352	0.008208
2^{-16}	0.000974	0.003861	0.014676	0.038952	**0.047085**	0.025730
2^{-18}	0.000244	0.000974	0.003861	0.014655	0.038285	**0.043835**
2^{-20}	0.000061	0.000244	0.000974	0.003861	0.014569	0.035787

The second numerical method is a nonlinear analogue of Method 10.2. It is composed of a fitted nonlinear finite difference operator of the form (10.7) on a uniform mesh. Again, in order to keep the required fitting factor simple, we introduce a method which is applicable to problems in the subclass of Problem Class 10.2 which have a boundary layer only at the boundary point $x = 0$.

Method 10.5.   Fitted finite difference operator with frozen fitting factor on uniform mesh for subclass of Problem Class 10.2.

$$\varepsilon \gamma_i \delta^2 U_\varepsilon(x_i) - c(U_\varepsilon(x_i))U_\varepsilon(x_i) = 0, \quad x_i \in \Omega_u^N, \tag{10.15a}$$
$$U_\varepsilon(0) = u_\varepsilon(0), \quad U_\varepsilon(1) = u_\varepsilon(1) = 0, \tag{10.15b}$$

with the frozen fitting factor

$$\gamma_i = \left(\frac{\rho(U_\varepsilon(0))}{\sinh \rho(U_\varepsilon(0))} \right)^2, \quad 0 \le i \le N, \quad \text{where } \rho(y) = \frac{1}{2N}\sqrt{\frac{c(y)}{\varepsilon}}. \tag{10.15c}$$

In Table 10.5 we give the computed maximum pointwise error E_ε^N for Method 10.5 applied to problem (10.12) for various values of ε and N. Again, the occurrence of a ridge of persistent, essentially constant, error along a diagonal of this table shows that Method 10.5 is not an ε–uniform method for the semilinear problem (10.12) and therefore that it is not ε–uniform for Problem Class 10.2.

It is instructive to compare the numerical results in Table 10.5 for the semilinear problem (10.12) with those in Table 10.2 for the linear problem (10.3). Both problems have the same boundary conditions, their solutions have just one boundary layer at $x = 0$, and the values of the coefficient of the term $u_\varepsilon(x)$ at the end points is the same in both problems. Despite these similarities the numerical methods based on analogous fitted finite difference operators on the same uniform mesh behave quite differently. For the linear problem (10.3) the

Table 10.5 *Computed maximum pointwise error E_ε^N generated by Method 10.5 applied to problem (10.12) for various values of ε and N.*

ε	Number of intervals N					
	8	16	32	64	128	256
2^{-5}	0.003723	0.001429	0.000383	0.000098	0.000024	0.000006
2^{-7}	0.003568	0.003721	0.001428	0.000382	0.000097	0.000023
2^{-9}	**0.011452**	0.003568	0.003717	0.001423	0.000377	0.000092
2^{-11}	0.001994	**0.011450**	0.003566	0.003701	0.001405	0.000359
2^{-13}	0.000009	0.001992	**0.011442**	0.003558	0.003639	0.001332
2^{-15}	0.000000	0.000009	0.001983	**0.011412**	0.003528	0.003398
2^{-17}	0.000000	0.000000	0.000009	0.001948	**0.011282**	0.003398
2^{-19}	0.000000	0.000000	0.000000	0.000008	0.001811	**0.010673**

convergence is ε–uniform, while for the semilinear problem (10.12) it is not. This illustrates the fact that the natural generalization of a numerical method, which is ε-uniform for a linear problem class, is not necessarily ε-uniform for a nonlinear problem class.

The third numerical method is a semilinear analogue of Method 10.3. It consists of a fitted finite difference operator with a variable fitting factor on a uniform mesh.

Method 10.6. Fitted finite difference operator with variable fitting factor on uniform mesh for Problem Class 10.2.

$$\varepsilon\gamma_i\delta^2 U_\varepsilon(x_i) - c(U_\varepsilon(x_i))U_\varepsilon(x_i) = 0, \quad x_i \in \Omega_u^N, \qquad (10.16\text{a})$$
$$U_\varepsilon(0) = u_\varepsilon(0), \ U_\varepsilon(1) = u_\varepsilon(1), \qquad (10.16\text{b})$$

with the fitting factor

$$\gamma_i = \left(\frac{\rho(U_\varepsilon(x_i))}{\sinh \rho(U_\varepsilon(x_i))}\right)^2, \quad 0 \leq i \leq N, \quad \text{where } \rho(y) = \frac{1}{2N}\sqrt{\frac{c(y)}{\varepsilon}}. \quad (10.16\text{c})$$

The computed maximum pointwise errors E_ε^N obtained by applying this method to the semilinear problem (10.13) are given in Table 10.6 for various values of ε and N. Again we see that there is a ridge of essentially constant error along a diagonal of this table. This shows that Method 10.6 is not ε–uniform for problem (10.13), and hence that it is not an ε–uniform method for Problem Class 10.2.

The numerical results in Tables 10.4–6 show that, in contrast to the situation for the linear problems considered in §10.1, we have not succeeded in finding an ε-uniform method for typical semilinear problems of Problem Class 10.2. The reason for this failure is that, so far, all of the numerical methods are on a uniform mesh. We now explain, by a simple analytic argument, why we cannot

Table 10.6 *Computed maximum pointwise error E_ε^N generated by Method 10.6 applied to problem (10.13) for various values of ε and N.*

ε	Number of intervals N					
	8	16	32	64	128	256
2^{-3}	0.001037	0.000271	0.000069	0.000017	0.000004	0.000001
2^{-5}	0.003949	0.001106	0.000289	0.000073	0.000018	0.000004
2^{-7}	**0.007913**	0.003968	0.001110	0.000290	0.000072	0.000017
2^{-9}	0.004933	**0.007911**	0.003965	0.001107	0.000286	0.000069
2^{-11}	0.000426	0.004931	**0.007902**	0.003951	0.001093	0.000273
2^{-13}	0.000002	0.000426	0.004921	**0.007867**	0.003897	0.001039
2^{-15}	0.000000	0.000002	0.000423	0.004882	**0.007726**	0.003682
2^{-17}	0.000000	0.000000	0.000002	0.000414	0.004731	**0.007183**

expect any numerical method, composed of a fitted finite difference operator on a uniform mesh, to be ε–uniform for Problem Class 10.2. We begin by considering the one–parameter family of linear problems

$$(P_A) \qquad \begin{cases} \varepsilon u_\varepsilon''(x) - c(x)u_\varepsilon(x) = 0, & x \in \Omega, \\ u_\varepsilon(0) = A, & u_\varepsilon(1) = 0, \end{cases}$$

and we define $v_\varepsilon(x)$ to be the solution of the fundamental problem

$$\begin{cases} \varepsilon v_\varepsilon''(x) - c(x)v_\varepsilon(x) = 0, & x \in \Omega, \\ v_\varepsilon(0) = 1, & v_\varepsilon(1) = 0. \end{cases}$$

The solution of each problem in (P_A) is given by

$$u_\varepsilon(x; A) = Av_\varepsilon(x).$$

Thus the function v_ε can be viewed as the fundamental solution or a basis function for the solution set of the family of problems (P_A), in the sense that each solution u_ε of a problem in (P_A) is a constant multiple of the single basis function v_ε. This shows that the family of linear problems is one–dimensional in character. It is easy to see that a fitted operator method, such as Method 10.2, which is ε–uniform for the fundamental problem, is also ε–uniform for the entire family of linear problems (P_A).

We also consider the one–parameter family of semilinear problems

$$(P_A^*) \qquad \begin{cases} \varepsilon u_\varepsilon''(x) - (1 + u_\varepsilon)u_\varepsilon(x) = 0, & x \in \Omega, \\ u_\varepsilon(0) = A, & u_\varepsilon(1) = 0 \end{cases}$$

which is analogous to the family (P_A) of linear problems. We denote the solution of (P_A^*) by $u_\varepsilon^*(x; A)$, and we define v_ε^* to be a solution of the semilinear problem

$$\begin{cases} \varepsilon v_\varepsilon''(x) - (1 + v_\varepsilon)v_\varepsilon(x) = 0, & x \in \Omega, \\ v_\varepsilon(0) = 1, & v_\varepsilon(1) = 0. \end{cases}$$

If, as before, $v_\varepsilon^*(x)$ can be regarded as a fundamental solution of the family of problems (P_A^*), then the function $Av_\varepsilon^*(x)$ should be a solution of (P_A^*). But it is not hard to see that

$$\begin{cases} \varepsilon(Av_\varepsilon^*)''(x) - (1 + Av_\varepsilon^*)Av_\varepsilon^*(x) = A(1-A)(v_\varepsilon^*)^2, & x \in \Omega, \\ \qquad\qquad Av_\varepsilon^*(0) = A, \quad Av_\varepsilon^*(1) = 0 \end{cases}$$

which shows that, for any $A \neq 1$, the solutions $u_\varepsilon^*(x; A)$ and $v_\varepsilon^*(x)$ are linearly independent; and the same is true of $u_\varepsilon^*(x; A)$ and $u_\varepsilon^*(x; B)$ for any $A \neq B$. It follows that there is no finite set of basis functions that spans the solution set of (P_A^*). Each change in the boundary condition gives rise to a totally different solution. This shows that this family of semilinear problems is infinite–dimensional in character. Therefore there is no fundamental problem in the family as in the linear case. An ε–uniform fitted operator method can be constructed for any given finite number of such problems, but not for the entire family.

This explains why fitted operator methods are less suitable for nonlinear than for linear problems. In fact, it is proved in Farrell et al. (1998) that it is impossible to construct a numerical method, composed of a fitted finite difference operator on a uniform mesh, which is ε–uniform for Problem Class 10.2. Therefore, if we want to construct an ε–uniform method for a class of singularly perturbed nonlinear problems it is essential to use non–uniform meshes. This is the topic of the next section.

10.5   Numerical methods on piecewise–uniform meshes

For linear problems we know from the results in previous chapters, that a piecewise–uniform fitted mesh suffices for the construction of an ε–uniform numerical method. In this section we show that the same approach is successful for the semilinear problems of Problem Class 10.2. To do this we introduce the following piecewise–uniform fitted mesh method.

Method 10.7.   Centred finite difference operator on piecewise–uniform fitted mesh for Problem Class 10.2.

$$\varepsilon\delta^2 U_\varepsilon(x_i) - c(U_\varepsilon(x_i))U_\varepsilon(x_i) = 0, \quad x_i \in \Omega_\varepsilon^N, \qquad (10.17\text{a})$$
$$U_\varepsilon(0) = u_\varepsilon(0), \quad U_\varepsilon(1) = u_\varepsilon(1), \qquad (10.17\text{b})$$

where

$$\overline{\Omega}_\varepsilon^N = \{x_i | x_i = 2i\sigma/N, \ i \le N/2; \ x_i = x_{i-1} + 2(1-\sigma)/N, \ i > N/2\} \quad (10.17\text{c})$$

with

$$\sigma = \min\{\frac{1}{2}, \sqrt{\varepsilon}\ln N\}. \qquad (10.17\text{d})$$

Table 10.7 *Computed maximum pointwise errors E_ε^N and E^N generated by Method 10.7 applied to problem (10.12) for various values of ε and N.*

	Number of intervals N						
ε	8	16	32	64	128	256	512
2^{-3}	0.004592	0.001192	0.000303	0.000076	0.000019	0.000004	0.000001
2^{-5}	0.016732	0.004679	0.001209	0.000306	0.000076	0.000018	0.000004
2^{-7}	0.026626	0.016162	0.004676	0.001206	0.000303	0.000072	0.000014
2^{-9}	0.026612	0.015549	0.006537	0.002496	0.000838	0.000243	0.000058
2^{-11}	0.056123	0.018139	0.006418	0.002480	0.000855	0.000254	0.000121
2^{-13}	0.075417	0.030679	0.010518	0.002646	0.000852	0.000254	0.000121
2^{-15}	0.086399	0.038817	0.016094	0.005624	0.001460	0.000263	0.000121
2^{-17}	0.092257	0.043418	0.019722	0.008229	0.002896	0.000759	0.000131
2^{-19}	0.095282	0.045861	0.021775	0.009929	0.004146	0.001451	0.000354
.	.	.	.	.	.	.	.
2^{-25}	0.097980	0.048072	0.023702	0.011652	0.005638	0.002595	0.000986
E^N	0.097980	0.048072	0.023702	0.011652	0.005638	0.002595	0.000986

We use the continuation algorithm in §10.3 to solve the nonlinear finite difference method (10.7). Just as in §10.4 we choose the tolerance *tol* to be 0.00001 and the initial guess to be $u_{\text{init}}(x) = u_\varepsilon(0) + x(u_\varepsilon(1) - u_\varepsilon(0))$, $0 \leq x \leq 1$.

In Tables 10.7 and 10.8 we give the computed maximum pointwise errors E_ε^N and E^N for Method 10.7 applied to problems (10.12) and (10.13), respectively, for various values of ε and N. In contrast to the methods on uniform meshes, considered in the previous section, we see from these tables that Method 10.7 is ε-uniform for these two semilinear problems. Comparing the results in Tables 10.4 and 10.5 with those in Table 10.7, and the results in Table 10.6 with those in Table 10.8, we conclude that, for a given value of N, the use of an ε-uniform method does not necessarily guarantee that the maximum pointwise error is smaller than when a uniform mesh is used, but it does guarantee that each

Table 10.8 *Computed maximum pointwise errors E_ε^N and E^N generated by Method 10.7 applied to problem (10.13) for various values of ε and N.*

	Number of intervals N						
ε	8	16	32	64	128	256	512
2^{-3}	0.003020	0.000780	0.000196	0.000049	0.000012	0.000003	0.000001
2^{-5}	0.010080	0.002769	0.000711	0.000179	0.000044	0.000011	0.000002
2^{-7}	0.016079	0.009553	0.002724	0.000698	0.000174	0.000041	0.000008
2^{-9}	0.020848	0.009062	0.003872	0.001458	0.000484	0.000143	0.000033
2^{-11}	0.042606	0.013832	0.003732	0.001438	0.000498	0.000149	0.000076
2^{-13}	0.056721	0.023153	0.007944	0.001998	0.000495	0.000149	0.000076
2^{-15}	0.064699	0.029175	0.012112	0.004233	0.001099	0.000198	0.000076
2^{-17}	0.068934	0.032568	0.014818	0.006186	0.002177	0.000570	0.000099
.	.	.	.	.	.	.	.
2^{-25}	0.073057	0.035988	0.017779	0.008748	0.004235	0.001950	0.000741
E^N	0.073057	0.035988	0.017779	0.008748	0.004235	0.001950	0.000741

boundary layer is resolved. It can be shown that the piecewise linear interpolant of the numerical solutions, and its discrete derivatives, also converge ε–uniformly to the exact solution and its derivative at each point of the domain. The ε–uniform error can be made as small as desired simply by taking a sufficiently large value of N.

We now consider two further problems belonging to Problem Class 10.2. The first is

$$\varepsilon u_\varepsilon''(x) - (1 - u_\varepsilon^2(x))u_\varepsilon(x) = 0, \; x \in (0,1), \tag{10.18a}$$
$$u_\varepsilon(0) = A, \quad u_\varepsilon(1) = B, \tag{10.18b}$$

and the second is

$$\varepsilon u_\varepsilon''(x) - (1 - u_\varepsilon(x))u_\varepsilon(x) = 0, \; x \in (0,1), \tag{10.19a}$$
$$u_\varepsilon(0) = A, \quad u_\varepsilon(1) = B. \tag{10.19b}$$

We apply Method 10.7 to these problems for various appropriate choices of the boundary values A and B. We observe that condition (10.6c), in the definition of Problem Class 10.2, imposes restrictions on the allowable choices of these boundary values. For example, to guarantee existence and uniqueness for the solution of the specific problem (10.18), we must have

$$|A| < 1 \quad \text{and} \quad |B| < 1$$

and for the specific problem (10.19), the boundary values must satisfy

$$A < 1 \quad \text{and} \quad B < 1.$$

We note that the reduced problems corresponding to problems (10.18) and (10.19) have more than one solution. Indeed, problem (10.18) has the three reduced solutions

$$v_0 = 0, +1, -1,$$

but it can be shown (Chang and Howes (1984)) that $v_0 = 0$ is the only stable solution.

In the case when $A = B = 1$, there are two solutions to problem (10.18). One is simply $u_\varepsilon = 1$, which is an unstable solution. The other is a stable solution, with boundary layers at $x = 0$ and $x = 1$, which converges to the stable reduced solution $v_0 = 0$ as $\varepsilon \to 0$. Numerical solutions of this special case are presented later in this chapter.

In Tables 10.9 and 10.10 we present numerical results obtained when Method 10.7 is applied to problem (10.18) for various values of ε and N. In Table 10.9 we give the computed maximum pointwise errors E_ε^N and E^N for problem (10.18)

Table 10.9 *Computed maximum pointwise error E_ε^N and E^N generated by Method 10.7 applied to problem (10.18) for various values of ε and N.*

	Boundary Conditions: $u_\varepsilon(0) = 1$, $u_\varepsilon(1) = 1$						
	Initial Guess : $\quad u_{\text{init}} = 0$						
	Number of intervals N						
ε	8	16	32	64	128	256	512
2^{-3}	0.006885	0.001755	0.000440	0.000110	0.000027	0.000006	0.000001
2^{-5}	0.011231	0.003309	0.000878	0.000219	0.000054	0.000013	0.000003
2^{-7}	0.016902	0.011381	0.003402	0.000899	0.000223	0.000053	0.000011
2^{-9}	0.056008	0.012978	0.005182	0.001950	0.000673	0.000226	0.000043
2^{-11}	0.094695	0.031616	0.007446	0.001971	0.000679	0.000230	0.000081
2^{-13}	0.118767	0.049207	0.016901	0.004247	0.000736	0.000231	0.000081
2^{-15}	0.131978	0.060343	0.025123	0.008769	0.002272	0.000409	0.000082
2^{-17}	0.138874	0.066527	0.030410	0.012707	0.004469	0.001170	0.000202
2^{-19}	0.142394	0.069775	0.033379	0.015264	0.006380	0.002232	0.000544
2^{-21}	0.144171	0.071436	0.034943	0.016704	0.007614	0.003131	0.000985
2^{-23}	0.145062	0.072274	0.035742	0.017458	0.008297	0.003688	0.001321
2^{-25}	0.145508	0.072693	0.036143	0.017840	0.008649	0.003985	0.001514
E^N	0.145508	0.072693	0.036143	0.017840	0.008649	0.003985	0.001514

Table 10.10 *Computed ε–uniform maximum pointwise error E^N and ε–uniform order of convergence p^N generated by Method 10.7 applied to problem (10.18) for various values of N.*

N	8	16	32	64	128	256
	Boundary Conditions: $u_\varepsilon(0) = 1$, $u_\varepsilon(1) = 1$					
	Initial Guess : $\quad u_{\text{init}} = 0$					
E^N	0.145508	0.072693	0.036143	0.017840	0.008649	0.003985
p^N	0.94	1.02	1.02	1.03	1.04	1.09
	Boundary Conditions: $u_\varepsilon(0) = .5$, $u_\varepsilon(1) = .7$					
	Initial Guess : $\quad u_{\text{init}} = 0$					
E^N	0.093222	0.046517	0.023123	0.011413	0.005533	0.002549
p^N	0.97	1.03	1.02	1.03	1.04	1.09
	Initial Guess : $\quad u_{\text{init}} = u_\varepsilon(0) + (u_\varepsilon(1) - u_\varepsilon(0))x$					
E^N	0.093257	0.046535	0.023132	0.011417	0.005536	0.002550
p^N	0.97	1.03	1.02	1.03	1.04	1.09

corresponding to the boundary conditions $A = 1$, $B = 1$, and the initial guess $u_{\text{init}} = 0$ in the continuation method, for various values of ε and N. In Table 10.10 we give the computed ε–uniform maximum pointwise error E^N and the computed ε–uniform order of convergence p^N for problem (10.18) with the same boundary conditions and initial guess as in Table 10.9, and also for problem (10.18) with the boundary conditions $A = 0.5, B = 0.7$ and two different choices of the initial guess, namely $u_{\text{init}} = 0$ and $u_{\text{init}} = u_\varepsilon(0) + (u_\varepsilon(1) - u_\varepsilon(0))x$, for various values of ε and N.

It is clear from these tables that, for these problems, Method 10.7, based on the centred finite difference operator on a piecewise–uniform fitted mesh, is an

Table 10.11 *Computed ε–uniform maximum pointwise error E^N and ε–uniform order of convergence p^N generated by Method 10.7 applied to problem (10.19) for various values of N.*

N	8	16	32	64	128	256
	Boundary Conditions: $u_\varepsilon(0)=1$, $u_\varepsilon(1)=1$					
	Initial Guess :		$u_{\text{init}}=0$			
E^N	0.097937	0.048050	0.023691	0.011648	0.005637	0.002595
p^N	1.06	1.08	1.04	1.04	1.05	1.09
	Initial Guess :		$u_{\text{init}}=.5$			
E^N	0.097873	0.048014	0.023672	0.011638	0.005632	0.002593
p^N	1.06	1.08	1.04	1.04	1.05	1.09
	Boundary Conditions: $u_\varepsilon(0)=.5$, $u_\varepsilon(1)=.7$					
	Initial Guess :		$u_{\text{init}}=0$			
E^N	0.073090	0.036008	0.017791	0.008755	0.004239	0.001952
p^N	1.05	1.07	1.04	1.03	1.05	1.09
	Initial Guess :		$u_{\text{init}}=u_\varepsilon(0)+(u_\varepsilon(1)-u_\varepsilon(0))x$			
E^N	0.073063	0.035993	0.017782	0.008751	0.004237	0.001951
p^N	1.05	1.07	1.04	1.03	1.05	1.09

ε–uniform method. Moreover, the computed order of ε–uniform convergence is approximately 1, and it is independent of the choice of the initial guess u_{init} in the continuation algorithm and of the chosen boundary conditions A and B.

In Table 10.11 we give the corresponding numerical results for problem (10.19), for various choices of the boundary conditions and the initial guess. It is remarkable that, for each choice of the boundary conditions and of the initial guess, the numerical solutions converge ε–uniformly to the exact solution and that the computed ε–uniform order of convergence of the method is essentially independent of the initial guess. These numerical experiments suggest that, once again, a simple recipe for the construction of an ε–uniform method for the semilinear problems in Problem Class 10.2 is a standard finite difference operator on an appropriately fitted piecewise–uniform mesh.

Indeed, assuming the additional condition

$$\frac{\partial}{\partial u}(c(u(x))u) > 0 \tag{10.20}$$

on the data of problems from Problem Class 10.2, it is proved in Shishkin (1991b) that the numerical solutions obtained from Method 10.7, with the same continuation algorithm as in §10.3, satisfy an ε–uniform error estimate of the form

$$\|U_\varepsilon(x,t_j) - u_\varepsilon(x)\|_{\Omega_\varepsilon^N} \le C(N^{-1}\ln N + q^{t_j/h_t}), \tag{10.21}$$

where $q < 1$ is a measure of the order of convergence of the continuation method. The error estimate (10.21) implies that the iterates of the continuation algorithm converge ε–uniformly to the solution $u_\varepsilon(x)$. That this is in fact the case for problem (10.12) is shown in Table 10.12, in which we give the number of iterations

Table 10.12 *Number of solver iterations in the continuation algorithm required to satisfy condition (10.11) with tol = 0.00001 for Method 10.7 applied to problem (10.12).*

ε	Number of intervals N						
	8	16	32	64	128	256	512
2^{-1}	5	5	5	5	5	5	5
2^{-2}	7	7	7	7	7	7	7
2^{-3}	8	8	8	8	8	8	8
2^{-4}	10	10	10	10	10	10	10
2^{-5}	12	12	12	12	12	12	12
2^{-6}	14	14	14	14	14	14	14
2^{-7}	15	15	15	15	15	15	15
2^{-8}	15	15	15	15	15	15	15
2^{-9}	15	16	16	16	16	16	16
2^{-10}	16	16	16	16	16	16	16
.	.	.	.	.	.	.	.
.	.	.	.	.	.	.	.
.	.	.	.	.	.	.	.
2^{-25}	16	16	16	16	16	16	16

required for the difference between successive iterates to satisfy condition (10.11) with *tol* = 0.00001.

It is easy to see that problems (10.12) and (10.13) satisfy condition (10.20). But for problem (10.18) to satisfy this condition, we need to impose the further constraint

$$\max\{|u_\varepsilon(0)|, |u_\varepsilon(1)|\} < 1/\sqrt{3}$$

on the boundary values. The numerical results presented in Table 10.9 show that Method 10.7 is ε–uniform for some problems that do not satisfy condition (10.20). This illustrates the obvious observation that a numerical method may, in practice, be ε–uniform for a much wider class of problems than that for which we can prove such behaviour theoretically by rigorous mathematical analysis.

10.6   An alternative stopping criterion

In this section, for various choices of the tolerance *tol*, we apply Method 10.7 to problem (10.18) with the boundary conditions $A = B = 1$ and the continuation algorithm of §10.3 with the initial guess $u_{\text{init}} = 0$. The computed ε–uniform maximum pointwise error and order of convergence corresponding to *tol* = 10^{-3} and *tol* = 10^{-6} are given in Table 10.13 for various values of N. We see that there is little difference between the entries in these tables, and so we conclude that the larger value *tol* = 10^{-3} is a satisfactory choice for this particular problem.

Instead of this stopping criterion , which is based on an arbitrary user–chosen tolerance *tol*, we can stop the iterative process when the two terms on the right hand side of (10.21) are balanced for all t_j, in the sense that their magnitudes are approximately equal:

Table 10.13 *Computed ε–uniform maximum pointwise error E^N and ε–uniform order of convergence p^N generated by Method 10.7 applied to problem (10.18) for two values of tol and various values of N.*

N	8	16	32	64	128	256
			$tol = 10^{-3}$			
E^N	0.145508	0.072693	0.036143	0.017840	0.008649	0.003985
p^N	0.94	1.02	1.02	1.03	1.04	1.09
			$tol = 10^{-6}$			
E^N	0.145605	0.072743	0.036168	0.017852	0.008655	0.003988
p^N	0.94	1.02	1.02	1.03	1.04	1.09

$$N^{-1} \ln N \approx q^{T/h_t}. \tag{10.22}$$

Since $h_t = T/K$, this is equivalent to requiring that

$$K \approx (\ln(1/q))^{-1} \ln(N(\ln N)^{-1})) \leq M \ln N,$$

where an appropriate value of the constant $M > 1$ must be chosen for each specific problem and each numerical method. This may be determined experimentally, by performing the computations for $K = \ln N, \, 2\ln N, \, 4\ln N, \, 8\ln N, ...$ and then choosing M to be the first value for which no difference is observed between the computed ε–uniform maximum pointwise errors E^N for $K = M \ln N$ and $K = 2M \ln N$. To illustrate this procedure we present in Table 10.14, for various values of N, the computed ε–uniform maximum pointwise error E^N and the computed ε–uniform order of convergence p^N for $K = \ln N, \, 2\ln N, \, 4\ln N, \, 8\ln N$ when Method 10.7 is applied to problem (10.18), with the boundary conditions $A = B = 1$ and the initial guess $u_{init} = 0$. Comparing the values of E^N in the four rows of Table 10.14, we see that the row corresponding to

$$K = 4\ln N$$

is the first row that is unchanged when K is increased by a factor of 2. We therefore take $M = 4$ as the appropriate value in this specific case.

When the number of iterations K is chosen according to this alternative stopping criterion, we see from (10.22) that

$$N^{-1} \ln N \approx q^K.$$

It follows that the theoretical error estimate (10.21) can be written in the form

$$\|U_\varepsilon(x,T) - u_\varepsilon(x)\|_{\Omega^N} \leq C_p N^{-p}.$$

We can therefore apply the experimental techniques of chapter 8 to compute realistic approximations to the error parameters p and C_p. In the particular case when Method 10.7 is applied to problem (10.18), the general algorithm of §8.3, generates the computed estimates of p and C_p shown in Table 10.15. This leads to the following computed explicit ε–uniform error bound for Method 10.7 applied

Table 10.14 *Computed ε–uniform maximum pointwise error E^N and order of ε–uniform convergence p^N generated by Method 10.7 applied to problem (10.18) with boundary conditions $u_\varepsilon(0) = 1$, $u_\varepsilon(1) = 1$ for four values of K and various values of N.*

N	8	16	32	64	128	256
			$K = \ln N$			
E^N	0.145599	0.101038	0.060681	0.033645	0.033614	0.014354
p^N	0.94	1.01	1.02	-0.51	0.42	3.24
			$K = 2\ln N$			
E^N	0.145605	0.072743	0.043904	0.024097	0.017150	0.006948
p^N	0.94	1.02	1.02	0.41	1.43	0.27
			$K = 4\ln N$			
E^N	0.145605	0.072743	0.036168	0.017852	0.008655	0.003988
p^N	0.94	1.02	1.02	1.03	1.04	1.09
			$K = 8\ln N$			
E^N	0.145605	0.072743	0.036168	0.017852	0.008655	0.003988
p^N	0.94	1.02	1.02	1.03	1.04	1.09

Table 10.15 *Computed ε–uniform mesh difference D^N, ε–uniform order of convergence p^N and error constant C_p^N generated by Method 10.7 with $K = 4\ln N$ applied to problem (10.18) for various values of N.*

	Number of intervals N						
	8	16	32	64	128	256	512
D^N	0.106648	0.055469	0.027427	0.013569	0.006663	0.003231	0.001515
p^N	0.94	1.02	1.02	1.03	1.04	1.09	
$C_{0.94}^N$	1.57	1.57	1.49	1.41	1.33	1.24	1.11

to problem (10.18), for all $N \geq 8$,

$$\|U_\varepsilon(x,T) - u_\varepsilon(x)\|_{\Omega_\varepsilon^N} \leq 1.57N^{-0.94}.$$

The possibility of obtaining an explicit ε–uniform error bound of this kind is a major motivation for using this alternative stopping criterion in the continuation algorithm.

CHAPTER 11

Prandtl flow past a flat plate - Blasius' method

11.1 Prandtl boundary layer equations

Up to this point in the book we have dealt mostly with problems and numerical methods for which mathematically rigorous ε–uniform error estimates are available. In the final two chapters we step beyond this to show that the techniques, which we have developed and validated in the previous chapters, can be used in an experimental manner to develop robust numerical methods for which no relevant rigorous mathematical results are at present known. We demonstrate this by applying our techniques to the steady laminar flow of an incompressible fluid past a thin flat semi–infinite plate $P = \{(x,0) \in \Re^2 : x \geq 0\}$, as shown in Fig. 11.1. Our goal is to model the flow for all Reynolds numbers for which the flow remains laminar and no separation occurs on the plate. In the present chapter we use a variant of the semi–analytic approach of Blasius to generate numerical approximations of guaranteed accuracy to the flow variables and their scaled derivatives. Then, in the final chapter, we construct a new numerical method for solving the problem. Although there are at present no theoretical error estimates for the resulting numerical solutions, we determine their accuracy by means of extensive numerical experiments and comparison with the

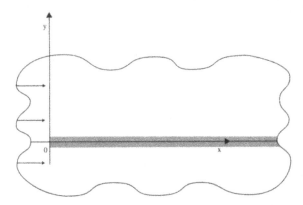

FIG. 11.1. Laminar flow past a semi–infinite flat plate.

previously determined semi–analytic approximations. We show that these numerical approximations are pointwise accurate and that they satisfy pointwise error estimates that are uniform with respect to the Reynolds number.

These results demonstrate that the techniques developed in the previous chapters enable us to obtain better numerical approximations than heretofore, even in cases where no theoretical error estimates are available. Moreover, we firmly believe that this approach is capable of further development and that, in the future, it will yield improved numerical solutions to many problems in computational fluid dynamics.

Incompressible flow past the plate P in the domain $D = \Re^2 \setminus P$ is governed by the non–dimensional Navier–Stokes equations, which can be written in the form

$$(P_{\text{NS}}) \begin{cases} \text{Find } \mathbf{u}_{\text{NS}} = (u_{\text{NS}}, v_{\text{NS}}), \ p_{\text{NS}} \text{ such that for all } (x,y) \in D, \\ \mathbf{u}_{\text{NS}} \text{ satisfies the differential equations} \\[2mm] -\frac{1}{Re}\Delta \mathbf{u}_{\text{NS}} + \mathbf{u}_{\text{NS}} \cdot \nabla \mathbf{u}_{\text{NS}} = -\frac{1}{\rho}\nabla p_{\text{NS}} \\[2mm] \nabla \cdot \mathbf{u}_{\text{NS}} = 0 \\[2mm] \text{with the boundary conditions} \\[2mm] \mathbf{u}_{\text{NS}}(x,0) = \mathbf{0} \quad \text{for all } x \geq 0 \\[2mm] \lim_{|y|\to\infty} \mathbf{u}_{\text{NS}}(x,y) = \lim_{x\to-\infty} \mathbf{u}_{\text{NS}}(x,y) = (1,0), \quad \text{for all } x \in \Re. \end{cases}$$

This is a nonlinear system of three equations for the three unknowns $\mathbf{u}_{\text{NS}}, p_{\text{NS}}$. The approach of Prandtl (1904), described for example in Acheson (1990, p. 271), simplifies (P_{NS}) to the following Prandtl problem in the domain D

$$(P_{\text{P}}) \begin{cases} \text{Find } \mathbf{u}_{\text{P}} = (u_{\text{P}}, v_{\text{P}}) \text{ such that for all } (x,y) \in D \\ \mathbf{u}_{\text{P}} \text{ satisfies the differential equations} \\[2mm] -\frac{1}{Re}\frac{\partial^2 u_{\text{P}}(x,y)}{\partial y^2} + \mathbf{u}_{\text{P}} \cdot \nabla u_{\text{P}}(x,y) = 0 \\[2mm] \nabla \cdot \mathbf{u}_{\text{P}}(x,y) = 0 \\[2mm] \text{with the boundary conditions} \\[2mm] \mathbf{u}_{\text{P}}(x,0) = \mathbf{0} \quad \text{for all } x \geq 0 \\[2mm] \lim_{|y|\to\infty} \mathbf{u}_{\text{P}}(x,y) = \lim_{x\to-\infty} \mathbf{u}_{\text{P}}(x,y) = (1,0), \quad \text{for all } x \in \Re. \end{cases}$$

This is a nonlinear system of two equations for the two unknown components u_P, v_P of the velocity $\mathbf{u}_P$. Since there is only one second order derivative in the first differential equation in (P_P), this is a parabolic equation in contrast to the elliptic equation in (P_{NS}). From Prandtl's work it is known that the solution of (P_P) is a good approximation to the solution of (P_{NS}) in a subdomain excluding the leading edge region, provided that the flow remains laminar and that no separation occurs.

There is no mathematical theory for the simultaneous existence and uniqueness of solutions $\mathbf{u}_{NS}, p_{NS}$, for arbitrary values of Re and for realistic data (see Doering and Gibbon (1995) for a discussion), for a class of problems of a form related to (P_{NS}). The class of problems for which solutions are known to exist does not intersect with any realistic class of problems known to have unique solutions. By a realistic class of problems we mean problems for which the solutions are of finite magnitude on a finite domain. Furthermore, there is no known maximum principle for (P_{NS}). On the other hand, if the Reynolds number Re is not too large, then it is known that a smooth unique solution exists for a class of problems of a form related to (P_{NS}). Furthermore, for moderately large values of Re, asymptotic analysis leads to the conjecture that, in any domain $D' = D \backslash \overline{D}_1$, where D_1 is an open neighbourhood of the leading edge of the plate, the solution $\mathbf{u}_P$ of (P_P) is the leading term in the asymptotic expansion of the solution $\mathbf{u}_{NS}$ of (P_{NS}). Indeed, in the laminar case it is known from classical analysis that the velocities $\mathbf{u}_{NS}$ and $\mathbf{u}_P$ satisfy a pointwise inequality of the form

$$\|\mathbf{u}_P - \mathbf{u}_{NS}\|_{D'} \leq C(Re)^{-p}, \qquad (11.1)$$

where p and C are positive constants independent of Re.

It is well known that problem (P_P) has a self-similar solution in the open subdomain $\{(x,y) : x > 0, |y| > 0\}$. By symmetry, it suffices to study the problem in the open quarter plane $\{(x,y) : x > 0, y > 0\}$. To avoid its more complicated behaviour near the leading edge of the plate, we compute numerical approximations to the solution of (P_P) on a finite rectangle, which does not contain the leading edge. More precisely, we take as the computational domain Ω a finite rectangle, which is a fixed distance to the right of the leading edge of the plate. We can take this domain to be both as close to the leading edge and as large as desired, provided that its location and size are independent of the Reynolds number. An appropriate domain $\Omega = (a, A) \times (0, B)$ is shown in Fig. 11.2, where a, A and B are fixed and independent of Re, with a arbitrarily small and A and B arbitrarily large. In all of the numerical computations in this chapter and the next we use the specific values

$$a = 0.1, \quad A = 1.1, \quad B = 1 \quad \text{and} \quad Re \in [1, \infty).$$

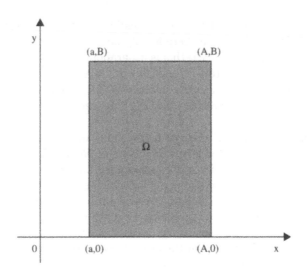

FIG. 11.2. Computational domain Ω for the Prandtl problem (P_P).

11.2   Blasius' solution

The classical approach of Blasius (1908), described for example in Schlichting (1979, p. 135–139), shows that in the open quarter plane $\{(x,y) : x > 0, y > 0\}$ a self–similar solution $\mathbf{u}_B = (u_B, v_B)$ of (P_P) can be written in the form

$$u_B(x,y) = f'(\eta) \tag{11.2a}$$

$$v_B(x,y) = \frac{1}{2}\sqrt{\frac{2}{xRe}}(\eta f'(\eta) - f(\eta)) \tag{11.2b}$$

where

$$\eta = y\sqrt{Re/2x} \tag{11.2c}$$

and the function f is the solution of the Blasius problem

(P_B)
$$\begin{cases} \text{Find a function } f \in C^3([0,\infty)) \text{ such that for all } \eta \in (0,\infty) \\[2mm] f'''(\eta) + f(\eta)f''(\eta) = 0 \\[2mm] \text{with the boundary conditions} \\[2mm] f(0) = f'(0) = 0, \quad \lim_{\eta \to \infty} f'(\eta) = 1. \end{cases}$$

The equation in (P_B) is known as the Blasius equation and, in what follows, we refer to $\mathbf{u}_B$ as the Blasius solution of (P_P). It is a third order nonlinear ordinary differential equation on the semi–infinite domain $(0, \infty)$. In Schlichting (1979) the existence and uniqueness of a solution to this problem is discussed. The exact solution of (P_B) can be obtained in the form of a series expansion. It can also be solved numerically on any bounded subdomain I of $[0, \infty)$, the standard approach being to use a Runge–Kutta method on a uniform mesh in the variable η.

In what follows we solve the Blasius problem (P_B) numerically for the function f and its derivatives, and then we use the analytic relations (11.2) to construct the Blasius solution u_B of (P_P) and its scaled derivatives. Our approach differs from the standard one, in that we construct Re–uniform analytic approximations, with guaranteed accuracy, to the solution u_P of (P_P) and its scaled first derivatives for all relevant values of Re at all points of the domain Ω. To achieve this it is clear that we need Re–uniform pointwise approximations with guaranteed accuracy to $f(\eta)$, $f'(\eta)$ and $f''(\eta)$ for all $\eta \in (0, \infty)$. We know of no standard numerical method for computing approximations of f, f' and f'' that fulfill these requirements. Therefore, in the following sections we construct a new numerical method, using the techniques discussed in earlier chapters for singular perturbation problems in one dimension, which enables us to generate such Re–uniform approximations.

11.3   Singularly perturbed nature of Blasius' problem

It is not well known that the Blasius problem (P_B) is a singularly perturbed problem, and so we give a brief explanation of this in the present section. In the next section we use this observation to motivate the construction of our new numerical method for solving problem (P_B).

We observe that, while we need the solution of (P_B) on an infinite interval, in practice we can find numerical solutions only on a finite interval. For this reason we introduce a one–parameter family of problems related to (P_B) on the finite interval $(0, L)$, where the length L of the interval is taken as the parameter of the family. The typical problem in this family is defined, for each value of L in the range $1 \leq L < \infty$, by

$$f_L'''(\eta) + f_L(\eta)f_L''(\eta) = 0, \quad \eta \in (0, L), \tag{11.3a}$$
$$f_L(0) = f_L'(0) = 0, \quad f_L'(L) = 1. \tag{11.3b}$$

We can reformulate problem (11.3) as a singularly perturbed convection–diffusion problem. It turns out that $1/L$ is the singular perturbation parameter, and so it is appropriate to introduce the temporary notation $\varepsilon = 1/L$. Then problem (11.3) can be written in the form

$$f_\varepsilon'''(\eta) + f_\varepsilon(\eta)f_\varepsilon''(\eta) = 0, \quad f_\varepsilon(0) = f_\varepsilon'(0) = 0, \ f_\varepsilon'(1/\varepsilon) = 1.$$

Putting $g_\varepsilon = f_\varepsilon'$, this problem becomes

$$g_\varepsilon''(\eta) + f_\varepsilon(\eta)g_\varepsilon'(\eta) = 0, \quad g_\varepsilon(0) = 0, \ g_\varepsilon(1/\varepsilon) = 1.$$

Changing variables from η to $\xi = \varepsilon\eta$, which is equivalent to a mapping of the interval $[0, 1/\varepsilon]$ to the unit interval $[0, 1]$, and writing $k_\varepsilon(\xi) = g_\varepsilon(\eta)$, $h(\xi) = f_\varepsilon(\eta)$, we obtain

$$\varepsilon k_\varepsilon''(\xi) + h(\xi)k_\varepsilon'(\xi) = 0, \quad k_\varepsilon(0) = 0, \ k_\varepsilon(1) = 1, \tag{11.4}$$

where $h(0) = 0$ and $h(\xi) = O(\xi)$ for all $\xi \geq \varepsilon$. This is a singularly perturbed problem for k_ε of standard form with a boundary layer at $\xi = 0$. Consequently, to construct an ε–uniform method for problem (11.4), the appropriate transition parameter σ_ξ in the piecewise–uniform fitted mesh is $\sigma_\xi = \sigma_\xi(\varepsilon, N) = \varepsilon\psi(N)$, where $\psi(N) \to \infty$ as $N \to \infty$. As usual we take $\psi(N) = \ln N$, which in terms of the original variable η gives $\sigma_\eta(N) = \ln N$.

In what follows it is natural to refer to L–uniform methods for (11.3). These are defined in an analogous way to the ε–uniform methods for problems in previous chapters.

11.4   Robust layer–resolving method for Blasius' problem

Our strategy for computing L–uniform approximations to the solution of (P_B) is to obtain a numerical solution F_L for problem (11.3), on the interval $[0, L]$, for an increasing sequence of values of the length L. Since, for each L, we need the values of f_L, f_L' and f_L'' at all points of the interval $[0, \infty)$, we extend the domain of definition of f_L, f_L' and f_L'' from $[0, L]$ to the semi–infinite interval $[0, \infty)$ by defining the following extrapolations

$$f_L''(\eta) = 0, \quad \text{for all} \quad \eta \geq L \tag{11.5a}$$
$$f_L'(\eta) = 1, \quad \text{for all} \quad \eta \geq L \tag{11.5b}$$
$$f_L(\eta) = (\eta - L) + f_L(L), \quad \text{for all} \quad \eta \geq L \tag{11.5c}$$

where (11.5c) is obtained by integrating both sides of (11.5b) from L to η.

We now describe our numerical method for finding approximations to the solution and its derivatives of problem (P_B). For each fixed N we write $L_N = \ln N$ and we divide the interval $[0, \infty)$ into the two subintervals $[0, L_N]$ and $[L_N, \infty)$. We construct a uniform mesh $\overline{I}_u^N = \{\eta_i : \eta_i = iN^{-1}\ln N, 0 \leq i \leq N\}_0^N$ on the subinterval $[0, L_N]$, and we determine numerical approximations F, D^+F and D^+D^+F to f_L, f_L' and f_L'' respectively, at the mesh points $\overline{I}_u^N$ using the nonlinear finite difference method

$$(P_B^N) \begin{cases} \text{Find } F \text{ on } \overline{I}_u^N \text{ such that , for all } \eta_i \in I_u^N, 2 \leq i \leq N - 1, \\[2mm] \delta^2(D^-F)(\eta_i) + F(\eta_i)D^+(D^-F)(\eta_i) = 0 \\[2mm] F(0) = D^+F(0) = 0, \quad \text{and} \quad D^0F(\eta_{N-1}) = 1. \end{cases}$$

It is worth observing that the use of the backward finite difference D^-F in the finite difference equations (P_{B}^N) ensures that this numerical method is monotone and that, with the given boundary conditions, its solution exists and is unique. We also note the use of the centred finite difference operator D^0, in the right–hand boundary condition, which is a higher order approximation to the first derivative, and it can be shown that monotonicity of the numerical method is preserved.

In practice, since (P_{B}^N) is nonlinear, we need a nonlinear solver to compute its solution. One possibility is the following continuation algorithm with the continuation parameter m.

$$(A_{\mathrm{B}}^N) \begin{cases} \text{For each integer } m,\ 1 \le m \le M \text{ find } F^m \text{ on } I_u^N \\ \text{such that , for all } \eta_i \in \overline{I}_u^N,\ 2 \le i \le N-1 \\[4pt] \delta^2(D^-F^m)(\eta_i) + F^{m-1}(\eta_i)D^+(D^-F^m)(\eta_i) - D^-(F^m - F^{m-1})(\eta_i) = 0 \\[4pt] F^m(0) = D^+F^m(0) = 0, \quad \text{and} \quad D^0 F^m(\eta_{N-1}) = 1 \\ \text{with the starting values for all mesh points } \eta_i \in \overline{I}_u^N \\ F^0(\eta_i) = \eta_i. \end{cases}$$

Again the use of the backward finite difference $D^-(F^m - F^{m-1})(\eta_i)$ ensures that the numerical method is monotone. We use the alternative stopping criterion described in the final section of chapter 10. Using the experimental technique described in §10.6, we find that for this problem the appropriate choice for the number of iterations is $M = 8 \ln N$. Note that the continuation algorithm (A_{B}^N) involves the solution of a sequence of *linear* problems, one linear problem corresponding to each value of m. To avoid cumbersome notation we suppress explicit mention of N and M, and we denote the final output of algorithm (A_{B}^N) simply by F. In what follows the finest mesh is taken to be $I_u^{N^*}$, with N^* chosen sufficiently large so that the asymptotic order of convergence of the numerical solutions is observed on several meshes I_u^N for which $N < N^*$. For our purposes $N^* = 65,536$ suffices and we use the numerical solution F on the mesh I_u^{65536} to replace the unknown exact solution in the expression for the pointwise error.

We assign the values $D^+F(\eta_N) = 1$ and $D^+D^+F(\eta_N) = D^+D^+F(\eta_{N-1}) = 0$, so that F, D^+F and D^+D^+F are defined at all points of the mesh $\overline{I}_u^N$. We then use piecewise linear interpolation to interpolate from $\overline{I}_u^N$ to each point of the subinterval $[0, L_N]$. We denote the corresponding interpolants by $\overline{F}, \overline{D^+F}$ and $\overline{D^+D^+F}$. We extend these three functions to the whole of the semi–infinite interval $[0, \infty)$ in an analogous way to the extensions (11.5) of their continuous counterparts, that is

$$\overline{F}(\eta) = \overline{F}(L_N) + (\eta - L_N), \quad \text{for all} \quad \eta \in [L_N, \infty) \qquad (11.6a)$$

$$\overline{D^+F}(\eta) = 1, \quad \text{for all} \quad \eta \in [L_N, \infty), \qquad (11.6b)$$

$$\overline{D^+D^+F}(\eta) = 0, \quad \text{for all} \quad \eta \in [L_N, \infty). \qquad (11.6c)$$

We take the values of $\overline{F}$, $\overline{D^+F}$, $\overline{D^+D^+F}$, respectively, to be the required numerical approximations to the exact values f, f', f'' of the Blasius solution and its derivatives on the semi–infinite interval $[0, \infty)$.

11.5   Numerical solution of Blasius' problem

The computed maximum pointwise errors E^N with respect to the finest mesh I_u^{65536}, the pointwise two–mesh differences D^N and the two–mesh orders of convergence p^N for $\overline{F}$, $\overline{D^+F}$, $\overline{D^+D^+F}$, respectively, are given in Tables 11.1, 11.2 and 11.3. The main conclusion to be drawn from the numerical results in these tables is that method (P_B^N), in conjunction with algorithm (A_B^N), is, in practice, a robust layer–resolving method for problem (11.3) in the sense that it is L–uniform and that the L–uniform order of convergence on $[0, L_N]$ of the numerical solution F to the exact solution f_L and of the discrete derivatives D^+F and D^+D^+F to the derivatives f'_L and f''_L, respectively, is, in practice, better than 0.8 for all $N \geq 512$.

It is important to observe that, when $N > 2048$, the effect of rounding error is significant in these calculations. This is due, in part, to the fact that the discretization matrix is lower Hessenberg with two lower diagonals, unlike the tridiagonal matrix encountered in the one–dimensional problems considered in previous chapters. In addition, we use much finer meshes here than in previous cases. Consequently, we need quadruple precision to achieve the L–uniform convergence behaviour achieved in Tables 11.1–11.3.

We recall that the order of local convergence p^N, defined in (8.17), corresponding to the theoretical behaviour $N^{-1} \ln N$ and $N^{-1} (\ln N)^2$, is given in Table 8.4 for various values of N. Comparison of the entries in the last rows of Tables 11.1–11.3 with those in the second row of Table 8.4 suggests strongly that the computed order of L–uniform convergence p^N observed in Tables 11.1–11.3, corresponds to the theoretical behaviour $N^{-1} \ln N$. Note that quadruple precision is also essential for the estimation of the error parameters, which we

Table 11.1 *Computed maximum pointwise error E^N, computed two–mesh difference D^N and computed order of convergence p^N in quadruple precision arithmetic for F on $\overline{I}_u^N$ generated by method (A_B^N) applied to problem (P_B) with $M = 8 \ln N$ and various values of N.*

N	128	256	512	1024	2048	4096	8192	16384
E^N	0.018727	0.011030	0.006181	0.003393	0.001826	0.000956	0.000478	0.000217
D^N	0.007741	0.004851	0.002788	0.001567	0.000870	0.000478	0.000261	0.000141
p^N	0.67	0.80	0.83	0.85	0.86	0.87	0.88	0.89

Table 11.2 *Computed maximum pointwise error E^N, two–mesh difference D^N and order of convergence p^N in quadruple precision arithmetic for D^+F on $\overline{I}_u^N \setminus \{\eta_N\}$ generated by method (A_B^N) with $M = 8 \ln N$ applied to problem (P_B) for various values of N.*

N	128	256	512	1024	2048	4096	8192	16384
E^N	0.001266	0.000679	0.000386	0.000212	0.000114	0.000060	0.000030	0.000014
D^N	0.000607	0.000296	0.000174	0.000098	0.000054	0.000030	0.000016	0.000009
p^N	1.04	0.77	0.83	0.85	0.86	0.87	0.88	0.89

Table 11.3 *Computed maximum pointwise error E^N, two–mesh difference D^N and order of convergence p^N in quadruple precision arithmetic for D^+D^+F on $\overline{I}_u^N \setminus \{\eta_N, \eta_{N-1}\}$ generated by method (A_B^N) with $M = 8 \ln N$ applied to problem (P_B) for various values of N .*

N	128	256	512	1024	2048	4096	8192	16384
E^N	0.006279	0.003608	0.002023	0.001112	0.000599	0.000314	0.000157	0.000071
D^N	0.002669	0.001585	0.000910	0.000513	0.000285	0.000157	0.000086	0.000046
p^N	0.75	0.80	0.83	0.85	0.86	0.87	0.88	0.89

undertake in the next section. We deduce this from the double precision values for p^N shown in Table 11.4, which could lead to the erroneous conclusion that method (P_B^N), in conjunction with algorithm (A_B^N), is not L–uniform for problem (11.3). Comparing the values in Tables 11.4 and 11.1 for the computed maximum pointwise error E^N, we see that there is a persistent difference in the entries of approximately 0.000046. This suggests that the lack of precision in the entries of Table 11.4 occurs because the fine mesh solution $F^{M,65536}$ is calculated in double rather than quadruple precision.

Graphs of $\overline{F}$ for $N = 8192$ on $[0, L_N]$, and a zoom to a neighbourhood of the boundary point $\eta = 0$, are given in Fig. 11.3. Graphs of $\overline{D^+F}$ and $\overline{D^+D^+F}$ on $[0, L_N]$ are displayed in Fig. 11.4.

11.6 Computed error estimates for Blasius' problem

Our preliminary theoretical investigations, which are not discussed in this book, indicate that, for all $N \geq N_0$, the approximations $\overline{F}, \overline{D^+F}, \overline{D^+D^+F}$, respec-

Table 11.4 *Computed maximum pointwise error E^N, two–mesh difference D^N and order of convergence p^N in double precision arithmetic for F on I_u^N generated by method (A_B^N) with $M = 8 \ln N$ applied to problem (P_B) for various values of N.*

N	128	256	512	1024	2048	4096	8192	16384
E^N	0.018773	0.011076	0.006226	0.003438	0.001872	0.001002	0.000524	0.000276
D^N	0.007741	0.004851	0.002788	0.001567	0.000870	0.000479	0.000249	0.000178
p^N	0.67	0.80	0.83	0.85	0.86	0.94	0.49	0.79

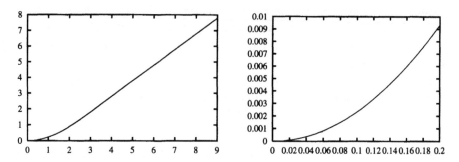

FIG. 11.3. Solution $\overline{F}$ generated by method (A_B^N) applied to problem (P_B) with $M = 8 \ln N$ and $N = 8192$.

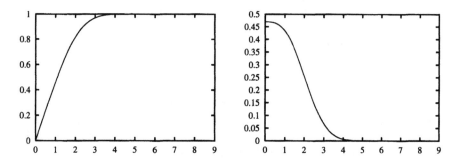

FIG. 11.4. First and second order discrete derivatives $\overline{D^+F}$ and $\overline{D^+D^+F}$ generated by method (A_B^N) applied to problem (P_B) with $M = 8 \ln N$ and $N = 8192$.

tively, to the exact values f, f', f'' of the solution of the Blasius problem (P_B) and its derivatives satisfy error bounds of the form

$$\|\overline{F} - f\|_{[0,\infty)} \leq C_p N^{-p}, \quad p > 0 \tag{11.7a}$$

$$\|\overline{D^+F} - f'\|_{[0,\infty)} \leq C_p N^{-p}, \quad p > 0 \tag{11.7b}$$

$$\|\overline{D^+D^+F} - f''\|_{[0,\infty)} \leq C_p N^{-p}, \quad p > 0. \tag{11.7c}$$

where the error parameters p and C_p are unknown. We now use the experimental technique of §8.6 to obtain computed parameter–uniform global error parameters for the function $\overline{F}$ on the semi–infinite interval $[0, \infty)$. All of our computations are carried out in quadruple precision for the reasons explained in the previous section. To determine the computed two–mesh differences $\bar{D}^N$, and hence the computed order of convergence $\bar{p}^N$ for each pair of meshes with N and $2N$ points, respectively, we need to consider the three sub–intervals $[0, L_N)$, $[L_N, L_{2N})$ and $[L_{2N}, \infty)$ separately. For $\eta \in [L_N, L_{2N})$, the two–mesh difference at η for $\overline{F}$ is

Table 11.5 *Computed two–mesh difference $\bar{D}^N$ and order of convergence $\bar{p}^N$ for $\overline{F}$ on $[0,\infty)$ generated by method (A_B^N) with $M = 8\ln N$ applied to problem (P_B) for various values of N.*

N	128	256	512	1024	2048	4096	8192	16384
$\bar{D}^N$	0.007741	0.004851	0.002788	0.001567	0.000870	0.000478	0.000261	0.000141
$\bar{p}^N$	0.67	0.80	0.83	0.85	0.86	0.87	0.88	0.89

Table 11.6 *Computed two–mesh difference $\bar{D}^N$ and order of convergence $\bar{p}^N$ for $\overline{D^+F}$ on $[0,\infty)$ generated by method (A_B^N) with $M = 8\ln N$ applied to problem (P_B) for various values of N.*

N	128	256	512	1024	2048	4096	8192	16384
$\bar{D}^N$	0.000607	0.000303	0.000176	0.000099	0.000055	0.000030	0.000016	0.000009
$\bar{p}^N$	1.01	0.78	0.84	0.85	0.87	0.88	0.89	0.89

$F^{2N}(\eta) - F^N(L_N) - (\eta - L_N)$, for $\overline{D^+F}$ it is $\overline{D^+F^{2N}}(\eta) - 1$, and for $\overline{D^+D^+F}$ it is $\overline{D^+D^+F^{2N}}(\eta)$. In the sub–interval $[L_{2N}, \infty)$, the two–mesh difference at η for $\overline{F}$ is $F^{2N}(L_{2N}) - F^N(L_N) - \ln 2$, and for $\overline{D^+F}$ and $\overline{D^+D^+F}$ it is zero. The resulting computed global two–mesh difference $\bar{D}^N$ and the computed global order of convergence $\bar{p}^N$ are given in Tables 11.5, 11.6 and 11.7, respectively, for various values of N.

Using these tables and the experimental technique of §8.6 we obtain the following error bounds for all $N \geq 256$

$$\|\overline{F} - f\|_{[0,\infty)} \leq 0.962N^{-0.80}, \tag{11.8a}$$

$$\|\overline{D^+F} - f'\|_{[0,\infty)} \leq 0.055N^{-0.78}, \tag{11.8b}$$

$$\|\overline{D^+D^+F} - f''\|_{[0,\infty)} \leq 0.315N^{-0.80}. \tag{11.8c}$$

In (11.8) we take the lower bound 256 for N, because comparison of the computed global order of convergence $\bar{p}^N$, in Tables 11.5–7, with the theoretical order of convergence $N^{-1}\ln N$, in Table 8.4, suggests that the asymptotic order of convergence is reached when $N = 256$.

The computed global error bounds (11.8a–11.8c) can be used for all values of $N \geq 256$, but, in the next chapter, we need only the value $N = 8192$. For this number of nodes, in accordance with the discussion in §8.1, we obtain in a

Table 11.7 *Computed two–mesh difference $\bar{D}^N$ and order of convergence $\bar{p}^N$ for $\overline{D^+D^+F}$ on $[0,\infty)$ generated by method (A_B^N) with $M = 8\ln N$ applied to problem (P_B) for various values of N.*

N	128	256	512	1024	2048	4096	8192	16384
$\bar{D}^N$	0.002675	0.001587	0.000911	0.000513	0.000285	0.000157	0.000086	0.000046
$\bar{p}^N$	0.75	0.80	0.83	0.85	0.86	0.87	0.88	0.89

Table 11.8 *Computed two–mesh difference $\bar{D}^N$ and order of convergence $\bar{p}^N$ of $\eta \overline{D^+ F} - \overline{F}$ to $\eta f'(\eta) - f(\eta)$ on $[0, \infty)$ generated by method (A_B^N) with $M = 8 \ln N$ applied to problem (P_B) for various values of N.*

N	128	256	512	1024	2048	4096	8192	16384
$\bar{D}^N$	0.008438	0.004893	0.002789	0.001567	0.000870	0.000478	0.000261	0.000141
$\bar{p}^N$	0.79	0.81	0.83	0.85	0.86	0.87	0.88	0.89

Table 11.9 *Computed two–mesh difference $\bar{D}^N$ and order of convergence $\bar{p}^N$ of $\eta \overline{D^+ D^+ F}$ to $\eta f''(\eta)$ on $[0, \infty)$ generated by method (A_B^N) with $M = 8 \ln N$ applied to problem (P_B) for various values of N.*

N	128	256	512	1024	2048	4096	8192	16384
$\bar{D}^N$	0.005374	0.003192	0.001837	0.001037	0.000577	0.000318	0.000174	0.000094
$\bar{p}^N$	0.75	0.80	0.82	0.84	0.86	0.87	0.88	0.89

similar manner to that for (11.8), the following sharper ε–uniform global error bounds for all $N \geq 2048$

$$\|\overline{F} - f\|_{[0,\infty)} \leq 1.364 N^{-0.86}, \tag{11.9a}$$

$$\|\overline{D^+ F} - f'\|_{[0,\infty)} \leq 0.092 N^{-0.87}, \tag{11.9b}$$

$$\|\overline{D^+ D^+ F} - f''\|_{[0,\infty)} \leq 0.447 N^{-0.86}. \tag{11.9c}$$

We deduce from (11.2b) that u_B, v_B, ∇u_B and ∇v_B involve the expressions $\eta f'(\eta) - f(\eta)$, $\eta f''(\eta)$ and $\eta^2 f''(\eta)$. Therefore, we compute estimates of the global error parameters for the corresponding approximations $\eta \overline{D^+ F}(\eta) - \overline{F}(\eta)$, $\eta \overline{D^+ D^+ F}(\eta)$ and $\eta^2 \overline{D^+ D^+ F}(\eta)$, respectively, using the techniques of §8.6 and the data in Tables 11.8, 11.9 and 11.10. The resulting computed global error bounds, which hold for all $N \geq 2048$, are

$$\|(\eta \overline{D^+ F} - \overline{F}) - (\eta f' - f)\|_{[0,\infty)} \leq 1.364 N^{-0.86}, \tag{11.10}$$

$$\|\eta(\overline{D^+ D^+ F} - f'')\|_{[0,\infty)} \leq 0.905 N^{-0.86}, \tag{11.11}$$

$$\|\eta^2(\overline{D^+ D^+ F} - f'')\|_{[0,\infty)} \leq 2.023 N^{-0.86}. \tag{11.12}$$

11.7 Computed global error estimates for Blasius' solution

We now obtain approximate expressions $\mathbf{U_B} = (U_B, V_B)$ for the Blasius solution u_B of (P_P) by substituting into the relations (11.2a–11.2b) the approximate

Table 11.10 *Computed two–mesh difference $\bar{D}^N$ and order of convergence $\bar{p}^N$ of $\eta^2 \overline{D^+ D^+ F}$ to $\eta^2 f''(\eta)$ on $[0, \infty)$ generated by method (A_B^N) with $M = 8 \ln N$ applied to problem (P_B) for various values of N.*

N	256	512	1024	2048	4096	8192	16384
$\bar{D}^N$	0.007102	0.004094	0.002315	0.001290	0.000710	0.000388	0.000210
$\bar{p}^N$	0.79	0.82	0.84	0.86	0.87	0.88	0.89

expressions $\overline{F}$ and $\overline{D^+F}$ for f and f', computed in §11.5. Thus, for each (x, y) in the rectangle $\overline{\Omega}$, we define

$$U_B(x, y) = \overline{D^+F}(\eta) \tag{11.13a}$$

and

$$V_B(x, y) = \frac{1}{2}\sqrt{\frac{2}{xRe}}(\eta\overline{D^+F}(\eta) - \overline{F}(\eta)) \tag{11.13b}$$

where $\eta \in [0, \infty)$ is given by (11.2c).

Graphs of the resulting approximate solution U_B are displayed in Figs. 11.5, 11.6, and 11.7 for $Re = 1.0$, 100, and 100,000 respectively. These graphs are constructed from the subset of computed values taken on a piecewise–uniform mesh of 17×17 points, using the data of the Blasius solution F corresponding to $M = 8\ln N$ and $N = 8192$. We observe from these figures that, while the maximum of u_P is of order 1, the maximum of v_P is of order $1/\sqrt{Re}$. Moreover, particularly in Fig. 11.7, the successful resolution of the boundary layer at the surface of the plate is apparent.

Our preliminary analytical investigations also indicate that, for all $N \geq N_0$ and some unknown constants p_1, p_2, C_{p_1} and C_{p_2}, the following theoretical error bounds hold for the approximations $(U_B, \sqrt{Re}V_B)$ to the scaled velocity $(u_B, \sqrt{Re}v_B)$:

$$\|U_B - u_B\|_{\overline{\Omega}} \leq C_{p_1} N^{-p_1}, \quad p_1 > 0 \tag{11.14}$$

$$\sqrt{Re}\,\|V_B - v_B\|_{\overline{\Omega}} \leq C_{p_2} N^{-p_2}, \quad p_2 > 0 \tag{11.15}$$

where U_B is the computed Blasius solution of the Prandtl problem on $\overline{\Omega}$ given by (11.13) and u_B is the exact solution of (P_P). Here N_0 and the error parameters

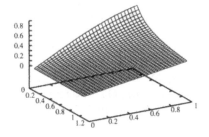
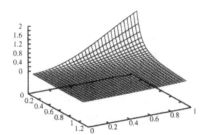

FIG. 11.5. Components of the semi–analytic solution (U_B, V_B) of problem (P_P) for $Re = 1.0$ generated from the numerical solution $\overline{F}$ of problem (P_B) with $M = 8\ln N$ and $N = 8192$.

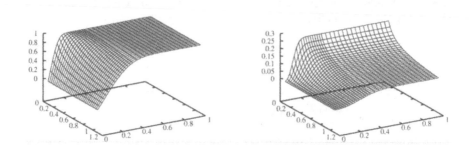

FIG. 11.6. Components of the semi–analytic solution (U_B, V_B) of problem (P_P) for $Re = 100$ generated from the numerical solution $\overline{F}$ of problem (P_B) with $M = 8 \ln N$ and $N = 8192$.

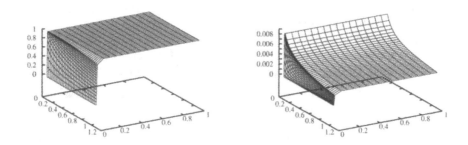

FIG. 11.7. Components of the semi–analytic solution (U_B, V_B) of problem (P_P) for $Re = 100,000$ generated from the numerical solution $\overline{F}$ of problem (P_B) with $M = 8 \ln N$ and $N = 8192$.

p_1, p_2, C_{p_1} and C_{p_2} are independent of M and N, provided that the number of iterations is sufficiently large, as discussed in §11.4.

In what follows, we again use the experimental techniques from §8.6 to determine realistic estimates of these global error parameters and, consequently, realistic error bounds of the form (11.14) and (11.15).

For the x–component U_B of the velocity we obtain, from (11.13a) and (11.9b), the following computed error bound for all $M \geq 8 \ln N$ and all $N \geq 2048$

$$\|U_B - u_B\|_{\overline{\Omega}} = u_\infty \|\overline{D^+ F} - f'\|_{[0,\infty)}$$
$$\leq 0.092 N^{-0.87}. \tag{11.16}$$

Similarly for the scaled y–component $\sqrt{Re}V_B$ of the velocity we obtain, from (11.13b) and (11.10), for all $M \geq 8 \ln N$ and all $N \geq 2048$

$$\sqrt{Re}\|V_{\rm B} - v_{\rm B}\|_{\overline{\Omega}} = \sqrt{Re}\|\frac{1}{2}\sqrt{\frac{2}{xRe}}[(\eta\overline{D^+F}(\eta) - \overline{F}(\eta)) - (\eta f' - f)]\|_{[0,\infty)}$$
$$\leq \sqrt{5}\,\|(\eta\overline{D^+F}(\eta) - \overline{F}(\eta)) - (\eta f' - f)\|_{[0,\infty)}$$
$$\leq \sqrt{5}(1.364N^{-0.86})$$
$$= 3.05N^{-0.86}. \tag{11.17a}$$

In the specific case $N = 8192$ the computed error bounds (11.16) and (11.17a) become

$$\|U_{\rm B} - u_{\rm B}\|_{\overline{\Omega}} \leq 0.092(8192)^{-0.87} = 0.362 \times 10^{-4}, \tag{11.18}$$
$$\sqrt{Re}\|V_{\rm B} - v_{\rm B}\|_{\overline{\Omega}} \leq 3.05(8192)^{-0.86} = 0.120 \times 10^{-2}. \tag{11.19}$$

In a similar manner to the above, we calculate computed error bounds for the approximations of the scaled first derivatives by scaled discrete derivatives. In order to estimate error bounds for the derivatives of $u_{\rm B}$ and $v_{\rm B}$, we introduce the expressions

$$\partial_y U_{\rm B}(\eta(x,y)) = \frac{\eta}{y}\overline{D_\eta^+ D_\eta^+ F}(\eta) \tag{11.20a}$$
$$\partial_y V_{\rm B}(\eta(x,y)) = \frac{\eta}{2x}\overline{D_\eta^+ D_\eta^+ F}(\eta) \tag{11.20b}$$
$$\partial_x U_{\rm B}(\eta(x,y)) = -\partial_y V_{\rm B}(\eta(x,y)) \tag{11.20c}$$
$$\partial_x V_{\rm B}(\eta(x,y)) = -\frac{1}{2x}(V_{\rm B} + \sqrt{\frac{1}{2xRe}}\eta^2\overline{D_\eta^+ D_\eta^+ F}(\eta)). \tag{11.20d}$$

In terms of these quantities the computed global error bounds for the scaled discrete derivatives of the velocity components, for all $N \geq 2048$, are

$$\frac{1}{\sqrt{Re}}\,\|\partial_y U_{\mathrm{B}} - \frac{\partial u_{\mathrm{B}}}{\partial y}\|_{\overline{\Omega}} = \frac{1}{\sqrt{Re}}\,\sqrt{\frac{Re}{2x}}\|\overline{D_\eta^+ D_\eta^+ F(\eta)} - f''(\eta)\|_{[0,\infty)}$$
$$\leq \sqrt{5}\|\overline{D_\eta^+ D_\eta^+ F(\eta)} - f''(\eta)\|_{[0,\infty)}$$
$$\leq 1.00 N^{-0.86}, \tag{11.21a}$$

$$\|\partial_y V_{\mathrm{B}} - \frac{\partial v_{\mathrm{B}}}{\partial y}\|_{\overline{\Omega}} = \|D_x U_{\mathrm{B}} - \frac{\partial u_{\mathrm{B}}}{\partial x}\|_{\overline{\Omega}}$$
$$= \frac{\eta}{2x}\|\overline{D_\eta^+ D_\eta^+ F}(\eta) - f''(\eta)\|$$
$$\leq 4.53 N^{-0.86}, \tag{11.21b}$$

$$\sqrt{Re}\|\partial_x V_{\mathrm{B}} - \frac{\partial v_{\mathrm{B}}}{\partial x}\|_{\overline{\Omega}} = \sqrt{Re}\frac{1}{2x}(\|V_{\mathrm{B}} - v_{\mathrm{B}}\| + \sqrt{\frac{1}{2xRe}}\eta^2\|\overline{D_\eta^+ D_\eta^+ F}(\eta) - f''(\eta)\|)$$
$$\leq \frac{1}{2x}(\sqrt{Re}\|V_{\mathrm{B}} - v_{\mathrm{B}}\| + \sqrt{\frac{1}{2x}}\eta^2\|\overline{D_\eta^+ D_\eta^+ F}(\eta) - f''(\eta)\|)$$
$$\leq 15.25 N^{-0.86} + 22.62 N^{-0.86}$$
$$\leq 37.9 N^{-0.86} \tag{11.21c}$$

where we have used (11.9c), (11.11), (11.12), (11.19) and the fact that $x \geq 0.1$. In the specific case $N = 8192$ these computed global error bounds become

$$\frac{1}{\sqrt{Re}}\|\partial_y U_{\mathrm{B}} - \frac{\partial u_{\mathrm{B}}}{\partial y}\|_{\overline{\Omega}} \leq 0.431 \times 10^{-3}, \tag{11.22a}$$

$$\|\partial_y V_{\mathrm{B}} - \frac{\partial v_{\mathrm{B}}}{\partial y}\|_{\overline{\Omega}} = \|\partial_x U_{\mathrm{B}} - \frac{\partial u_{\mathrm{B}}}{\partial x}\|_{\overline{\Omega}} \leq 0.195 \times 10^{-2}, \tag{11.22b}$$

$$\sqrt{Re}\|\partial_x V_{\mathrm{B}} - \frac{\partial v_{\mathrm{B}}}{\partial x}\|_{\overline{\Omega}} \leq 0.163 \times 10^{-1}. \tag{11.22c}$$

CHAPTER 12

Prandtl flow past a flat plate - direct method

12.1 Prandtl problem in a finite domain

This chapter is both the culmination of and different from the previous ones. In it we construct a robust layer–resolving method to compute pointwise–accurate and parameter–uniform numerical approximations to the solution $\mathbf{u}_P$ of the Prandtl problem (P_P) in the same finite rectangular computational domain Ω as in the previous chapter. Here the numerical method is direct in the sense that (P_P) is replaced by a nonlinear finite difference discretization, which is then solved by an appropriate nonlinear solver, in contrast to the indirect semi–analytic approach of Blasius. We use the computed Blasius solution $\mathbf{U}_B^{8192}$, found in the previous chapter, to give the required boundary conditions on the top edge Γ_T and inflow conditions on the left–hand edge Γ_L of Ω. Having found a numerical solution with this direct method, we investigate its error in two different ways. In the first, the unknown exact solution in the expression for the error is replaced by the numerical solution generated by our direct numerical method on the finest available mesh, and in the second by the computed Blasius solution $\mathbf{U}_B^{8192}$. It is convenient to introduce the notation $\varepsilon = \frac{1}{Re}$, which casts the problems considered here into a form similar to that in previous chapters.

We introduce the following class of nonlinear problems, which is considered in detail in Oleinik and Samokhin (1999).

Problem Class 12.1. Oleinik and Samokhin nonlinear system in two dimensions.

Find $\mathbf{u}_\varepsilon = (u_\varepsilon, v_\varepsilon)$ *such that for all* $(x, y) \in \Omega$

$$-\varepsilon \frac{\partial^2 u_\varepsilon(x, y)}{\partial y^2} + \mathbf{u}_\varepsilon \cdot \nabla u_\varepsilon(x, y) = 0 \qquad (12.1a)$$

$$\nabla \cdot \mathbf{u}_\varepsilon(x, y) = 0 \qquad (12.1b)$$

$$\mathbf{u}_\varepsilon = \mathbf{0} \quad \text{on} \quad \Gamma_B \qquad (12.1c)$$

$$u_\varepsilon = g, \quad \text{on} \quad \Gamma_L \cup \Gamma_T \qquad (12.1d)$$

For a typical problem from Problem Class 12.1, with sufficiently smooth data g and sufficient compatibility, the existence and uniqueness of a solution $\mathbf{u}_\varepsilon$ can

be established by the techniques in Oleinik and Samokhin (1999). We denote by
Γ' that part of the boundary on which boundary conditions must be specified.
From the parabolic nature of the problem, and the fact that the variable x is the
time–like variable, it is clear that $\Gamma' = \Gamma_L \cup \Gamma_T \cup \Gamma_B$.

A particular problem from this class is the following

$$(P_\varepsilon) \begin{cases} \text{Find } \mathbf{u}_\varepsilon = (u_\varepsilon, v_\varepsilon) \text{ such that for all } (x,y) \in \Omega \\[2mm] -\varepsilon \frac{\partial^2 u_\varepsilon(x,y)}{\partial y^2} + \mathbf{u}_\varepsilon \cdot \nabla u_\varepsilon(x,y) = 0 \\[2mm] \nabla \cdot \mathbf{u}_\varepsilon(x,y) = 0 \\ \mathbf{u}_\varepsilon = \mathbf{0} \quad \text{on} \quad \Gamma_B \\ u_\varepsilon = u_P \quad \text{on} \quad \Gamma_L \cup \Gamma_T \end{cases}$$

where u_P is the exact solution of (P_P). Since the boundary conditions in (P_ε)
are the appropriate values of u_P on Γ', it follows that on the finite rectangle $\overline{\Omega}$
the solution u_ε of (P_ε) is equal to the solution u_P of (P_P).

As far as we are aware, there are no rigorous theoretical error estimates
available in the literature for any numerical method for solving problems from
Problem Class 12.1. Likewise, we make no attempt to establish theoretical error
estimates. Instead we present extensive experimental evidence, based on the
numerical and experimental techniques of the previous chapters, which strongly
suggests that the direct numerical method described in the following sections is
a robust layer–resolving method for (P_ε).

We are convinced that the techniques used in this chapter to construct a
robust layer–resolving numerical method for laminar flow past a flat plate can
also aid the construction of robust layer–resolving methods for more complicated
problems. Furthermore, the tables of numerical results contained in this chapter
will serve as useful reference results for future work.

12.2 Nonlinear finite difference method

We now construct our direct numerical method for solving (P_ε). This is composed
of a standard upwind finite difference operator on an appropriate piecewise–
uniform fitted mesh. Because the computational domain is rectangular, the
piecewise–uniform fitted rectangular mesh Ω_ε^N is the tensor product of one dimensional meshes (see Fig. 12.1). Since the parabolic boundary layer on Γ_B is
the only boundary layer in the solution, we can take a uniform mesh in the
direction of the time–like x–axis and a piecewise–uniform fitted mesh in the di-
rection of the y–axis. The tensor product of these meshes is $\overline{\Omega}_\varepsilon^N = \overline{\Omega}_u^{N_x} \times \overline{\Omega}_\varepsilon^{N_y}$,
where $\mathbf{N} = (N_x, N_y)$, $\overline{\Omega}_u^{N_x}$ is a uniform mesh with N_x mesh intervals on the
interval $[a, A]$ of the x–axis, and $\overline{\Omega}_\varepsilon^{N_y}$ is a piecewise–uniform fitted mesh with
N_y mesh intervals on the interval $[0, B]$ of the y–axis, such that the subinterval

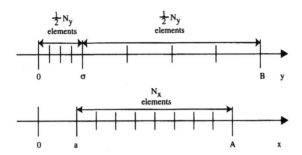

FIG. 12.1. Meshes $\Omega_u^{N_x}$ and $\Omega_\varepsilon^{N_y}$ in the case $N_x = N_y = 8$.

$[0, \sigma]$ and the subinterval $[\sigma, B]$ are both subdivided into $\frac{1}{2}N_y$ uniform mesh intervals. To determine the correct choice of the transition parameter σ we note that the parabolic boundary layer in the solution of (P_ε) is degenerate in nature. However, this degeneracy is not of the power form discussed in chapter 7, where the velocity component v_ε has the power behaviour y^α. In the present case the velocity component v_ε has an exponential form closely related to $1 - e^{-y/\sqrt{\varepsilon}}$. This motivates us to define a similar σ to that for non–degenerate parabolic boundary layers, namely

$$\sigma = \min\{\frac{1}{2}B, \ \sqrt{\varepsilon}\ln N\},$$

and the choice of $\sqrt{\varepsilon}$ can be motivated either from *a priori* estimates of the derivatives of the solution $\mathbf{u}_\varepsilon$ or from the asymptotic analysis in Schlichting (1979). The resulting piecewise–uniform mesh $\overline{\Omega}_\varepsilon^{\mathbf{N}}$ for the case $\mathbf{N} = (8,8)$ is illustrated in Fig. 12.2.

Using the above piecewise–uniform fitted mesh $\Omega_\varepsilon^{\mathbf{N}}$, the problem (P_ε) is discretized by the following nonlinear system of upwind finite difference equations for the approximate velocity $\mathbf{U}_\varepsilon = (U_\varepsilon, V_\varepsilon)$

$(P_\varepsilon^{\mathbf{N}})$
$\begin{cases}
\text{Find } \mathbf{U}_\varepsilon = (U_\varepsilon, V_\varepsilon) \text{ such that for all mesh points } (x_i, y_j) \in \Omega_\varepsilon^{\mathbf{N}} \\[2mm]
-\varepsilon\delta_y^2 U_\varepsilon(x_i, y_j) + (\mathbf{U}_\varepsilon \cdot \mathbf{D}^-)U_\varepsilon(x_i, y_j) = 0 \\[2mm]
(\mathbf{D}^- \cdot \mathbf{U}_\varepsilon)(x_i, y_j) = 0 \\[2mm]
\text{with the boundary conditions} \\[2mm]
\mathbf{U}_\varepsilon = \mathbf{0} \quad \text{on} \quad \Gamma_B \\
U_\varepsilon = U_B^{8192} \quad \text{on} \quad \Gamma_L \cup \Gamma_T
\end{cases}$

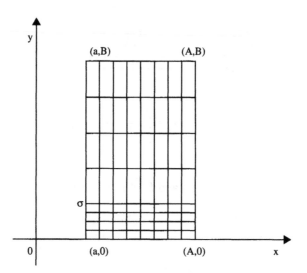

FIG. 12.2. Piecewise–uniform mesh Ω_ε^N in the case $N_x = N_y = 8$.

where $\mathbf{D}^- = (D_x^-, D_y^-)$. We note that in (P_ε^N) we use the known approximate boundary values U_B^{8192} on $\Gamma_L \cup \Gamma_T$ to replace the unknown exact boundary values u_P, where U_B^{8192} is the computed Blasius solution of the Prandtl problem (P_P) from the previous chapter. Since (P_ε^N) is a nonlinear finite difference method, it is necessary to prescribe an ε–uniform nonlinear solver for computing ε–uniformly convergent approximations to its solution. This is the topic of the next section.

We note also that, because there are no known theoretical results concerning the convergence of the solutions $\mathbf{U}_\varepsilon$ of (P_ε^N) to the solution $\mathbf{u}_\varepsilon$ of (P_ε), we have no theoretical estimate for the pointwise error $(\mathbf{U}_\varepsilon - \mathbf{u}_\varepsilon)(x_i, y_j)$. It is for this reason that we are forced to adopt an experimental technique.

12.3   Solution of the nonlinear finite difference method

Finding the solution $\mathbf{U}_\varepsilon$ of (P_ε^N) involves the solution of a nonlinear system of finite difference equations. We employ the following iterative method which involves solving successively a nonlinear system for U_ε and a linear system for V_ε. We replace (P_ε^N) by an algorithm, which sweeps across the domain $\overline{\Omega}$ from left to right, that is, in the same direction as the physical flow. At the ith stage of the sweep, we compute the values of $\mathbf{U}_\varepsilon$ on the x–level $X_i = \{(x_i, y_j), 1 \leq j \leq N\}$ assuming that values of $\mathbf{U}_\varepsilon$ are known on X_{i-1}. For each i, $1 \leq i \leq N$, we linearize the nonlinear tridiagonal system for U_ε on X_i by introducing the sequence of linear problems

$$(-\varepsilon \delta_y^2 U_\varepsilon^m + (\mathbf{U}_\varepsilon^{m-1} \cdot \mathbf{D}^-) U_\varepsilon^m)(x_i, y_j) = 0, \quad 1 \leq j \leq N - 1$$

where for each $m, 1 \leq m \leq M$, $\{U_\varepsilon^m(x_i, y_j),\ 1 \leq j \leq N - 1\}$ are the unknown values of the mth iterate. In order to solve this system on X_i we need values of U_ε on X_{i-1}, boundary values for U_ε^m at points of $\Gamma_B \cup \Gamma_T$ and an initial guess U_ε^0 on X_i. The required values on $\Gamma_B \cup \Gamma_T$ are taken to be the given boundary conditions for U_ε on $\Gamma_B \cup \Gamma_T$, and in general, we take as the initial guess U_ε^0 on X_i the value of the final iterate on X_{i-1}. The initial guess U_ε^0 on X_1 is taken to be the prescribed boundary conditions for U_ε on Γ_L and V_ε^0 is taken to be zero. We then find V_ε by solving the linear system

$$\mathbf{D}^- \cdot U_\varepsilon^m(x_i, y_j) = 0, \quad 1 \leq j \leq N$$

for the unknown values $\{V_\varepsilon(x_i, y_j),\ 1 \leq j \leq N\}$. Note that this requires only the given initial condition for V_ε on Γ_B.

We continue this process until the change between two successive iterates for the scaled velocity $(U_\varepsilon^m, \frac{1}{\sqrt{\varepsilon}} V_\varepsilon^m)$ is less than a specified tolerance tol, that is $\max(|U_\varepsilon^m - U_\varepsilon^{m-1}|_{\overline{X}_i}, \frac{1}{\sqrt{\varepsilon}}|V_\varepsilon^m - V_\varepsilon^{m-1}|_{\overline{X}_i}) \leq tol$, and we denote by M_i the value of m for which this occurs. This completes the ith stage of the sweep. The complete algorithm can be expressed in the following form

$$(A_\varepsilon^N) \begin{cases} \text{With the boundary condition } U_\varepsilon^M = U_B^{8192} \text{ on } \Gamma_L, \\[2mm] \text{for each } i, 1 \leq i \leq N, \text{ use the initial guess } U_\varepsilon^0|_{X_i} = U_\varepsilon^{M_{i-1}}|_{X_{i-1}} \\[2mm] \text{and for } m = 1, \dots, M_i \text{ solve the following} \\ \text{two point boundary value problem for } U_\varepsilon^m(x_i, y_j) \\[2mm] (-\varepsilon \delta_y^2 + U_\varepsilon^{m-1} \cdot \mathbf{D}^-)U_\varepsilon^m(x_i, y_j) = 0, \quad 1 \leq j \leq N - 1 \\[2mm] \text{with the boundary conditions } U_\varepsilon^m = U_B \text{ on } \Gamma_B \cup \Gamma_T, \\[2mm] \text{and the initial guess for } V_\varepsilon^0|_{X_1} = 0. \\[2mm] \text{Also solve the initial value problem for } V_\varepsilon^m(x_i, y_j) \\[2mm] (\mathbf{D}^- \cdot U_\varepsilon^m)(x_i, y_j) = 0, \\[2mm] \text{with initial condition } V_\varepsilon^m = 0 \text{ on } \Gamma_B. \\[2mm] \text{Continue to iterate between the equations for } U_\varepsilon^m \text{ until } m = M_i, \\ \text{where } M_i \text{ is such that} \\[2mm] \max(|U_\varepsilon^{M_i} - U_\varepsilon^{M_i-1}|_{\overline{X}_i}, \frac{1}{\sqrt{\varepsilon}}|V_\varepsilon^{M_i} - V_\varepsilon^{M_i-1}|_{\overline{X}_i}) \leq tol. \end{cases}$$

For notational simplicity, we suppress explicit mention of the iteration super-script M_i henceforth, and we write simply $\mathbf{U}_\varepsilon$ for the solution generated by (A_ε^N). We take $tol = 10^{-6}$ in the following computations.

Graphs of the scaled components $(U_\varepsilon, \varepsilon^{-1/2}V_\varepsilon)$ of the solution $\mathbf{U}_\varepsilon$ generated by the direct method (A_ε^N) with $N = 32$ for $\varepsilon = 1, 0.01$ and 0.00001 respectively are shown in Figs. 12.3, 12.4 and 12.5. We observe that these graphs are qual-itatively similar to the corresponding graphs of the computed Blasius solution $\mathbf{U}_B^{8192}$ in Figs. 11.5, 11.6 and 11.7.

In the next section we show, computationally, that the numerical solutions and their discrete derivatives generated by the algorithm (A_ε^N) are pointwise–accurate ε–uniform approximations to the scaled velocity $(u_\varepsilon, \varepsilon^{-1/2}v_\varepsilon)$ and its scaled derivatives.

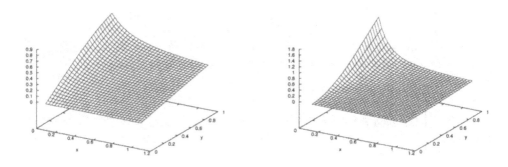

FIG. 12.3. Components of $(U_\varepsilon, \varepsilon^{-1/2}V_\varepsilon)$ for $\varepsilon = 1.0$ with $\mathbf{U}_\varepsilon$ generated by method (A_ε^N) applied to problem (P_ε) and $N = 32$.

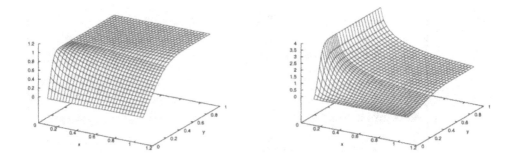

FIG. 12.4. Components of $(U_\varepsilon, \varepsilon^{-1/2}V_\varepsilon)$ for $\varepsilon = 0.01$ with $\mathbf{U}_\varepsilon$ generated by method (A_ε^N) applied to problem (P_ε) and $N = 32$.

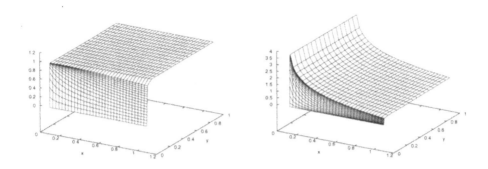

FIG. 12.5. Components of $(U_\varepsilon, \varepsilon^{-1/2}V_\varepsilon)$ for $\varepsilon = 0.00001$ with $\mathbf{U}_\varepsilon$ generated by method (A_ε^N) applied to problem (P_ε) and $N = 32$.

12.4   Error analysis based on the finest mesh solution

In this section we estimate, computationally, the maximum pointwise error in the numerical solution and its discrete derivatives generated by the algorithm (A_ε^N) of the previous section. Since the exact solution in the expression for the computed maximum pointwise error is unknown, we replace it by the solution $\mathbf{U}_\varepsilon^{512}$ generated by (A_ε^N) on the finest available mesh Ω_ε^{512}. Thus, for U_ε and $\varepsilon^{-1/2}V_\varepsilon$, we define the computed maximum pointwise errors

$$E_\varepsilon^N(U_\varepsilon) = \|U_\varepsilon - \overline{U}_\varepsilon^{512}\|_{\overline{\Omega}_\varepsilon^N} \quad \text{and} \quad E_\varepsilon^N(\varepsilon^{-1/2}V_\varepsilon) = \varepsilon^{-1/2}\|V_\varepsilon - \overline{V}_\varepsilon^{512}\|_{\overline{\Omega}_\varepsilon^N \backslash \Gamma_\mathrm{L}}.$$

The values of these errors generated by method (A_ε^N) applied to problem (P_ε) are given in Tables 12.1 and 12.2 respectively for various values of ε and N. The corresponding computed orders of convergence p_ε^N and p^N, defined in (8.21) and (8.17) respectively, for U_ε and $\varepsilon^{-1/2}V_\varepsilon$ are given in Tables 12.3 and 12.4 respectively for various values of ε and N. The results in these tables suggest that (A_ε^N) is an ε–uniform numerical method and that its ε–uniform order of convergence is at least 0.8 for all $N \geq 8$. We note that the irregular pattern in the order of convergence, for small values of N, in Tables 12.3 and 12.4 is due to the interaction of the errors in the numerical approximations U_ε^N and U_ε^{2N}.

We also approximate the scaled partial derivatives $\dfrac{\partial u_\varepsilon}{\partial x}$, $\varepsilon^{1/2}\dfrac{\partial u_\varepsilon}{\partial y}$, $\varepsilon^{-1/2}\dfrac{\partial v_\varepsilon}{\partial x}$

and $\dfrac{\partial v_\varepsilon}{\partial y}$ by the correspondingly scaled discrete derivatives $D_x^- U_\varepsilon$, $\varepsilon^{1/2}D_y^- U_\varepsilon$, $\varepsilon^{-1/2}D_x^- V_\varepsilon$ and $D_y^- V_\varepsilon$. Graphs of these computed scaled discrete derivatives generated by method (A_ε^N) applied to problem (P_ε) are given for $N = 32$ and two values of ε in Figs. 12.6, 12.7, 12.8 and 12.9. From the definition of the algorithm (A_ε^N) we know that $D_x^- U_\varepsilon = -D_y^- V_\varepsilon$, which is reflected in the graphs in Figs. 12.6 and 12.9.

Table 12.1 *Computed maximum pointwise errors $E_\varepsilon^N(U_\varepsilon)$ generated by method (A_ε^N) applied to problem (P_ε) for various values of ε and N.*

ε	8	16	32	64	128	256
2^0	0.416D-02	0.442D-02	0.266D-02	0.146D-02	0.688D-03	0.239D-03
2^{-2}	0.501D-01	0.240D-01	0.116D-01	0.544D-02	0.233D-02	0.776D-03
2^{-4}	0.205D+00	0.768D-01	0.332D-01	0.147D-01	0.615D-02	0.202D-02
2^{-6}	0.217D+00	0.111D+00	0.579D-01	0.288D-01	0.118D-01	0.384D-02
2^{-8}	0.209D+00	0.109D+00	0.558D-01	0.282D-01	0.130D-01	0.467D-02
.	.	.	.	.	.	.
.	.	.	.	.	.	.
2^{-20}	0.204D+00	0.108D+00	0.558D-01	0.282D-01	0.130D-01	0.467D-02

Table 12.2 *Computed maximum pointwise errors $E_\varepsilon^N(\varepsilon^{-1/2}V_\varepsilon)$ generated by method (A_ε^N) applied to problem (P_ε) for various values of ε and N.*

ε	8	16	32	64	128	256
2^0	0.527D+00	0.363D+00	0.200D+00	0.947D-01	0.404D-01	0.136D-01
2^{-2}	0.105D+01	0.660D+00	0.351D+00	0.172D+00	0.776D-01	0.277D-01
2^{-4}	0.395D+01	0.160D+01	0.731D+00	0.344D+00	0.154D+00	0.556D-01
2^{-6}	0.454D+01	0.266D+01	0.149D+01	0.778D+00	0.333D+00	0.118D+00
2^{-8}	0.444D+01	0.263D+01	0.146D+01	0.785D+00	0.394D+00	0.156D+00
.	.	.	.	.	.	.
.	.	.	.	.	.	.
2^{-20}	0.420D+01	0.261D+01	0.146D+01	0.785D+00	0.394D+00	0.156D+00

Since the scaled derivatives of the exact solution $\mathbf{u}_\varepsilon$ are unknown, we replace them in the expression for the error by the appropriate scaled discrete derivatives of the computed solution $\mathbf{U}_\varepsilon^{512}$, generated by method (A_ε^N) on the finest available mesh. The resulting computed maximum pointwise errors E_ε^N of the computed scaled discrete derivatives $D_x^- U_\varepsilon$, $\sqrt{\varepsilon}D_y^- U_\varepsilon$ and $\varepsilon^{-1/2}D_x^- V_\varepsilon$ are given in Tables

Table 12.3 *Computed orders of convergence $p_\varepsilon^N(U_\varepsilon)$ and $p^N(U_\varepsilon)$ generated by method (A_ε^N) applied to problem (P_ε) for various values of ε and N.*

ε	8	16	32	64	128
2^0	-0.01	0.59	0.65	0.82	0.91
2^{-2}	1.05	1.02	1.00	1.00	1.00
2^{-4}	1.62	1.25	1.11	1.05	1.03
2^{-6}	1.10	0.92	0.76	1.10	1.05
2^{-8}	0.95	0.98	0.87	0.84	0.84
2^{-10}	0.94	0.97	0.87	0.84	0.84
2^{-12}	0.92	0.97	0.87	0.84	0.84
2^{-14}	0.91	0.97	0.87	0.84	0.84
2^{-16}	0.74	1.13	0.87	0.84	0.84
2^{-18}	0.56	1.24	0.94	0.84	0.84
2^{-20}	0.48	1.20	1.06	0.84	0.85
$p^N(U_\varepsilon)$	0.91	1.20	0.87	1.03	0.84

Table 12.4 *Computed orders of convergence $p_\varepsilon^N(\varepsilon^{-1/2}V_\varepsilon)$ and $p^N(\varepsilon^{-1/2}V_\varepsilon)$ generated by method (A_ε^N) applied to problem (P_ε) for various values of ε and N.*

ε	8	16	32	64	128
2^0	0.65	0.84	0.96	1.01	1.00
2^{-2}	0.79	0.94	0.96	0.96	0.94
2^{-4}	1.52	1.25	1.09	1.01	0.95
2^{-6}	0.94	0.94	0.87	1.13	1.02
2^{-8}	0.93	0.96	0.91	0.85	0.82
.	.	.	.	.	.
.	.	.	.	.	.
.	.	.	.	.	.
2^{-20}	0.86	0.95	0.91	0.85	0.82
$p^N(\varepsilon^{-1/2}V_\varepsilon)$	0.94	0.94	0.87	0.93	0.82

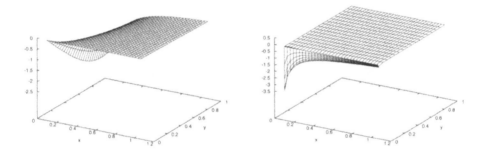

FIG. 12.6. Graphs of $D_x^- U_\varepsilon$ for $\varepsilon = 1.0$ and 0.00001 with $\mathbf{U}_\varepsilon$ generated by method (A_ε^N) applied to problem (P_ε) and $N = 32$.

12.5, 12.6 and 12.7 respectively. Of course, the errors for $D_y^- V_\varepsilon$ are identical to those for $D_x^- U_\varepsilon$, since $D_x^- U_\varepsilon = -D_y^- V_\varepsilon$, and so we omit a separate table for them.

The results in Tables 12.5 and 12.6 suggest that the method is ε-uniform for $D_x^- U_\varepsilon$ and $\sqrt{\varepsilon} D_y^- U_\varepsilon$, and the corresponding computed orders of convergence in Tables 12.8 and 12.9 indicate that the ε-uniform order of convergence is at least 0.6 for all $N \geq 32$. On the other hand, the results in Table 12.7 show that the method is not ε-uniform for $\varepsilon^{-1/2} D_x^- V_\varepsilon$, which is apparent too from the computed orders of convergence given in Table 12.10. We discuss this phenomenon at the end of the next section.

It is also of interest to investigate whether or not the number of iterations required for convergence of the approximations generated by of the nonlinear solver in (A_ε^N) is independent of ε. The experimentally determined results in Table 12.11 show the total number of one-dimensional linear solver iterations $\sum_{i=1}^{N-1} M_i$, required to achieve the chosen tolerance for each value of ε. We see

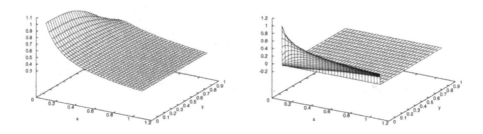

FIG. 12.7. Graphs of $\varepsilon^{1/2} D_y^- U_\varepsilon$ for $\varepsilon = 1.0$ and 0.00001 with $\mathbf{U}_\varepsilon$ generated by method $(A_\varepsilon^{\mathbf{N}})$ applied to problem (P_ε) and $N = 32$.

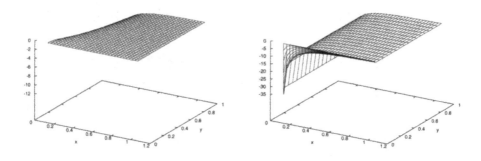

FIG. 12.8. Graphs of $\varepsilon^{-1/2} D_x^- V_\varepsilon$ for $\varepsilon = 1.0$ and 0.00001 with $\mathbf{U}_\varepsilon$ generated by method $(A_\varepsilon^{\mathbf{N}})$ applied to problem (P_ε) and $N = 32$.

that this number grows as expected as N increases, and also that it stabilizes to an ε–independent limit in each column. Dividing each entry in Table 12.11 by the value of N corresponding to its column leads to Table 12.12, in which the numbers tend to a small constant as N increases. This indicates that the method is equivalent essentially to a direct, rather than an iterative, method when N is sufficiently large.

12.5 Error analysis based on the Blasius solution

In this section we use the computed error estimates, obtained in §11.7 for the quantities $(U_{\mathrm{B}}^{8192}, \sqrt{Re} V_{\mathrm{B}}^{8192})$ and their scaled discrete derivatives, to estimate the error in the numerical approximations $(U_\varepsilon, \sqrt{Re} V_\varepsilon)$ generated by the direct

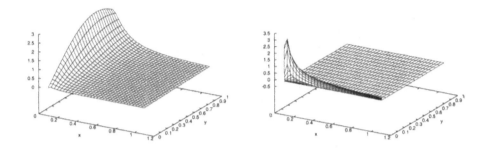

FIG. 12.9. *Graphs of* $D_y^- V_\varepsilon$ *for* $\varepsilon = 1.0$ *and* 0.00001 *with* $\mathbf{U}_\varepsilon$ *generated by method* $(A_\varepsilon^\mathbf{N})$ *applied to problem* (P_ε) *and* $N = 32$.

Table 12.5 *Computed maximum pointwise error* $E_\varepsilon^N(D_x^- U_\varepsilon)$ *generated by method* $(A_\varepsilon^\mathbf{N})$ *applied to problem* (P_ε) *for various values of* ε *and* N.

ε	8	16	32	64	128	256
2^0	0.607D+00	0.429D+00	0.239D+00	0.113D+00	0.492D-01	0.166D-01
2^{-2}	0.891D+00	0.616D+00	0.350D+00	0.176D+00	0.788D-01	0.277D-01
2^{-4}	0.188D+01	0.112D+01	0.619D+00	0.321D+00	0.149D+00	0.543D-01
2^{-6}	0.197D+01	0.164D+01	0.116D+01	0.692D+00	0.315D+00	0.114D+00
2^{-8}	0.195D+01	0.163D+01	0.115D+01	0.700D+00	0.371D+00	0.150D+00
$\cdot$	$\cdot$	$\cdot$	$\cdot$	$\cdot$	$\cdot$	$\cdot$
$\cdot$	$\cdot$	$\cdot$	$\cdot$	$\cdot$	$\cdot$	$\cdot$
$\cdot$	$\cdot$	$\cdot$	$\cdot$	$\cdot$	$\cdot$	$\cdot$
2^{-20}	0.193D+01	0.163D+01	0.115D+01	0.700D+00	0.371D+00	0.150D+00

method $(A_\varepsilon^\mathbf{N})$. The resulting estimates are independent of those obtained in the previous section.

First, we use the triangle inequality to obtain

$$\|U_\varepsilon - u_P\|_{\Omega_\varepsilon^N} = \|U_\varepsilon - u_B\|_{\Omega_\varepsilon^N}$$
$$\leq \|U_\varepsilon - U_B^{8192}\|_{\Omega_\varepsilon^N} + \|U_B^{8192} - u_B\|_{\Omega_\varepsilon^N} \qquad (12.2)$$

$$\varepsilon^{-1/2}\|V_\varepsilon - v_P\|_{\overline{\Omega}_\varepsilon^N \setminus \Gamma_L} = \varepsilon^{-1/2}\|V_\varepsilon - v_B\|_{\overline{\Omega}_\varepsilon^N \setminus \Gamma_L}$$
$$\leq \varepsilon^{-1/2}(\|V_\varepsilon - V_B^{8192}\|_{\overline{\Omega}_\varepsilon^N \setminus \Gamma_L} + \|V_B^{8192} - v_B\|_{\overline{\Omega}_\varepsilon^N}) \; (12.3)$$

where $\mathbf{U}_\varepsilon = (U_\varepsilon, V_\varepsilon)$ is the solution generated by the direct algorithm $(A_\varepsilon^\mathbf{N})$ on the mesh $\Omega_\varepsilon^\mathbf{N}$ with $\mathbf{N} = (N, N)$, $\mathbf{u}_B = (u_B, v_B)$ is the exact solution of the Prandtl problem constructed from the Blasius formulae (11.2) and $\mathbf{U}_B^{8192} = (U_B^{8192}, V_B^{8192})$ is the computed Blasius solution generated in the previous chapter on a mesh with 8192 intervals. We then observe that the first term on the right–hand side of (12.2) and (12.3) involves the computable quantities $U_\varepsilon, U_B^{8192}$ and

Table 12.6 *Computed maximum pointwise error $E_\varepsilon^N(\sqrt{\varepsilon}D_y^- U_\varepsilon)$ generated by method (A_ε^N) applied to problem (P_ε) for various values of ε and N.*

ε	8	16	32	64	128	256
2^0	0.689D-01	0.343D-01	0.166D-01	0.775D-02	0.332D-02	0.111D-02
2^{-2}	0.189D+00	0.107D+00	0.562D-01	0.274D-01	0.120D-01	0.405D-02
2^{-4}	0.262D+00	0.136D+00	0.657D-01	0.310D-01	0.133D-01	0.443D-02
2^{-6}	0.271D+00	0.183D+00	0.109D+00	0.614D-01	0.266D-01	0.885D-02
2^{-8}	0.266D+00	0.179D+00	0.105D+00	0.593D-01	0.291D-01	0.107D-01
.	.	.	.	.	.	.
.	.	.	.	.	.	.
.	.	.	.	.	.	.
2^{-20}	0.266D+00	0.179D+00	0.105D+00	0.593D-01	0.291D-01	0.107D-01

Table 12.7 *Computed maximum pointwise error $E_\varepsilon^N(\varepsilon^{-1/2}D_x^- V_\varepsilon)$ generated by method (A_ε^N) applied to problem (P_ε) for various values of ε and N.*

ε	8	16	32	64	128	256
2^0	0.362D+01	0.395D+01	0.279D+01	0.158D+01	0.890D+00	0.543D+00
2^{-2}	0.788D+01	0.634D+01	0.453D+01	0.296D+01	0.191D+01	0.123D+01
2^{-4}	0.277D+02	0.141D+02	0.955D+01	0.647D+01	0.440D+01	0.286D+01
2^{-6}	0.311D+02	0.279D+02	0.232D+02	0.179D+02	0.114D+02	0.710D+01
2^{-8}	0.320D+02	0.279D+02	0.229D+02	0.187D+02	0.149D+02	0.108D+02
.	.	.	.	.	.	.
.	.	.	.	.	.	.
.	.	.	.	.	.	.
2^{-20}	0.341D+02	0.281D+02	0.229D+02	0.187D+02	0.149D+02	0.108D+02

$V_\varepsilon, V_B^{8192}$ respectively. Furthermore, the second term on each right–hand side involves the scaled pointwise errors $U_B^{8192} - u_B$, $\varepsilon^{-1/2}(V_B^{8192} - v_B)$, which have already been estimated in §11.7 of the previous chapter. This shows that we can estimate the errors in the scaled numerical solutions and their scaled discrete derivatives, generated by the numerical method (A_ε^N) applied to problem (P_ε), even though no theoretical error analysis is available for this numerical method.

Table 12.8 *Computed orders of convergence $p_\varepsilon^N(D_x^- U_\varepsilon)$ and $p^N(D_x^- U_\varepsilon)$ generated by method (A_ε^N) applied to problem (P_ε) for various values of ε and N.*

ε	8	16	32	64	128
2^0	0.62	0.81	0.97	0.98	0.97
2^{-2}	0.67	0.83	0.92	0.96	0.96
2^{-4}	0.97	0.96	0.94	0.95	0.94
2^{-6}	0.40	0.62	0.72	1.04	0.98
2^{-8}	0.38	0.68	0.71	0.80	0.78
.	.	.	.	.	.
.	.	.	.	.	.
.	.	.	.	.	.
2^{-20}	0.36	0.68	0.71	0.80	0.78
$p^N(D_x^- U_\varepsilon)$	0.40	0.62	0.72	0.85	0.78

Table 12.9 *Computed orders of convergence $p_\varepsilon^N(\sqrt{\varepsilon}D_y^- U_\varepsilon)$ and $p^N(\sqrt{\varepsilon}D_y^- U_\varepsilon)$ generated by method (A_ε^N) applied to problem (P_ε) for various values of ε and N.*

ε	8	16	32	64	128
2^0	0.99	1.00	1.00	1.00	1.00
2^{-2}	0.68	0.82	0.91	0.95	0.97
2^{-4}	0.98	0.99	0.99	1.00	1.00
2^{-6}	0.52	0.54	0.49	0.99	1.00
2^{-8}	0.50	0.63	0.62	0.72	0.77
2^{-10}	0.48	0.63	0.62	0.72	0.77
2^{-12}	0.47	0.63	0.62	0.72	0.77
2^{-14}	0.46	0.63	0.62	0.72	0.77
2^{-16}	0.46	0.48	0.73	0.77	0.77
2^{-18}	0.45	0.36	0.64	0.98	0.77
2^{-20}	0.45	0.30	0.59	1.08	0.77
$p^N(\sqrt{\varepsilon}D_y^- U_\varepsilon)$	0.93	0.30	0.59	1.08	0.77

Table 12.10 *Computed order of convergence $p_\varepsilon^N(\varepsilon^{-1/2}D_x^- V_\varepsilon)$ generated by method (A_ε^N) applied to problem (P_ε) for various values of ε and N.*

ε	8	16	32	64	128
2^0	0.16	0.53	0.62	0.56	0.29
2^{-2}	0.49	0.51	0.52	0.43	0.24
2^{-4}	1.15	0.71	0.55	0.43	0.27
2^{-6}	0.26	0.34	0.35	0.60	0.40
2^{-8}	0.29	0.37	0.31	0.24	0.16
2^{-10}	0.31	0.38	0.31	0.24	0.16
.	.	.	.	.	.
2^{-20}	0.34	0.39	0.31	0.24	0.16

We now compare the magnitudes of the two terms on the right–hand side of both (12.2) and (12.3). The first terms are the scaled maximum pointwise differences $\|U_\varepsilon - U_B^{8192}\|_{\overline{\Omega}_\varepsilon^N}$ and $\varepsilon^{-1/2}\|V_\varepsilon - V_B^{8192}\|_{\overline{\Omega}_\varepsilon^N \setminus \Gamma_L}$. These quantities are found immediately from the solutions U_ε of (A_ε^N) and the solution U_B^{8192}, respectively,

Table 12.11 *Number of one dimensional linear solver iterations required for convergence for method (A_ε^N) applied to problem (P_ε) for various values of ε and N.*

ε	8	16	32	64	128	256	512
2^0	53	124	283	622	1295	2498	4597
2^{-2}	135	263	497	913	1676	3056	5554
2^{-4}	271	377	608	1049	1863	3333	6012
2^{-6}	288	445	729	1215	2035	3526	6228
2^{-8}	279	443	730	1229	2107	3664	6460
.	.	.	.	.	.	.	.
2^{-20}	286	456	739	1232	2107	3664	6460

Table 12.12 *Number of one dimensional linear solver iterations per level X_i required for convergence for method (A_ε^N) applied to problem (P_ε) for various values of ε and N.*

ε	8	16	32	64	128	256	512
2^0	7	8	9	10	10	10	9
2^{-2}	17	16	16	14	13	12	11
2^{-4}	34	24	19	16	15	13	12
2^{-6}	36	28	23	19	16	14	12
2^{-8}	35	28	23	19	16	14	13
.	.	.	.	.	.	.	.
2^{-20}	36	29	23	19	16	14	13

computed in the previous section and chapter. Their numerical values are given in Tables 12.13 and 12.14 for various values of ε and N. The second terms on the right–hand side of (12.2) and (12.3) are the scaled maximum pointwise errors $\|U_B^{8192} - u_B\|_{\Omega_\varepsilon^N}$ and $\varepsilon^{-1/2}\|V_B^{8192} - v_B\|_{\Omega_\varepsilon^N}$ in the computed Blasius solution. The corresponding error bounds (11.18) and (11.19), respectively, show that the second terms are bounded above by 0.362×10^{-4} and 0.120×10^{-2}.

Table 12.13 *Computed maximum pointwise difference $\|U_\varepsilon - U_B^{8192}\|_{\overline{\Omega}_\varepsilon^N}$ generated by method (A_ε^N) applied to problem (P_ε) for various values of ε and N.*

ε	8	16	32	64	128	256	512
2^0	0.420D-02	0.459D-02	0.287D-02	0.166D-02	0.898D-03	0.450D-03	0.211D-03
2^{-2}	0.509D-01	0.248D-01	0.124D-01	0.622D-02	0.312D-02	0.157D-02	0.792D-03
2^{-4}	0.207D+00	0.787D-01	0.352D-01	0.167D-01	0.817D-02	0.404D-02	0.202D-02
2^{-6}	0.220D+00	0.115D+00	0.616D-01	0.326D-01	0.156D-01	0.762D-01	0.378D-02
2^{-8}	0.213D+00	0.114D+00	0.616D-01	0.340D-01	0.189D-01	0.105D-01	0.581D-02
.	.	.	.	.	.	.	.
2^{-20}	0.208D+00	0.113D+00	0.616D-01	0.340D-01	0.189D-01	0.105D-01	0.581D-02

Table 12.14 *Computed maximum pointwise difference $\varepsilon^{-1/2}\|V_\varepsilon - V_B^{8192}\|_{\overline{\Omega}_\varepsilon^N \backslash \Gamma_L}$ generated by method (A_ε^N) applied to problem (P_ε) for various values of ε and N.*

ε	8	16	32	64	128	256	512
2^0	0.533D+0	0.374D+0	0.213D+0	0.108D+0	0.536D-01	0.268D-01	0.142D-01
2^{-2}	0.106D+1	0.677D+0	0.371D+0	0.194D+0	0.101D+0	0.531D-01	0.287D-01
2^{-4}	0.396D+1	0.163D+1	0.763D+0	0.382D+0	0.197D+0	0.104D+0	0.562D-01
2^{-6}	0.457D+1	0.271D+1	0.154D+1	0.849D+0	0.416D+0	0.215D+0	0.114D+0
2^{-8}	0.448D+1	0.269D+1	0.154D+1	0.893D+0	0.523D+0	0.309D+0	0.183D+0
.	.	.	.	.	.	.	.
2^{-20}	0.424D+1	0.267D+1	0.154D+1	0.893D+0	0.523D+0	0.309D+0	0.183D+0

It follows that the first term dominates the second term on the right–hand side of both (12.2) and (12.3). Indeed, from Table 12.13, we see that the smallest value of the first term is 0.211×10^{-3} and so the ratio of the second term to the first term on the right–hand side of (12.2) is at most $36.2/211 \approx 0.17$. In other words the magnitude of the second term is at most 17% that of the first term. Similarly, for (12.3) the corresponding ratio is at most $12.0/142 \approx 0.09$ or 9%. Therefore each entry in Table 12.13 is within at worst 17% of the true error in the numerical solution U_ε, and the corresponding figure for each entry in Table 12.14 is 9%. In other words, each entry in Table 12.13 differs from the true value of the maximum pointwise error by at most 0.4×10^{-4}, and in Table 12.14 by at most 0.2×10^{-2}. We conclude that the first term on the right–hand side of both (12.2) and (12.3) approximates the true error at least to experimental accuracy. This means that the error in the numerical approximations generated by the direct numerical method $(A_\varepsilon^{\mathbf{N}})$ may be found, at least to experimental accuracy, simply by comparing them with appropriate semi–analytic approximations generated by the Blasius technique.

Graphs of the differences $U_\varepsilon - U_{\mathrm{B}}^{8192}$ and $\varepsilon^{-1/2}(V_\varepsilon - V_{\mathrm{B}}^{8192})$ between the scaled components of the Prandtl and Blasius solutions, respectively, are shown in Figs. 12.10, 12.11 and 12.12 for $N = 32$ and $\varepsilon = 1.0, 0.01$ and 0.00001. Since the scaled semi–analytic approximations $(U_{\mathrm{B}}^{8192}, \varepsilon^{-1/2}V_{\mathrm{B}}^{8192})$ generated by the Blasius technique are representative approximations to the scaled exact solution, these graphs provide accurate representations of the scaled errors $\overline{U}_\varepsilon - u_{\mathrm{P}}$ and $\varepsilon^{-1/2}(\overline{V}_\varepsilon - v_{\mathrm{P}})$.

The situation is even better if we restrict the range of ε, by excluding its larger values. For example, if we take $\varepsilon \leq 0.25$ we see, from the entries in Tables 12.13 and 12.14, that the magnitude of the second term on the right–hand side of both (12.2) and (12.3) is at most 5% that of the first. It is worth noting

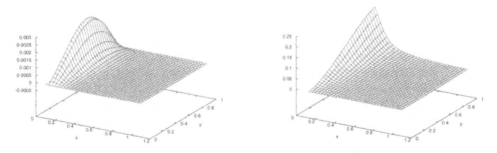

FIG. 12.10.   Graphs of $U_\varepsilon - U_{\mathrm{B}}^{8192}$ and $\varepsilon^{-1/2}(V_\varepsilon - V_{\mathrm{B}}^{8192})$ for $\varepsilon = 1.0$ with $\mathbf{U}_\varepsilon$ generated by method $(A_\varepsilon^{\mathbf{N}})$ applied to problem (P_ε) and $N = 32$.

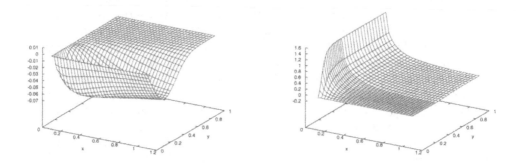

FIG. 12.11. Graphs of $U_\varepsilon - U_B^{8192}$ and $\varepsilon^{-1/2}(V_\varepsilon - V_B^{8192})$ for $\varepsilon = 0.01$ with $\mathbf{U}_\varepsilon$ generated by method $(A_\varepsilon^{\mathbf{N}})$ applied to problem (P_ε) and $N = 32$.

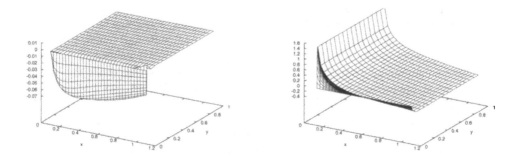

FIG. 12.12. Graphs of $U_\varepsilon - U_B^{8192}$ and $\varepsilon^{-1/2}(V_\varepsilon - V_B^{8192})$ for $\varepsilon = 0.00001$ with $\mathbf{U}_\varepsilon$ generated by method $(A_\varepsilon^{\mathbf{N}})$ applied to problem (P_ε) and $N = 32$.

also that the ε–uniform behaviour of the method is determined essentially by the maximum entry in each column of these tables. Observe that the relative magnitude of the first and second terms on the right–hand sides of (12.2) and (12.3), respectively, corresponding to the entries in each column, is at most 1% in both tables.

We now estimate the order of convergence of the numerical approximations $\mathbf{U}_\varepsilon$, generated by the direct numerical method $(A_\varepsilon^{\mathbf{N}})$, by introducing the computed orders of convergence $p_{\varepsilon,\mathrm{comp}}^N$ and p_{comp}^N, which are defined for the first component by

$$p_{\varepsilon,\text{comp}}^N(U_\varepsilon) = \log_2 \frac{\|U_\varepsilon^N - U_B^{8192}\|_{\overline{\Omega}_\varepsilon^N}}{\|U_\varepsilon^{2N} - U_B^{8192}\|_{\overline{\Omega}_\varepsilon^{2N}}} \tag{12.4a}$$

$$p_{\text{comp}}^N(U_\varepsilon) = \log_2 \frac{\max_\varepsilon \|U_\varepsilon^N - U_B^{8192}\|_{\overline{\Omega}_\varepsilon^N}}{\max_\varepsilon \|U_\varepsilon^{2N} - U_B^{8192}\|_{\overline{\Omega}_\varepsilon^{2N}}} \tag{12.4b}$$

with corresponding definitions for the scaled second component. The values of the $p_{\varepsilon,\text{comp}}^N$ and p_{comp}^N for the scaled components of $\mathbf{U}_\varepsilon$ are given in Tables 12.15 and 12.16 respectively, from which we conclude that, in practice, the method $(A_\varepsilon^{\mathbf{N}})$ is ε-uniform of order at least 0.8 for U_ε and 0.7 for $\varepsilon^{-1/2}V_\varepsilon$. Comparing the entries in these tables with the corresponding entries in Tables 12.3 and 12.4, we see that they are comparable. This is significant in view of the fact that they are generated by completely independent methods.

We now estimate the maximum pointwise error in the scaled components of the discrete derivatives of $\mathbf{U}_\varepsilon$, in an analogous manner to the above estimates

Table 12.15 *Computed orders of convergence $p_{\varepsilon,\text{comp}}^N$ and p_{comp}^N of U_ε to U_B^{8192} generated by method $(A_\varepsilon^{\mathbf{N}})$ applied to problem (P_ε) for various values of ε and N.*

ε	8	16	32	64	128	256
2^0	-0.13	0.68	0.78	0.89	1.00	1.09
2^{-2}	1.03	1.00	0.99	1.00	0.99	0.98
2^{-4}	1.39	1.16	1.07	1.03	1.01	1.00
2^{-6}	0.94	0.90	0.92	1.07	1.03	1.01
2^{-8}	0.90	0.89	0.86	0.85	0.85	0.85
.	.	.	.	.	.	.
.	.	.	.	.	.	.
.	.	.	.	.	.	.
2^{-20}	0.88	0.88	0.86	0.85	0.85	0.85
p_{comp}^N	0.94	0.90	0.86	0.85	0.85	0.85

Table 12.16 *Computed orders of convergence $p_{\varepsilon,\text{comp}}^N$ and p_{comp}^N of $\varepsilon^{-1/2}V_\varepsilon$ to $\varepsilon^{-1/2}V_B^{8192}$ generated by method $(A_\varepsilon^{\mathbf{N}})$ applied to problem (P_ε) for various values of ε and N.*

ε	8	16	32	64	128	256
2^0	0.51	0.81	0.98	1.01	1.00	0.92
2^{-2}	0.65	0.87	0.94	0.95	0.92	0.89
2^{-4}	1.28	1.09	1.00	0.96	0.91	0.89
2^{-6}	0.76	0.81	0.86	1.03	0.95	0.92
2^{-8}	0.73	0.80	0.79	0.77	0.76	0.76
.	.	.	.	.	.	.
.	.	.	.	.	.	.
2^{-20}	0.67	0.79	0.79	0.77	0.76	0.76
p_{comp}^N	0.76	0.81	0.79	0.77	0.76	0.76

of the pointwise errors in U_ε and $\varepsilon^{-1/2}V_\varepsilon$. As before, we examine the relative magnitudes of both terms on the right–hand sides of the analogous expressions to (12.2) and (12.3), respectively. The values of the computed scaled discrete derivatives $\|D_x^- U_\varepsilon - \partial_x U_B^{8192}\|_{\overline{\Omega}_\varepsilon^N \backslash \Gamma_L}$ and $\sqrt{\varepsilon}\|D_y^- U_\varepsilon - \partial_y U_B^{8192})\|_{\overline{\Omega}_\varepsilon^N \backslash \Gamma_B}$ are given in Tables 12.17 and 12.18 for various values of ε and N. We observe that the values of $\|D_x^- U_\varepsilon - \partial_x U_B^{8192}\|_\Omega$ are identical to those of $\|D_y^- V_\varepsilon - \partial_y V_B^{8192}\|_\Omega$, and so no separate table is required. From Tables 12.17 and 12.18 we see that the smallest values of $\|D_x^- U_\varepsilon - \partial_x U_B^{8192}\|_{\overline{\Omega}_\varepsilon^N \backslash \Gamma_L}$ and $\sqrt{\varepsilon}\|D_y^- U_\varepsilon - \partial_y U_B^{8192}\|_{\overline{\Omega}_\varepsilon^N \backslash \Gamma_B}$, respectively, are 0.178×10^{-1} and 0.139×10^{-2}. Recalling (11.22a) and (11.22b) we see that the magnitudes of $\|\partial_x U_B^{8192} - \frac{\partial u_B}{\partial x}\|_{\overline{\Omega}}$ and $\sqrt{\varepsilon}\|\partial_y U_B^{8192} - \frac{\partial u_B}{\partial y}\|_{\overline{\Omega}}$ are bounded by, respectively, 0.195×10^{-2} and 0.431×10^{-3}. This shows that, as before, the maximum pointwise errors are determined essentially by the first term on each right–hand side. The corresponding computed orders of convergence $p_{\varepsilon,\text{comp}}^N$ and p_{comp}^N of $D_x^- U_\varepsilon$ to $\partial_x U_B^{8192}$ and $\sqrt{\varepsilon}D_y^- U_\varepsilon$ to $\sqrt{\varepsilon}\partial_y U_B^{8192}$, defined analogously to (12.4a) and (12.4a), are given in Tables 12.19 and 12.20 respectively. The entries in these tables indicate that the ε–uniform orders of convergence are at least

Table 12.17 *Computed maximum pointwise difference* $\|D_x^- U_\varepsilon - \partial_x U_B^{8192}\|_{\overline{\Omega}_\varepsilon^N \backslash \Gamma_L}$ *generated by method* (A_ε^N) *applied to problem* (P_ε) *for various values of* ε *and* N.

ε	8	16	32	64	128	256	512
2^0	0.614D+0	0.444D+0	0.256D+0	0.130D+0	0.668D-01	0.344D-01	0.178D-01
2^{-2}	0.900D+0	0.633D+0	0.372D+0	0.201D+0	0.105D+0	0.556D-01	0.307D-01
2^{-4}	0.189D+1	0.114D+1	0.650D+0	0.360D+0	0.194D+0	0.105D+0	0.573D-01
2^{-6}	0.198D+1	0.167D+1	0.121D+1	0.759D+0	0.397D+0	0.210D+0	0.113D+0
2^{-8}	0.197D+1	0.167D+1	0.121D+1	0.798D+0	0.496D+0	0.300D+0	0.180D+0
.	.	.	.	.	.	.	.
2^{-20}	0.195D+1	0.167D+1	0.121D+1	0.798D+0	0.496D+0	0.300D+0	0.180D+0

Table 12.18 *Computed maximum pointwise difference* $\sqrt{\varepsilon}\|D_y^- U_\varepsilon - \partial_y U_B^{8192}\|$ *in the domain* $\overline{\Omega}_\varepsilon^N \backslash \Gamma_B$ *generated by method* (A_ε^N) *applied to problem* (P_ε) *for various values of* ε *and* N.

ε	8	16	32	64	128	256	512
2^0	0.703D-01	0.357D-01	0.180D-01	0.914D-02	0.471D-02	0.249D-02	0.139D-02
2^{-2}	0.193D+00	0.111D+00	0.603D-01	0.315D-01	0.162D-01	0.819D-02	0.414D-02
2^{-4}	0.266D+00	0.140D+00	0.703D-01	0.357D-01	0.180D-01	0.914D-02	0.471D-02
2^{-6}	0.279D+00	0.192D+00	0.118D+00	0.703D-01	0.357D-01	0.180D-01	0.914D-02
2^{-8}	0.279D+00	0.192D+00	0.118D+00	0.733D-01	0.432D-01	0.248D-01	0.141D-01
.	.	.	.	.	.	.	.
2^{-20}	0.279D+00	0.192D+00	0.118D+00	0.733D-01	0.432D-01	0.248D-01	0.141D-01

Table 12.19 *Computed orders of convergence $p^N_{\varepsilon,\text{comp}}$ and p^N_{comp} for $D^-_x U_\varepsilon$ to $\partial_x U^{8192}_{\text{B}}$ generated by method (A^N_ε) applied to problem (P_ε) for various values of ε and N.*

ε	8	16	32	64	128	256
2^0	0.47	0.79	0.98	0.96	0.96	0.95
2^{-2}	0.51	0.76	0.89	0.94	0.92	0.86
2^{-4}	0.73	0.81	0.85	0.89	0.89	0.87
2^{-6}	0.25	0.46	0.68	0.93	0.92	0.90
2^{-8}	0.24	0.46	0.60	0.69	0.72	0.74
$\cdot$	$\cdot$	$\cdot$	$\cdot$	$\cdot$	$\cdot$	$\cdot$
2^{-20}	0.22	0.46	0.60	0.69	0.72	0.74
p^N_{comp}	0.25	0.46	0.60	0.69	0.72	0.74

Table 12.20 *Computed orders of convergence $p^N_{\varepsilon,\text{comp}}$ and p^N_{comp} of $\sqrt{\varepsilon}D^-_y U_\varepsilon$ to $\sqrt{\varepsilon}\partial_y U^{8192}_{\text{B}}$ generated by method (A^N_ε) applied to problem (P_ε) for various values of ε and N.*

ε	8	16	32	64	128	256
2^0	0.98	0.99	0.98	0.96	0.92	0.85
2^{-2}	0.79	0.88	0.94	0.96	0.98	0.99
2^{-4}	0.93	1.00	0.98	0.99	0.98	0.96
2^{-6}	0.54	0.71	0.74	0.98	0.99	0.98
2^{-8}	0.54	0.71	0.68	0.77	0.80	0.82
$\cdot$	$\cdot$	$\cdot$	$\cdot$	$\cdot$	$\cdot$	$\cdot$
2^{-20}	0.54	0.71	0.68	0.77	0.80	0.82
p^N_{comp}	0.54	0.71	0.68	0.77	0.80	0.82

0.6 for all $N \geq 32$. This is the same result as that determined independently in the previous section from Tables 12.8 and 12.9. Again, this is a significant confirmation of the soundness of our approach. It should be observed that the computed orders of convergence given in Tables 12.19 and 12.20 appear to follow a more regular pattern than those in Tables 12.8 and 12.9.

The qualitative behaviour of the errors in the computed scaled discrete derivatives can be seen from the graphs of $D^-_x U_\varepsilon - \partial_x U^{8192}_{\text{B}}$, $\sqrt{\varepsilon}(D^-_y U_\varepsilon - \partial_y U^{8192}_{\text{B}})$, $D^-_y V_\varepsilon - \partial_y V^{8192}_{\text{B}}$ and $\varepsilon^{-1/2}(D^-_x V_\varepsilon - \partial_x V^{8192}_{\text{B}})$ displayed in Figs. 12.13, 12.14, 12.15 and 12.16. We see from the graph of $\varepsilon^{-1/2}(D^-_x V_\varepsilon - \partial_x V^{8192}_{\text{B}})$ in Fig. 12.16, and also from the entries in Table 12.21, that there is a singularity in the scaled derivative $\varepsilon^{-1/2}\frac{\partial v_\varepsilon}{\partial x}$ in a neighbourhood of the edge Γ_{L}. This singularity is not resolved by the numerical method (A^N_ε). Nevertheless, the computed scaled discrete derivatives $\varepsilon^{-1/2}D^-_x V_\varepsilon$ appear to converge ε-uniformly to the corresponding scaled exact derivative $\varepsilon^{-1/2}\frac{\partial v_\varepsilon}{\partial x}$ in subdomains which exclude a sufficiently large neighbourhood of this edge. This may be seen from the numerical results

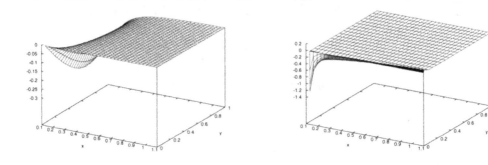

FIG. 12.13. Graphs of $D_x^- U_\varepsilon - \partial_x U_B^{8192}$ for $\varepsilon = 1.0$ and 0.00001 with U_ε generated by method (A_ε^N) applied to problem (P_ε) and $N = 32$.

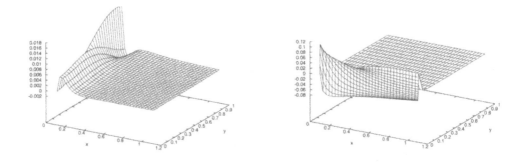

FIG. 12.14. Graphs of $\sqrt{\varepsilon}(D_y^- U_\varepsilon - \partial_y U_B^{8192})$ for $\varepsilon = 1.0$ and 0.00001 with U_ε generated by method (A_ε^N) applied to problem (P_ε) and $N = 32$.

in Tables 12.22 and 12.23, respectively, in which the errors and orders of convergence corresponding to the subdomain $[0.2, 1.1] \times [0, 1]$ are given.

In summary, the numerical results of this section demonstrate, experimentally, that the numerical method (A_ε^N) generates ε–uniform approximations to the scaled components of $\mathbf{u}_P$ and its scaled derivatives. This is a separate and independent confirmation of the same experimental conclusion reached in the previous section.

Since $\mathbf{U}_B^{8192}$ is a good approximation to the exact solution $\mathbf{u}_\varepsilon$, we can use it to compute a good approximation to the error at the mesh points. Therefore, we do not need to compute the error parameters in the above analysis of the errors at the mesh points. However, if we want to estimate the global errors, then we need to use the experimental technique of §8.6, in the usual way, to determine

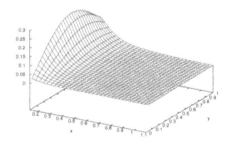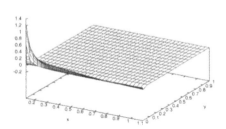

FIG. 12.15. Graphs of $D_y^- V_\varepsilon - \partial_y V_B^{8192}$ for $\varepsilon = 1.0$ and 0.00001 with V_ε generated by method (A_ε^N) applied to problem (P_ε) and $N = 32$.

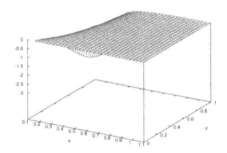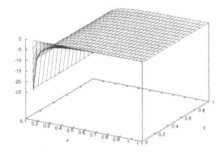

FIG. 12.16. Graphs of $\varepsilon^{-1/2}(D_x^- V_\varepsilon - \partial_x V_B)$ for $\varepsilon = 1.0$ and 0.00001 with V_ε is generated by method (A_ε^N) applied to problem (P_ε) and $N = 32$.

the computed global error parameters associated with the convergence of the scaled components of the numerical solution $\mathbf{U}_\varepsilon$ of (A_ε^N) and their scaled discrete derivatives to the scaled components of the exact solution $\mathbf{u}_\varepsilon$ of (P_ε) and their scaled derivatives. This experimental technique is based on the hypothesis that the numerical solutions are ε–uniformly convergent of some (usually unknown) order. In the present case we do not know that the method (A_ε^N) is ε–uniform, and so, in principle, we cannot use this technique. On the other hand, we can infer from the two experimental approaches, described in this and the previous section, that (A_ε^N) is ε–uniform. Therefore, in practice, we can apply the experimental technique as before to determine these global error parameters for (A_ε^N). We do not pursue this topic here.

Table 12.21 *Computed maximum pointwise difference* $\varepsilon^{-1/2}\|D_x^- V_\varepsilon - \partial_x V_B^{8192}\|$ *in the domain* $\overline{\Omega}_\varepsilon^N \setminus (X_1 \cup \Gamma_L)$ *generated by method* (A_ε^N) *applied to problem* (P_ε) *for various values of* ε *and* N.

ε	8	16	32	64	128	256	512
2^0	0.364D+1	0.406D+1	0.300D+1	0.180D+1	0.113D+1	0.886D+0	0.922D+0
2^{-2}	0.794D+1	0.647D+1	0.476D+1	0.328D+1	0.232D+1	0.191D+1	0.188D+1
2^{-4}	0.278D+2	0.143D+2	0.989D+1	0.700D+1	0.519D+1	0.428D+1	0.396D+1
2^{-6}	0.311D+2	0.281D+2	0.237D+2	0.189D+2	0.130D+2	0.101D+2	0.875D+1
2^{-8}	0.321D+2	0.282D+2	0.237D+2	0.202D+2	0.174D+2	0.158D+2	0.152D+2
.	.	.	.	.	.	.	.
.	.	.	.	.	.	.	.
.	.	.	.	.	.	.	.
2^{-20}	0.342D+2	0.284D+2	0.237D+2	0.202D+2	0.174D+2	0.158D+2	0.152D+2

Table 12.22 *Computed maximum pointwise difference* $\varepsilon^{-1/2}\|D_x^- V_\varepsilon - \partial_x V_B^{8192}\|$ *in the subdomain* $(\overline{\Omega}_\varepsilon^N \setminus (X_1 \cup \Gamma_L)) \cap [0.2, 1.1] \times [0, 1]$ *generated by method* (A_ε^N) *applied to problem* (P_ε) *for various values of* ε *and* N.

ε	8	16	32	64	128	256	512
2^0	0.364D+1	0.406D+1	0.189D+1	0.107D+1	0.578D+0	0.287D+0	0.142D+0
2^{-2}	0.794D+1	0.647D+1	0.248D+1	0.130D+1	0.672D+0	0.332D+0	0.166D+0
2^{-4}	0.278D+2	0.143D+2	0.353D+1	0.177D+1	0.894D+0	0.432D+0	0.214D+0
2^{-6}	0.311D+2	0.281D+2	0.547D+1	0.298D+1	0.163D+1	0.842D+0	0.447D+0
2^{-8}	0.321D+2	0.282D+2	0.547D+1	0.298D+1	0.163D+1	0.842D+0	0.447D+0
.	.	.	.	.	.	.	.
.	.	.	.	.	.	.	.
.	.	.	.	.	.	.	.
2^{-20}	0.342D+2	0.284D+2	0.547D+1	0.298D+1	0.163D+1	0.842D+0	0.447D+0

Table 12.23 *Computed orders of convergence* $p_{\varepsilon,\mathrm{comp}}^N$ *and* p_{comp}^N *of* $D_x V_\varepsilon$ *to* $\partial_x V_B^{8192}$ *in the subdomain* $(\overline{\Omega}_\varepsilon^N \setminus (X_1 \cup \Gamma_L)) \cap [0.2, 1.1] \times [0, 1]$ *generated by method* (A_ε^N) *applied to problem* (P_ε) *for various values of* ε *and* N.

ε	8	16	32	64	128	256
2^0	-0.15	1.10	0.81	0.89	1.01	1.01
2^{-2}	0.29	1.38	0.93	0.95	1.02	1.00
2^{-4}	0.96	2.01	1.00	0.99	1.05	1.02
2^{-6}	0.15	2.36	0.92	1.04	1.09	1.04
2^{-8}	0.19	2.37	0.88	0.87	0.95	0.91
2^{-10}	0.22	2.37	0.88	0.87	0.95	0.91
.	.	.	.	.	.	.
.	.	.	.	.	.	.
.	.	.	.	.	.	.
2^{-20}	0.27	2.38	0.88	0.87	0.95	0.91
p_{comp}^N	0.27	2.38	0.88	0.87	0.95	0.91

12.6   A benchmark solution for laminar flow

An interesting application of the numerical method (A_ε^N) is to use it to generate benchmark numerical solutions against which other numerical methods for solving the Prandtl boundary layer equations, and the Navier–Stokes equations in the laminar, case can be tested. The manner in which we can test the effectiveness of a proposed new numerical method is now outlined.

First, we use the new numerical method to solve the problem on the finite rectangle Ω defined in §11.2. We then compare the resulting numerical solution U_{new} with the Blasius solution U_B^{8192} (or with U_ε^{512} generated by (A_ε^N)) as follows. In the case when the proposed new method is applied to the Prandtl boundary layer equations, since the exact solutions u_B and u_P coincide on Ω, we observe that

$$\|U_{\text{new}} - u_P\|_{\Omega_\varepsilon^N} \leq \|U_{\text{new}} - U_B^{8192}\|_{\Omega_\varepsilon^N} + \|U_B^{8192} - u_B\|_{\Omega_\varepsilon^N}$$

and

$$\sqrt{Re}\|V_{\text{new}} - v_P\|_{\Omega_\varepsilon^N} \leq \sqrt{Re}\left(\|V_{\text{new}} - V_B^{8192}\|_{\Omega_\varepsilon^N} + \|V_B^{8192} - v_B\|_{\Omega_\varepsilon^N}\right).$$

The first term on each right–hand side may be determined from the tables of numerical results. Furthermore, the second term on each right–hand side is already known to be negligible, at least relative to the experimental error, because of the numerical results and bounds (11.18) and (11.19) in chapter 11 and the discusssion in the previous section.

Likewise, if the proposed new method is applied to the Navier–Stokes equations in the laminar case, we observe that

$$\|U_{\text{new}} - u_{\text{NS}}\|_{\Omega_\varepsilon^N} \leq \|U_{\text{new}} - U_B^{8192}\|_{\Omega_\varepsilon^N} + \|U_B^{8192} - u_B\|_{\Omega_\varepsilon^N} + \|u_B - u_{\text{NS}}\|_{\Omega_\varepsilon^N}$$

and

$$\sqrt{Re}\|V_{\text{new}} - v_{\text{NS}}\|_{\Omega_\varepsilon^N} \leq$$
$$\sqrt{Re}\left(\|V_{\text{new}} - V_B^{8192}\|_{\Omega_\varepsilon^N} + \|V_B^{8192} - v_B\|_{\Omega_\varepsilon^N} + \|v_B - v_{\text{NS}}\|_{\Omega_\varepsilon^N}\right).$$

The first two terms on each right–hand side may be determined, as before, from the tables of numerical results. Moreover, the final term on each right–hand side is negligible, because u_B and u_P coincide on $\overline{\Omega}$ and the classical estimate (11.1) applies.

We now describe in more detail the required computations for one possible test. We use the boundary conditions $u_\varepsilon = 0$ on the bottom edge Γ_B of the rectangle and $\mathbf{u}_\varepsilon = \mathbf{U}_B^{8192}$ on the top and inflow edges of the rectangle. On the meshes Ω_ε^N for $N = 8, 16, 32, \ldots$, we use the numerical method under test to solve either the Navier–Stokes or the Prandtl boundary layer equations for the values

$\varepsilon = 1, 2^{-1}, 2^{-3}, \ldots, 10^{-6}$. We then compute tables of the maximum pointwise error in the approximations to $\mathbf{u}_\varepsilon$ on the whole of $\overline{\Omega}_\varepsilon^N$, and also the corresponding tables for the maximum pointwise error in the approximations to $\varepsilon^{1/2} \frac{\partial u_\varepsilon}{\partial y}$ and $\frac{\partial v_\varepsilon}{\partial y}$ on Γ_B. Comparison of the resulting tables with the corresponding tables for the ε-uniform pointwise-accurate algorithm of chapter 11, in the same way as in the previous section for the numerical method (A_ε^N), determines experimentally whether or not the new numerical method is ε-uniform.

References

Acheson D. J. (1990). *Elementary Fluid Dynamics*. Oxford University Press.

Allen D. N. de G and Southwell R. V. (1955). Relaxation methods applied to determine the motion, in 2D, of a viscous fluid past a fixed cylinder. *Quart. J. Mech. Appl. Math.*, **VIII**, (2), 129–145.

Andreyev V. B. and Kopteva N. V. (1996). A study of difference schemes with the first derivative approximated by a central difference ratio. *Comput. Maths. Math. Phys.*, **36** (8) 1065–1078.

Bagaev B. M. and Shaidurov V. V. (1998). *Numerical Methods for the Solution of Problems with a Boundary Layer*. Nauka, Novosibirsk (in Russian).

Bakhvalov N. S. (1969). On the optimization of methods for boundary-value problems with boundary layers. *J. Numer. Meth. Math. Phys.*, **9**, (4), 841–859 (in Russian).

Berger A. E., Solomon J. M. and Ciment M. (1981). An analysis of a uniformly accurate difference method for a singular perturbation problem. *Math. Comp.*, **37**, 465–492.

Blasius H. (1908). Grenzschichten in Flüssigkten mit Kleiner Reibung. *Z. Math. u. Phys.*, **56**, 1–37; Engl. trans. in NACA TM 1256.

Chang K. W. and Howes F. A. (1984). *Nonlinear Singular Perturbation Phenomena*. Springer-Verlag, New York.

Doering C. R. and Gibbon J. D. (1995). *Applied Analysis of the Navier-Stokes Equations*. Cambridge University Press.

Doolan E. P., Miller J. J. H. and Schilders W. H. A. (1980). *Uniform Numerical Methods for Problems with Initial and Boundary Layers*. Boole Press, Dublin.

Eckhaus W. (1973). *Matched Asymptotic Expansions and Singular Perturbations*. North-Holland, Amsterdam.

Eckhaus W. (1979). *Asymptotic Analysis of Singular Perturbations*. North-Holland, Amsterdam.

El–Mistikawy T. M. and Werle M. J. (1978). Numerical method for boundary layers with blowing – the exponential box scheme. *AIAA J.*, **16**, 749–751.

Emel'ianov K. V. (1973). A difference scheme for a three dimensional elliptic equation with a small parameter multiplying the highest derivative. *Boundary Value Problems for Equations of Mathematical Physics*. USSR Academy of Sciences, Ural Scientific Centre, 30–42 (in Russian).

Farrell P. A., Hemker P. and Shishkin G. I. (1996). Discrete approximations for singularly perturbed boundary value problems with parabolic layers III *J. of Comput. Maths*, **14**, (3), 273–290.

Farrell P. A., Miller J. J. H., O'Riordan E. and Shishkin G.I. (1998). On the non-existence of ε-uniform finite difference methods on uniform meshes for semilinear two-point boundary value problems. *Math. Comp.*, **67**, (222), 603–617.

Gartland E. C. (1988). Graded-mesh difference schemes for singularly perturbed two-point boundary value problems. *Math. Comput.*, **51**, 631–657.

Han H. and Kellogg R. B. (1990). Differentiability properties of solutions of the equation $-\varepsilon^2 \Delta u + ru = f(x,y)$ in a square. *SIAM J. Math. Anal.*, **21**, 394–408.

Hegarty A. F. (1986). Analysis of finite difference methods for two-dimensional elliptic singular perturbation problems. Ph. D. Thesis, Trinity College, Dublin.

Hegarty A. F., Miller J. J. H., O'Riordan E. (1980). Uniform second order difference schemes for singular perturbation problems, *Proceedings BAIL I Conference, Dublin*. Boole Press, Dublin, 301–305.

Hegarty A. F., Miller J. J. H., O'Riordan E. and Shishkin G. I. (1995). On a novel mesh for the regular boundary layers arising in advection—dominated transport in two dimensions. *Comms. in Numer. Methods in Engineering*, **11**, 435–441.

Hegarty A. F., Miller J. J. H., O'Riordan E. and Shishkin G. I. (1997). Numerical solution of elliptic convection–diffusion problems on fitted meshes. *CWI Quarterly*, **10**, n. 3, 4 239–251.

Hemker P. W. (1977). *A Numerical Study of Stiff Two–Point Boundary Value Problems*. Mathematical Center, Amsterdam.

Hemker P. W., Shishkin G. I. and Shishkina L. P. (1997). The use of defect correction for the solution of parabolic singular perturbation problems. *Zeitschrift für Angewandte Mathematik und Mechanik*, **76**, 59–74.

Il'in A. M. (1969). Differencing scheme for a differential equation with a small parameter affecting the highest derivative. *Math. Notes*, **6**, (2), 596–602.

Ilin, A. M. (1992). *Matching of Asymptotic Expansions of Solutions of Boundary Value Problems*, American Math. Society Translations, Providence.

Ladyzhenskaya O. A. and Ural'tseva N. N. (1968). *Linear and Quasilinear Elliptic Equations*. Academic Press, New York and London.

Linss T. and Stynes M. (1998). Asymptotic analysis and Shishkin-type decomposition for an elliptic convection–diffusion problem. Preprint 1998-4, Department of Mathematics, University College Cork, Ireland.

Liseikin V. D. (1983). On the numerical solution of a two–dimensional second–order elliptic equation with a small parameter at the highest derivatives. *Numer. Meth. Mechs. Continuous Media*, **14**, (4), 110–115 (in Russian).

Miller J. J. H., O'Riordan E. and Shishkin G.I. (1996). *Fitted Numerical Methods for Singular Perturbation Problems*. World Scientific Publishing Co. Singapore.

Oleinik O. A. and Samokhin V. N. (1999). *Mathemathical Models in Boundary Layer Theory*. Chapman and Hall/CRC, Boca Raton.

O'Malley Jr R. E., (1991). *Singular Perturbation Methods for Ordinary Differential Equations*. Springer–Verlag, New York.

O'Riordan E. (1984). Singular perturbation finite element methods. *Numer. Math.*, **44**, 425–434.

O'Riordan E. and Stynes M. (1986). A uniformly accurate finite–element method for a singularly perturbed one–dimensional reaction–diffusion problem. *Math. Comput.*, **47**, (176), 555–570.

Prandtl L. (1904). Über Flussigkeitsbewegung bei sehr kleiner Reibung. *Proceedings III Intern. Congr. Math.*, Heidelberg.

Protter M. H. and Weinberger H. F. (1984). *Maximum Principles in Differential Equations*. Springer–Verlag, New York.

Roos H.-G., Stynes M. and Tobiska L. (1996). *Numerical Methods for Singularly Perturbed Differential Equations. Convection–Diffusion and Flow Problems*, Springer–Verlag, New York.

Schlichting H. (1979). *Boundary Layer Theory*, 7th edition, McGraw–Hill, New York.

Shih S. D. and Kellogg R. B. (1987). Asymptotic analysis of a singular perturbation problem. *SIAM J. Math. Anal.*, **18**, 1467–1511.

Shishkin G. I. (1983). A difference scheme on a non–uniform mesh for a differential equation with a small parameter multiplying the highest derivative. *USSR Comput. Maths. Math. Phys.*, **23**, 59–66.

Shishkin G. I. (1988a). A difference scheme for a singularly perturbed equation of parabolic type with a discontinuous initial condition. *Soviet Math. Dokl.*, **37**, 792–796.

Shishkin G. I. (1988b). Grid approximation of singularly perturbed parabolic equations with internal layers. *Sov. J. Numer. Anal. Math. Modelling*, **3**, (5), 392–407.

Shishkin G. I. (1989). Approximation of solutions of singularly perturbed boundary value problems with a parabolic boundary layer. *USSR Comput. Maths. Math. Phys.*, **29**, (4), 1–10.

Shishkin G. I. (1991a). Grid Approximation of singularly perturbed parabolic equations degenerating on the boundary. *USSR Comput. Maths. Math. Phys.*, **31**, (10), 53–63.

Shishkin G. I. (1991b). Grid approximation of singularly perturbed boundary value problem for a quasi–linear elliptic equation in the completely degenerate case. *USSR Comput. Maths. Math. Phys.*, **31**, 33–46.

Shishkin G. I. (1992a). A difference scheme for a singularly perturbed parabolic equation degenerating on the boundary. *USSR Comput. Maths. Math. Phys.*, **32**, 621–636.

Shishkin G. I. (1992b). Discrete approximation of singularly perturbed elliptic and parabolic equations. *Russian Academy of Sciences, Ural Section, Ekaterinburg*, (in Russian).

Shishkin G. I. (1993). Grid approximation of a singularly perturbed quasilinear equation in the presence of a transition layer. *Russian Acad. Sci. Dokl. Math.*, **47** (1), 83–88.

Shishkin G. I. (1994). Difference approximation of the Dirichlet problem for a singularly perturbed quasilinear parabolic equation in the presence of a transition layer. *Russian Acad. Sci. Dokl. Math.*, **48** (2), 346–352.

Shishkin G. I. (1995). On finite difference fitted schemes for singularly perturbed boundary value problems with a parabolic boundary layer. *INCA Preprint No. 4*, Dublin.

Shishkin G. I. (1996). Approximation of solutions and of diffusion flows in the case of singularly perturbed boundary value problems with discontinuous initial conditions. *Comput. Math. Math. Phys.*, **36**, 1233–1250.

Shishkin G. I. (1997). Singularly perturbed boundary value problems with concentrated sources and discontinuous initial conditions. *Comput. Math. Math. Phys.*, **37** (4), 417–434.

Shishkin G. I. (2000). Approximation of singularly perturbed convection-diffusion equations with low smoothness of the derivatives involved in the equation. Analytical and Numerical Methods for Convection-Dominated and Singularly Perturbed Problems, L.G. Vulkov, J.J.H. Miller and G.I. Shishkin eds., Nova Science, N.Y., 2000, 13 pp. (to appear)

Sleijpen G. L. G. and Fokkema D. R. (1993). BICGSTAB(*L*) for linear equations involving unsymmetric matrices with complex spectrum. *Electronic Transactions on Numerical Analysis*, **1**, 11–32.

Sonneveld P. (1989). CGS, a fast Lanczos–type solver for nonsymmetric linear systems. *SIAM J. Sci. Statist. Comput*, **10**, 36–52.

Tobiska L. (1976). Die asymptotische lösung von Wärmeleitungsproblemen. Ph. D. Thesis, Otto von Guericke University, Magdeburg.

Varga R. S. (1962). *Matrix Iterative Analysis*. Prentice–Hall, N.J..

Volkov E. A. (1965). Differentiability properties of solutions of boundary value problems for the laplace and poisson equations. *Proc. Steklov Inst. Math.*, **77**, 101–126.

van der Vorst H. A. (1992). Bi–CGSTAB: A fast and smoothly converging variant of Bi–CG for the solution of nonsymmetric linear systems. *SIAM J. Sci. Comput.*, **13**, 631–644.

Vulanović R. (1986). Mesh Construction for Discretization of Singularly Perturbed Boundary Value Problems. Doctoral Dissertation, Faculty of Sciences, University of Novi Sad.

Index